D1791139

SPINORS IN PHYSICS AND GEOMETRY

SPINORS IN PHYSICS AND GEOMETRY

Trieste, 11 – 13 September 1986

Editors

A Trautman

University of Warsaw

G Furlan

University of Trieste

Published by

World Scientific Publishing Co. Pte. Ltd.
P O Box 128, Farrer Road, Singapore 9128

USA office: World Scientific Publishing Co., Inc.
687 Hartwell Street, Teaneck, NJ 07666, USA

UK office: World Scientific Publishing Co. Pte. Ltd.
P O Box 379, London N12 7JS, England

SPINORS IN PHYSICS AND GEOMETRY

ISBN 9971-50-763-3

Printed in Singapore by Utopia Press.

Preface

Some time during the summer of 1985, in conversations with some colleagues, the idea of organizing a conference on spinors was put forward. We thought that a meeting of mathematicians and physicists working on the subject was appropriate, in view of the rôle played by spinors in recent research on supersymmetry, Kaluza-Klein theories, twistors and general relativity. Trieste seemed to be a good place for such a meeting, especially in view of Paolo Budinich's keen interest in the subject, and the encouragement he extended to his collaborators and students. With the warm support of Abdus Salam the preparation for the meeting went on under the joint sponsorship of the International School for Advanced Studies and the International Center for Theoretical Physics.

The Conference on Spinors in Physics and Geometry was held in Trieste from 11 to 13 September 1986. It gathered about 70 participants; there were 23 invited lectures. During the opening ceremony, Abdus Salam congratulated Paolo Budinich and André Lichnerowicz on the occasion of their anniversaries and briefly described their academic activity. All lecturers and participants, during their intervention at the meeting, joined in expressing their admiration and warm feelings to the two scientists.

We would like to dedicate the volume of the Proceedings of the First Conference on Spinors in Physics and Geometry to Paolo Budinich, on the occasion of his seventieth birthday, as a token of gratitude for his invaluable contributions to Science and to the development of Theoretical Physics in Trieste.

Andrzej Trautman and Giuseppe Furlan

I would like to add my tribute to Paolo Budinich on his 70th birthday. Paolo Budinich is the man who brought the International Centre for Theoretical Physics to Trieste. It was his vision and his imaginative and gracious approach which persuaded those from the international Physics community on the one hand, and the Italian Government and the Town of Trieste on the other, to create the Centre in Trieste.

Paolo Budinich's contributions to Physics are important. His love of Spinors is proverbial and legendary together with his veneration for Cartan.

It has been a pleasure to have known him as a person and to have his friendship. I wish him many happy returns on this occasion.

Abdus Salam

I would like to add my tribute to Paolo Budinich on his 70th birthday. Paolo Budinich is the man who brought the International Centre for Theoretical Physics to Trieste. It was his vision and his imaginative and generous approach which persuaded those from the International Atomic Energy Agency on the one hand and the Italian Government and the Commune of Trieste on the other, to create the Centre at Trieste.

Paolo Budinich's contributions to physics are important. His ideas of spinors in physics and geometry together with his dedication to Trieste.

It has been a pleasure to have known him as a friend and to have his friendship. I wish him many happy returns on this occasion.

Abdus Salam

CONTENTS

SPINORS IN PHYSICS AND GEOMETRY

Killing spinors according to O. Hijazi and applications

A. Lichnerowicz

Introduction.

Killing spinors were first introduced in Mathematical Physics : General Relativity, 11-dimensional (resp. 10-dimensional) supergravity theory, super symmetry, matter fields. This notion also appears in a purely geometrical way in direct relation with the Dirac operator P of a spin manifold [5], [6], [7]. A Killing spinor is automatically an eigenspinor of P and is a generalization of the notion of parallel spinor. I have studied many years ago properties of parallel spinors in the context of the harmonic spinors.

The most interesting step in this way has been recently obtained by O. Hijazi |7|. He gives a lower bound for the absolute values of the eigenvalues of the Dirac operator of a compact spin manifold in terms of conformal geometry. The limiting case is precisely the case where the manifold admits Killing spinors.

Other interesting results on the Killing spinors have been given recently by different authors and some of these results will appear certainly in this conference. I will give here a survey on the Killing spinors mainly connected with the results of Hijazi. I will give also some of my results.

The physical philosophy which appears can be the following : consider a compact spin manifold as giving a euclidean model for a fermionic field. If the manifold admits a Killing spinor, we can consider the corresponding eigenvalue of P as the main eigenvalue (ground state for the model); all the other eigenvalues are superior in absolute value.

I - Definitions and general formulas.

1 - Spin structure.

a) Let (W,g) be an oriented riemannian manifold of dimension $n \geq 2$, admitting the positive definite metric g; introduce the orthonormal frame bundle E of structural group S O(n). The manifold (W,g) admits a spin structure if and only if the second Stiefel-Whitney class $w_2 \in H^2(W;Z_2)$ is zero; there are then principal Spin(n)-bundles $\mathcal{E}$ on W which are 2-fold coverings of E such that the following diagram is commutative

$$\begin{array}{ccc} \mathrm{Spin}(n) & \longrightarrow & \mathcal{E} \\ \bar{p} \downarrow & & \downarrow p \\ SO(n) & \longrightarrow & E \\ & & \downarrow \\ & & W \end{array}$$

$\mathcal{E}$ defines on W a spin structure. For example (S^n,can.) is a spin manifold and (CP^n, can.) is not a spin manifold for even n.

A point z of $\mathcal{E}$ is called an orthonormal spin frame and $\pi : \mathcal{E} \to W$ is the orthonormal spin frame bundle. A tensor at $x = \pi z$ is said to be refered to z if it is to $y = p z \in E$.

b) Set $n = 2\nu$ or $2\nu + 1$ and introduce the $2^\nu \times 2^\nu$ Dirac matrices $\gamma_\alpha = (\gamma_\alpha{}^a{}_b)$ ($\alpha = 1,..,n;a,b,.. = 1,..,2^\nu$) satisfying :

(1-1) $$\gamma_\alpha \gamma_\beta + \gamma_\beta \gamma_\alpha = - 2 g_{\alpha\beta} e$$

These matrices can be also considered as defining linear operators on a complex vector space S of dimension 2^ν. The group Spin(n) can be considered as a group of linear operators on S or as a group of $2^\nu \times 2^\nu$-matrices $\Lambda = (\Lambda^{a'}_b)$ satisfying :

(1-2) $$\Lambda \gamma_\alpha \Lambda^{-1} = A^{\lambda'}_\alpha \gamma_{\lambda'}$$

where $A = (A^{\lambda'}_\alpha) = \bar{p} \Lambda \in SO(n)$.

A <u>contravariant 1-spinor</u> ψ at $x \in W$ is a map $z \to \psi(z)$ of $\pi^{-1}(x)$ in S such that :

$$\psi(z\,\Lambda^{-1}) = \Lambda\,\psi(z) \qquad\qquad (\Lambda \in Spin(n))$$

We define thus the 2^{ν}-dimensional complex vector space Σ_x of the contravariant 1-spinors at $x \in W$; Σ is the corresponding vector bundle associated with $\mathcal{E}$. We obtain by duality and tensor product the notion of tensor-spinor and the corresponding bundles.

The formula (1-2) expresses that the elements of the γ_α are the components of a vector-spinor γ of spinor type (1,1) with respect to a spinor frame; γ is said to be the fundamental vector-spinor of the spin manifold (W,g).

It is well known that the matrices γ_α can be choosed antihermitian :

$$\tilde{\gamma}_\alpha = -\,\gamma_\alpha \tag{1-3}$$

where $\sim$ denotes the adjunction. If such is the case, $\tilde{\Lambda} = \Lambda^{-1}$ and the representation of Spin(n) is given by unitary matrices. If ψ is a contravariant 1-spinor, $\tilde{\psi}$ is a covariant 1-spinor and the spinor space Σ_x admits a canonical structure of hermitian vector space given by the scalar product $(\psi^{(1)}, \psi^{(2)}) = \tilde{\psi}^{(2)}\,\psi^{(1)}$.

c) <u>Suppose that n is even ($n = 2\nu$)</u>. Consider the (1,1)-spinor $\mathcal{S}\eta$ associated with the volume element η of (W,g) in the classical isomorphism $\mathcal{S}$ between forms and (1,1) spinors. Set

$$\xi = \varepsilon\,\mathcal{S}\,\eta = \varepsilon\,\gamma^1 \ldots \gamma^n$$

where $\varepsilon = 1$ if ν is even and $\varepsilon = i$ if ν is odd; ξ is hermitian ($\tilde{\xi} = \xi$) and satisfies :

$$\xi\,\gamma_\alpha + \gamma_\alpha\,\xi = 0 \qquad\qquad \xi^2 = e \tag{1-4}$$

Associate to each contravariant 1-spinor ψ the 1-spinor $B\,\psi = \xi\,\psi$. We have $B^2 = Id$ and

(1-5) $$(B\,\psi^{(1)},\ \psi^{(2)}) = (\psi^{(1)},\ B\,\psi^{(2)})$$

We can decompose Σ_x in a direct sum $\Sigma_x^+ \oplus \Sigma_x^-$, where the elements of Σ_x^+ (resp. Σ_x^-) are the eigenspaces of B corresponding to the eigenvalue +1 (resp. - 1). It follows from (1-5) that positive spinors and negative spinors are orthogonal in our scalar product.

2 - Spinor connection.

a) The riemannian connection of (W,g) induces a connection on the principal bundle ξ by means of the isomorphism between the Lie algebras of Spin(n) and SO(n). Introduce local sections of the orthonormal frame bundle. If $\omega_U = (\omega_{\alpha\beta})$ $(\omega_{\alpha\beta} + \omega_{\beta\alpha} = 0)$ is for a domain U of W the 1-form defining the riemannian connection, the corresponding spinor connection is given by the 1-form :

$$\sigma_U = -\frac{1}{4}\,\omega_{\alpha\beta}\,\gamma^\alpha\,\gamma^\beta$$

We denote by ∇ the corresponding covariant differentiation. We have $\nabla\gamma = 0$ and the adjunction $\mathcal{A}$ commutes with ∇. If ψ is a 1-spinor field, we have on U :

(2-1) $$(\nabla_\alpha\,\nabla_\beta - \nabla_\beta\,\nabla_\alpha)\,\psi = -\frac{1}{4}\,R_{\lambda\mu,\alpha\beta}\,\gamma^\lambda\,\gamma^\mu\,\psi$$

where $RC = (R_{\lambda\mu,\alpha\beta})$ is the riemannian curvature tensor. Let $Ri = (R_{\alpha\beta})$ be the Ricci tensor and R the scalar curvature of (W,g). We have the classical formulas :

(2-2) $$R_{\alpha\beta,\,\lambda\mu}\,\gamma^\beta\,\gamma^\lambda\,\gamma^\mu = 2\,R_{\alpha\beta}\,\gamma^\beta \qquad R_{\alpha\beta}\,\gamma^\alpha\,\gamma^\beta = -R$$

It follows :

(2-3) $$\gamma^\beta(\nabla_\alpha\nabla_\beta - \nabla_\beta\nabla_\alpha)\,\psi = -\frac{1}{2}\,R_{\alpha\beta}\,\gamma^\beta\,\psi$$

The <u>Dirac operator</u> P on the contravariant 1-spinor fields is given by :

(2-4) $$P\,\psi = \gamma^\alpha\,\nabla_\alpha\,\psi$$

It satisfies for even n :

$$P\,B = -\,B\,P$$

b) Let f be a complex-valued function on W. We can introduce a corresponding <u>connection</u> on the vector bundle Σ by $\sigma_U^{(f)} = \sigma_U + \frac{f}{n}\, \gamma|_U$. We denote by $\nabla^{(f)}$ the corresponding covariant differentiation. For a contravariant 1-spinor field ψ we have :

(2-6) $$\nabla^{(f)}\psi = \nabla\psi + \frac{f}{n}\gamma\psi$$

and for a (1,1)-spinor φ :

$$\nabla^{(f)}\varphi = \nabla\varphi + \frac{f}{n}(\gamma\varphi - \varphi\gamma)$$

Similarly :

(2-7) $$\nabla_\alpha^{(f)}\gamma^\beta = \frac{f}{n}(\gamma_\alpha\gamma^\beta - \gamma^\beta\gamma_\alpha)$$

For a tensor T we have $\nabla^{(f)} T = \nabla T$; in particular $\nabla^{(f)} g = 0$. It follows from (2-6) :

(2-8) $$P^{(f)}\psi = \gamma^\alpha\nabla_\alpha^{(f)}\psi = P\psi - f\psi \quad .$$

The naturality of such a connection has been studied by M. Cahen - S. Gutt in a different context. By a straightforward computation using (2-3) and the definition of $\nabla^{(f)}$, we get :

(2-9) $$-\gamma^\beta(\nabla_\alpha^{(f)}\nabla_\beta^{(f)} - \nabla_\beta^{(f)}\nabla_\alpha^{(f)})\psi =$$
$$= \frac{1}{2}(R_{\alpha\beta} - \frac{4(n-1)}{n^2} f^2 g_{\alpha\beta})\gamma^\beta\psi + \nabla_\alpha f\psi + \frac{1}{n}\nabla_\beta f\gamma^\beta\gamma_\alpha\psi$$

It is easy to show that, for $P^{(f)} = P - f$, we have :

(2-10) $$(P^{(f)})^2\psi = -g^{\alpha\beta}\nabla_\alpha^{(f)}\nabla_\beta^{(f)}\psi + (\frac{R}{4} - \frac{n-1}{n}f^2)\psi - \frac{n-1}{n}(2fP^{(f)}\psi + \nabla_\alpha f\gamma^\alpha\psi)$$

<u>3 - Properties of spinors such that $\nabla^{(f)}\psi = 0$.</u>

In the following, we will study classes of spin manifolds which admit spinors ψ such that $\nabla^{(f)}\psi = 0$. If ψ is such a spinor, it follows from (2-9), (2-10) :

(3-1) $$\frac{1}{2}(R_{\alpha\beta} - \frac{4(n1)}{n^2} f^2 g_{\alpha\beta})\gamma^\beta\psi + \nabla_\alpha f\psi + \frac{1}{n}\nabla_\beta f\gamma^\beta\gamma_\alpha\psi = 0$$

and :

(3-2) $$\left(\frac{R}{4} - \frac{n-1}{n} f^2\right) \psi - \frac{n-1}{n} \nabla_\alpha f \gamma^\alpha \psi = 0$$

a) Set $f = a + ib$, where a,b are real-valued functions. We have :

$$\nabla_\alpha \psi = - \frac{a+ib}{n} \gamma_\alpha \psi \qquad \nabla_\alpha \tilde{\psi} = \frac{a-ib}{n} \tilde{\psi} \gamma_\alpha$$

Introduce the real scalar and vector

$$u = \tilde{\psi} \psi > 0 \quad (\text{for } \psi \neq 0) \qquad U_\alpha = i \tilde{\psi} \gamma_\alpha \psi$$

A straightforward computation gives :

(3-3) $$du = - \frac{db}{n} U \qquad \nabla_\alpha U_\beta + \nabla_\beta U_\alpha = - \frac{4b}{n} u g_{\alpha\beta}$$

Now taking the product of (3-2) by $\tilde{\psi}$ and separating real and purely imaginary parts we get :

(3-4) $$\left(\frac{R}{4} - \frac{n-1}{n} (a^2 - b^2)\right) u - \frac{n-1}{n} U^\alpha \nabla_\alpha b = 0$$

and

(3-5) $$2 a b u - U^\alpha \nabla_\alpha a = 0$$

b) <u>Suppose that n is even</u>. If $\psi = \psi^+ + \psi^-$, we have $B \psi = \psi^+ - \psi^-$. Introduce the real scalar and vector :

$$v = (B \psi)^{2} \psi = \tilde{\psi}^+ \psi^+ - \tilde{\psi}^- \psi^- \qquad V_\alpha = (B \psi)^2 \gamma_\alpha \psi$$

A straightforward computation gives :

(3-6) $$dv = - \frac{2a}{n} V \qquad \nabla_\alpha V_\beta + \nabla_\beta V_\alpha = \frac{4a}{n} v g_{\alpha\beta}$$

Taking the product of (3-2) by $(B \psi)^\sim$ and separating real and purely imaginary parts we get :

(3-7) $$\left(\frac{R}{4} - \frac{n-1}{n} (a^2 - b^2)\right) v - \frac{n-1}{n} V^\alpha \nabla_\alpha a = 0$$

and :

(3-8) $$2 ab v + V^\alpha \nabla_\alpha b = 0$$

II - Killing spinors and connection $\nabla^{(f)}$

4 - Killing spinors

If W is compact, introduce the global scalar product $\langle \psi^{(1)}, \psi^{(2)} \rangle = \int_W (\psi^{(1)}, \psi^{(2)}) \eta$. If we set $\Delta = P^2$ for spinors, we have $\langle \Delta\psi, \psi \rangle = \langle P\psi, P\psi \rangle \geqq 0$. It follows that <u>if W is compact the spectrum of P is real</u>.

a) (W,g) being an arbitrary spin manifold, a <u>Killing spinor</u> is a spinor ψ such that $\nabla^{(\lambda)} \psi = 0$, where λ is a complex constant. According to (2-8) we have $P\psi = \lambda\psi$ and λ is an eigenvalue of the Dirac operator. It follows from (3-5) that λ is either real or purely imaginary and we deduce from (3-1), (3-2) :

$$(4\text{-}1) \qquad R_{\alpha\beta} = (R/n)\, g_{\alpha\beta} \qquad\qquad \lambda^2 = (n/4(n-1))\, R$$

In particular if ψ is a parallel spinor field ($\lambda = 0$), the Ricci tensor of the spin manifold is zero. We say that a Killing spinor is <u>non trivial</u> if $\lambda \neq 0$.

If W is compact, λ is necessarily real. Conversely if λ is real, the Ricci tensor is positive definite and W is compact according to a classical theorem of Myiers. We have the following well-known proposition [2] [3]

<u>Proposition - If (W,g) admits a non zero Killing spinor, it is an Einstein space and $\lambda^2 = (n/4(n-1))R$. If $\lambda \neq 0$ is real, W is compact, if λ is purely imaginary, W is non compact</u>.

b) All the Killing spinors of a manifold (W,g) correspond to the constants λ and $-\lambda$. The Killing spinors of (W,g) define two complex vector spaces K_λ and $K_{-\lambda}$. If n is even, B defines an isomorphism between K_λ and $K_{-\lambda}$. If $n \equiv 1$ (mod. 4) for real λ or if $n \equiv 3$ (mod. 4) for purely imaginary λ, the charge conjugation defines an antiisomorphism between K_λ and $K_{-\lambda}$; in all these cases $\dim_R K_\lambda = \dim_R K_{-\lambda}$. For the sphere $S^{2\nu}$ we have $\dim_R K_\lambda = 2^\nu$.

If λ is <u>real</u>, it follows from (3-3) that u = const. and if $U \neq 0$, it defines an infinitesimal isometry, that is a Killing vector. It is the origin of the name of Killing spinor.

5 - The case where f is a real-valued or purely imaginary-valued function.

a) Consider a spinor $\psi \neq 0$ such that $\nabla^{(f)}\psi = 0$, where f is a real valued function (f = a). It follows from (3-1) by product by $\nabla^{\alpha} a . \tilde{\psi}$:

$$(5\text{-}1) \qquad -\frac{i}{2}(R_{\alpha\beta} - \frac{4(n-1)}{n^2} a^2 g_{\alpha\beta}) \nabla^{\alpha} a.U^{\beta} + (\nabla^{\alpha} a \nabla_{\alpha} a)(1 - \frac{1}{n})u = 0$$

Taking the real part of the left member of (5-1), we get :

$$(\nabla^{\alpha} a \nabla_{\alpha} a) = 0$$

and $a = \lambda$ = const; therefore ψ is a Killing spinor corresponding to a real λ [7] :

Theorem (Hijazi) - On the spin manifold (W,g), let $\psi \neq 0$ be a spinor such that $\nabla^{(f)}\psi = 0$, where f is a real-valued function. Then f is constant and ψ is a Killing spinor; the manifold W is compact.

b) Now suppose that n is even. Consider a spinor $\psi \neq 0$ such that $\nabla^{(f)}\psi = 0$ where f = a+ib. It follows from (3-5) that if a = const., either a = 0 or b = 0. We are led to study the case where f is a purely imaginary-valued function (f = ib). According to (3-6) we have v = const. and if $V \neq 0$, it defines an infinitesimal isometry.

It follows from (3-1) by product by $\nabla^{\alpha} b \ \widetilde{(B\psi)}$:

$$(5\text{-}2) \qquad \frac{1}{2}(R_{\alpha\beta} + \frac{4(n-1)}{n} b^2 g_{\alpha\beta}) \nabla^{\alpha} b\, V^{\beta} + i(\nabla^{\alpha} b \nabla_{\alpha} b)(1 - \frac{1}{n}) v = 0$$

Taking the purely imaginary part of the left member of (5-2), we get :

$$(\nabla^{\alpha} b \nabla_{\alpha} b) v = 0$$

Theorem - Suppose that (W,g) is even-dimensional. Let $\psi \neq 0$ be a spinor such that $\nabla^{(f)}\psi = 0$, where f is a purely imaginary-valued function. Theen either f = const. and ψ is a Killing spinor or v = 0.

III - Conformal bound for the smallest eigenvalue of the Dirac operator in the compact case.

6 - Conformal change of metric.

Now I will give results of Hijazi connecting the smallest eigenvalue of the Dirac operator and Killing spinors in a conformal way.

In this section we suppose that (W,g) is compact and $n \geq 3$.

a)) Consider a metric $\bar{g} = e^{2c} g$ for some real function c. It is convenient to write the conformal factor under the form :

(6-1) $$e^{2c} = h^{4/n-2} \qquad \text{(for a positive function } h)$$

The scalar curvature of $(W,\bar{g})$ is given by |10| :

(6-2) $$\bar{R}\, e^{2c} = \frac{4(n-1)}{n-2}\, h^{-1}\, \Delta\, h + R$$

where Δ is the positive laplacien. We are led to introduce the Yamabe operator on the functions :

(6-3) $$L = \frac{4(n-1)}{n-2}\, \Delta + R \; .$$

It is well known that if μ_1 is the first eigenvalue of L, there is a function $h_1 > 0$ such that $L\, h_1 = \mu_1\, h_1$, and so

(6-4) $$\mu_1 = \frac{4(n-1)}{n-2}\, h_1^{-1}\, \Delta\, h_1 + R$$

If $\mu_1 \neq 0$, $\bar{R}_1$ has everywhere the sign of μ_1 or is null if $\mu_1 = 0$. The spin of μ_1 is a conformally invariant [10]. Study the function :

$$Q(h) = \left(\frac{4(n-1)}{n-2}\, h^{-1}\, \Delta\, h + R\right) - \mu_1 = \frac{4(n-1)}{n-2}\, (h^{-1}\, \Delta\, h - h_1^{-1}\, \Delta\, h_1)$$

where h is > 0. We have trivially

$$\int_W (h_1\, \Delta\, h - h\, \Delta\, h_1)\eta = 0$$

Therefore $(h_1\, \Delta\, h - h\, \Delta\, h_1)$ and so the function $(h^{-1}\, \Delta\, h - h_1^{-1}\, \Delta\, h_1)$ take values of arbitrary signs. It follows :

$$\inf_W \left(\frac{4(n-1)}{n-2} h^{-1} \Delta h + R\right) \leq \mu_1 \leq \sup_W \left(\frac{4(n-1)}{n-2} h^{-1} \Delta h + R\right)$$

and we have :

(6-5) $$\sup_{h>0} \inf_W \left(\frac{4(n-1)}{n-2} h^{-1} \Delta h + R\right) \leq \mu_1 \leq \inf_{h>0} \sup_W \left(\frac{4(n-1)}{n-2} h^{-1} \Delta h + R\right)$$

Taking h = const., we have in particular :

(6-6) $$\inf_W R \leq \mu_1 \leq \sup_W R$$

If R = const., $\mu_1 = R$.

b) Let E_g be the g-orthonormal frame bundle of (W,g), $E_{\bar{g}}$ the $\bar{g}$-orthonormal frame bundle of $(W,\bar{g})$. We have an isomorphism G_c of E_g onto $E_{\bar{g}}$. Let γ be a spin structure on (W,g), $\mathcal{E}_g$ the corresponding spin frame bundle. One can define a spin structure $\bar{\gamma}$ on $(W,\bar{g})$ in such a way that the diagram :

$$\begin{array}{ccc} \mathcal{E}_g & \xrightarrow{\tilde{G}_c} & \mathcal{E}_{\bar{g}} \\ \downarrow & & \downarrow \\ E_g & \xrightarrow{G_c} & E_{\bar{g}} \end{array}$$

commutes. Taking the vector bundles associated with $\mathcal{E}_g$ and $\mathcal{E}_{\bar{g}}$, for a spinor ψ of (W,g) we write $\bar{\psi} = G_c\psi$. We have $\bar{\gamma}_\alpha = e^c \gamma_\alpha$. It is easy to prove that, for any spinor field ψ, we have :

(6-7) $$\bar{\nabla}_\alpha \bar{\psi} = \overline{\nabla_\alpha \psi} - \frac{1}{2} \partial_\beta c \, \overline{\gamma_\alpha \gamma^\beta \psi} - \frac{1}{2} \partial_\alpha c \, \bar{\psi}$$

It follows that :

(6-8) $$\bar{P} \bar{\psi} = e^{-c} \left(\overline{P \psi} + \frac{n-1}{2} \partial_\alpha c \, \overline{\gamma^\alpha \psi}\right)$$

7 - The inequality of Hijazy.

We will prove the following [7] :

Theorem (Hijazi) - Let (W,g) be a compact spin manifold of dimension $n \geq 3$. Any eigenvalue λ of the Dirac operator satisfies :

(7-1) $$\lambda^2 \geq (n/4(n-1)) \, \mu_1$$

where μ_1 is the first eigenvalue of the Yamabe operator. If R = const. we have $\lambda^2 \geq \frac{n}{4(n-1)} R$.

This statement is only of interest for the eigenvalue λ_1 of P with the smallest absolute value and on a manifold for which $\mu_1 \geq 0$. If such is the case, there is on a conformal metric with non negative scalar curvature (see also [5]).

Consider the metric $\bar{g} = e^{2c} g$. Let ψ be an eigenspinor of P ($P\psi = \lambda\psi$). Associate with ψ the spinor $\varphi = e^{-\frac{n-1}{2}} \psi$. We have :

$$P\varphi = \lambda\varphi - \frac{n-1}{2} \partial_\alpha c \gamma^\alpha \varphi$$

It follows from (6-8) :

$$\bar{P}\bar{\varphi} = e^{-c}(\overline{P\varphi} + \frac{n-1}{2} \overline{\partial_\alpha c \gamma^\alpha \varphi}) = e^{-c}(\lambda\bar{\varphi} - \frac{n-1}{2}\overline{\partial_\alpha c\gamma^\alpha\varphi} + \frac{n-1}{2}\overline{\partial_\alpha c\gamma^\alpha\varphi})$$

that is

$$\bar{P}\bar{\varphi} = \lambda e^{-c}\bar{\varphi} \tag{7-2}$$

Take (2-10) for $f = \lambda e^{-c}$ and for $(W,\bar{g})$ and $\bar{\varphi}$. We obtain by integration on W of the scalar product by $\bar{\varphi}$:

$$\int_W (\bar{\nabla}^{(f)}\bar{\varphi}, \bar{\nabla}^{(f)}\bar{\varphi})_{\bar{g}}\,\bar{\eta} + \int_W (\frac{\bar{R}}{4} - \frac{n-1}{n} f^2)(\bar{\varphi},\bar{\varphi})_{\bar{g}}\,\bar{\eta} - \frac{n-1}{n}\int_W (\nabla_\alpha f \bar{\gamma}^\alpha \bar{\varphi}, \bar{\varphi})_{\bar{g}}\,\bar{\eta} = 0 \tag{7-3}$$

The last term of (6-3) is purely imaginary and the other terms are real. Thus :

$$\int_W (\bar{\nabla}^{(f)}\bar{\varphi}, \bar{\nabla}^{(f)}\bar{\varphi})_{\bar{g}}\,\bar{\eta} + \frac{1}{4}\int_W e^{-2c}(\bar{R}e^{2c} - \frac{4(n-1)}{n}\lambda^2)(\bar{\varphi},\bar{\varphi})_{\bar{g}}\,\bar{\eta} = 0 \tag{7-4}$$

The second integral is non positive and $\bar{R} e^{2c} - (4(n-1)/n)\lambda^2$ takes values which are ≤ 0. We have according to (6-2) :

$$\lambda^2 \geq \frac{n}{4(n-1)} \inf_W (\bar{R} e^{2c}) = \frac{n}{4(n-1)} \inf_W (\frac{4(n-1)}{n} h^{-1} \Delta h + R) \qquad (\forall h > 0)$$

For $h = h_1$ we have $\lambda^2 \geq (n/4(n-1))\,\mu_1$ and the inequality is proved.

8 - The limiting case.

Now we prove the following theorem :

Theorem (Hijazy) - Let (W,g) be a compact spin manifold of dimension $n \geq 3$. If there exists an eigenspinor ψ of P for the eigenvalue λ_1, with $\lambda_1^2 = (n/4(n-1))\mu_1 > 0$, ψ is a Killing spinor corresponding to λ_1 ($\nabla^{(\lambda_1)}\psi = 0$). Conversely if (W,g) admits a non trivial Killing spinor ψ, with $\nabla^{(\lambda_1)}\psi = 0$, then the eigenvalue λ_1 of P satisfies $\lambda_1^2 = (n/4(n-1))\,\mu_1$ and is an eigenvalue of P with the smallest absolute value.

In fact introduce $h_1 > 0$ such that $\mu_1 = (4(n-1)/n-2)\, h_1^{-1}\,\Delta\, h_1 + R$ and set $e^{2c_1} = h_1^{4/n-2}$, $f_1 = \lambda_1 e^{-c_1}$. The scalar curvature $\bar{R}_1$ of the metric $\bar{g}$, is given by $\bar{R}_1\, e^{2c_1} = \mu_1$ according to (6-2).

If $\lambda_1^2 = (n/4(n-1))\,\mu_1$ we have :

$$\bar{R}_1\, e^{2c_1} - (4(n-1)/n)\,\lambda_1^2 = 0$$

It follows from (7-4) that $\bar{\nabla}^{(f)}\bar{\varphi} = 0$ and $f_1 = \text{const.}$, $c_1 = \text{const.}$ for $\lambda_1 \neq 0$.

We have :

$$\bar{\nabla}\bar{\varphi} + \frac{f_1}{h}\,\bar{\gamma}\,\bar{\varphi} = \bar{\nabla}\bar{\varphi} + \frac{\lambda_1}{n}\, e^{-c_1}\,\bar{\gamma}\,\bar{\varphi} = 0$$

According to (6-7) $\bar{\nabla}\bar{\varphi} = \overline{\nabla\varphi}$, where $\varphi = e^{-\frac{n-1}{2}c_1}\psi$. It follows :

$$\nabla\varphi + \frac{\lambda_1}{n}\,\gamma\,\varphi = 0$$

or $\nabla^{(\lambda_1)}\psi = 0$. Conversely if ψ is a non trivial Killing spinor with $\nabla^{(\lambda_1)}\psi = 0$, (W,g) is an Einstein manifold and $\mu_1 = R$. The conclusion follows from (4-1) and from the inequality.

Remark. If $\mu_1 = 0$, $\lambda_1 = 0$, the same argument gives $\bar{\nabla}\bar{\varphi} = 0$ and the manifold $(W, \bar{g}_1 = e^{2c_1} g)$ admits a parallel spinor. Therefore if $\mu_1 = 0$, there is a metric $\bar{g}_1$ such that all the harmonic spinors of $(W,\bar{g}_1)$ are parallel spinors.

9 - Harmonic forms and Killing spinor

We have the following :

Theorem - let (W,g) be a compact spin manifold of dimension $n \geq 3$ admitting a non trivial Killing spinor ψ. If β is a non trivial harmonic k-form ($k \neq 0,n$) of (W,g), the corresponding (1,1) spinor $\varphi(\beta)$ kills ψ.

In fact, let β be a harmonic k-form ($k \neq 0,n$), $\varphi(\beta)$ the corresponding (1,1-spinor :

$$\varphi(\beta) = \frac{1}{k!}\, \beta_{\lambda_1 \cdot\cdot \lambda_k}\, \gamma^{\lambda_1} \cdots \gamma^{\lambda_k}.$$

β being harmonic, $d\beta = \delta\beta = 0$ and thus $P\,\varphi(\beta) = 0$. If ψ is a non trivial Killing spinor ($\nabla^{(\lambda)}\psi = 0$), we have

$$P(\varphi(\beta)\,\psi) = -\frac{\lambda}{n}\gamma^{\alpha}\,\varphi(\beta)\,\gamma_{\alpha}\,\psi$$

A straightforward computation gives

$$\gamma^{\alpha}\,\varphi(\beta)\gamma_{\alpha} = (-1)^{k-1}(n - 2k)\varphi(\beta) \tag{9-1}$$

Therefore :

$$P(\varphi(\beta)\,\psi) = (-1)^{k}\,\lambda(1 - 2k/n)(\varphi(\beta)\,\psi)$$

If $\varphi(\beta)\,\psi \not\equiv 0$, P admits the eigenvalue $\lambda' = (-1)^k\,\lambda\,(1 - 2k/n)$. But $|\lambda'| < |\lambda|$ and we know that it is impossible. It follows $\varphi(\beta)\,\psi \equiv 0$.

IV - Parallel forms and Killing spinor in the general case.

10 - A parallel form kills a Killing spinor

Let (W,g) be a spin manifold of dimension $n \geq 2$ admitting a non trivial Killing spinor ψ ($\nabla^{(\lambda)}\psi = 0$). If β is a parallel k-form ($k \neq 0,n$), $\varphi(\beta)$ is also parallel and

$$\nabla_{\alpha}(\varphi(\beta)\,\psi) = -\frac{\lambda}{n}\,\varphi(\beta)\,\gamma_{\alpha}\,\psi$$

It follows :

$$-\nabla^{\alpha}\,\nabla_{\alpha}(\varphi(\beta)\,\psi) = \frac{\lambda^2}{n}\,\varphi(\beta)\,\psi$$

We set $\chi = \varphi(\beta)\,\psi$. If $\Delta = P^2$, we have

$$\Delta\,\chi = (\frac{\lambda^2}{n} + \frac{R}{4})\,\chi \qquad \lambda^2 = (n/4(n-1))\,R \tag{10-1}$$

But we deduce from (9-1) that $P\,\chi = \lambda'\,\chi$ and so $\Delta\,\chi = \lambda'^2\,\chi$ ($\lambda'^2 = (1 - 2k/n)^2\lambda^2$).

If $\chi \neq 0$ we have :

$$\lambda'^2 = \frac{\lambda^2}{n} + \frac{R}{4} = \frac{R}{4(n-1)} + \frac{R}{4} = \frac{n\,R}{4(n-1)} = \lambda^2$$

and so $(1-2k/n)^2 = 1$, impossible for $k \neq 0,n$. It follows $\varphi(\beta)\psi = 0$.

11 - Parallel forms and Killing spinor.

In our situation we have

$$(1/k!)\,\beta_{\lambda_1 .. \lambda_k}\,\gamma^{\lambda_1} .. \gamma^{\lambda_k}\,\psi = 0$$

It follows by differentiation :

$$\beta_{\lambda_1 ... \lambda_k}\,\gamma^{\lambda_1} ... \gamma^{\lambda_k}\gamma^{\alpha}\,\psi = 0$$

and so :

$$(-1)^k\,\beta_{\lambda_1 .. \lambda_k}\,\gamma^{\alpha}\,\gamma^{\lambda_1} .. \gamma^{\lambda_k}\,\psi - 2\,k\,\beta_{\lambda_1 .. \lambda_{k-1}\,\alpha}\,\gamma^{\lambda_1} ... \gamma^{\lambda_{k-1}}\psi = 0$$

where the first term of the left member is null. We have :

$$\beta_{\alpha_1 \lambda_2 .. \lambda_k}\,\gamma^{\lambda_2} .. \gamma^{\lambda_k}\,\psi = 0$$

After new differentiations, we get

$$\beta_{\alpha_1\,\alpha_2 ...\,\alpha_{k-1}\,\lambda_k}\,\gamma^{\lambda_k}\psi = 0$$

and so $\beta = 0$.

<u>Theorem - Let (W,g) be an arbitrary spin manifold of dimension $n \geq 2$ admitting a non trivial Killing spinor. There are not non trivial parallel k-forms ($k \neq 0,n$) on (W,g). In particular such a manfold is necessarily irreducible and non-kählerian.</u>

V - Isometries and Killing spinors

12 - A formula

a) Let (W,g) be a spin manifold admitting an infinitesimal isometry $X \neq 0$ and a Killing spinor $\psi \neq 0$. The Lie derivative of a spinor field by an infinitesimal isometry is given by :

$$(12\text{-}1) \qquad \mathcal{L}(X)\,\psi = X^{\rho}\,\nabla_{\rho}\,\psi - \frac{1}{4}\nabla_{\alpha}X_{\beta}\,\gamma^{\alpha}\,\gamma^{\beta}\,\psi$$

We set :

$$\chi = X^{\rho}\,\gamma_{\rho}\,\psi \qquad\qquad \varphi = \frac{1}{4}\nabla_{\alpha}\,X_{\beta}\,\gamma^{\alpha}\,\gamma^{\beta}\,\psi$$

If ψ is a Killing spinor ($\nabla^{(\lambda)}\psi = 0$), $\mathcal{L}(X)\,\psi$ is also a Killing spinor and we have :

$$(12\text{-}2) \qquad \mathcal{L}(X)\,\psi = -\,(\lambda/n)\,\chi - \varphi$$

A straightforward computation gives :

$$(12\text{-}3) \qquad P\,\chi = -\,\lambda(1 - \frac{2}{n})\chi + 4\,\varphi$$

We deduce from (12-2), (12-3) the formula :

$$(12\text{-}4) \qquad \mathcal{L}(X)\,\psi = -\,\frac{1}{4}\,(P - \nu)\,\chi \qquad\qquad \nu = -\,\lambda(1 + 2/n)$$

$\mathcal{L}(X)\psi$ being a Killing spinor, $(P - \lambda)\mathcal{L}(X)\,\psi = 0$ and we have :

$$(12\text{-}5) \qquad (P - \nu)(P - \lambda)\chi = 0 \qquad\qquad \chi = X^{\rho}\,\gamma_{\rho}\,\psi$$

b) Suppose that W is compact. It follows from (12-4) that if $\lambda = 0$, that is if ψ is a parallel spinor, it is invariant under X [9].

For any λ, consider the infinitesimal isometry $U^{\alpha} = i\,\tilde{\psi}\,\gamma^{\alpha}\,\psi$ and the corresponding $\chi = U^{\alpha}\,\gamma_{\alpha}\,\psi$; we have $i\,\tilde{\psi}\,\chi = U^{\alpha}\,U_{\alpha}$ and it follows from (12-4) :

$$\tilde{\psi}\,\mathcal{L}(U)\,\psi = -\,\frac{1}{4}\,\tilde{\psi}\,P\,\chi + \frac{i}{4}\,\lambda(1 + \frac{2}{n})\,U^{\alpha}\,U_{\alpha}$$

We obtain by integration :

$$<\mathcal{L}(U)\,\psi,\,\psi> = -\,\frac{1}{4} < P\chi,\psi > + \frac{i}{4}\,\lambda(1 + \frac{2}{n}) < U,U >$$

But $< P\chi,\psi > = < \chi,P\psi > = \lambda< \chi,\psi > = -\,i\,\lambda< U,U >$. It follows :

$$(12\text{-}6) \qquad <\mathcal{L}(U)\psi,\psi > = (i/2)\,\lambda(1 + \frac{1}{n})\,<U,U>$$

Therefore, if ψ is non trivial and for $U \neq 0$, ψ is not invariant by U.

13 - A new eigenvalue for P.

Let (W,g) be a compact spin manifold having the non trivial Killing spinor ψ. It follows from (12-5) that if $(P-\lambda)\chi \not\equiv 0$ ($\chi = X^\rho \gamma_\rho \psi$), ν is an eigenvalue of P.

Suppose $(P-\lambda)\chi \equiv 0$; we deduce from (12-4) :

(13-1) $$\mathcal{L}(X)\psi = \frac{1}{4}(\nu-\lambda)\chi = -\frac{1}{2}\lambda(1+\frac{1}{n})\chi$$

We know that :

$$(\mathcal{L}(X)\psi, \mathcal{L}(X)\psi) = \text{const.}$$

If X has a zero on W, χ and thus $\mathcal{L}(X)\psi$ have zeros.We see that $\mathcal{L}(X)\psi \equiv \chi \equiv 0$; therefore X is necessarily null. We have :

Theorem - Let (W,g) be a compact spin manifold of dimension $n \geq 2$ admitting a non trivial Killing spinor ψ ($\nabla^{(\lambda)}\psi = 0$) and a non trivial infinitesimal isometry X having a zero on W. The Dirac operator admits the eigenvalue $\nu = -\lambda(1+2/n)$. It is in particular the case if W has a Euler-Poincaré characteristic $\neq 0$.

For example the sphere $S^{2\nu}$ has a characteristic equal to 2 and non trivial Killing spinors corresponding to

$$\lambda = \pm (n/2\sqrt{n(n-1)})\sqrt{R}$$

The corresponding Dirac operator admits the eigenvalue λ and ν with

$$\nu = \pm((n+2)/2\sqrt{n(n-1)})\sqrt{R}$$

14 - Killing spinor and spheres

Let (W,g) be a compact spin manifold of dimension $n = 2\nu$ admitting a non trivial Killing spinor ψ. A straightforward computation shows that the corresponding scalar v satisfies :

(14-1) $$\Delta v = (R/(n-1))v$$

According to Yano-Obata or myself, any compact Einstein manifold admitting a function $v \not\equiv 0$ satisfying (14-1) coincides with (S^n, can). We thus have the well known proposition [3] [4] [7] :

<u>Proposition - Let (W,g) be a compact spin manifold of dimension $n = 2\nu$ admitting a non trivial Killing spinor ψ. Either $v = 0$ or $(W,g) = (S^{2\nu}, can.)$.</u>

In particular for $n = 4$, V is $\neq 0$ and so $v \not\equiv 0$; the spin manifold coincides with S^4. [4] [6] [7].

References

|1| T. AUBIN The scalar curvature, Cahen and Flato (eds) Differential Geometry and Relativity, Reidel Dordrecht Holland, p 5-8 (1976).

|2| R. D'AURIA, P. FRE, P. van NIEUWENHUIZEN, Phys. Lett. 122B, p 225, (1983).

|3| B. BRIAN, P. SPINDEL, C.E.R.N. preprint TH 4313/85 (to be published in Nucl. Phys.)(1985).

|4| M. CAHEN, preprint 1986 (in collab. with S. GUTT, L. LEMAIRE, P. SPINDEL).

|5| T. FRIEDRICH, Math. Nachr. 97, p 117-146, (1980).

|6| T. FRIEDRICH, Math. Nachr. 102, p 53-56, (1981).

|7| O. HIJAZY, A conformal lower bound for the smallest eigenvalue of the Dirac operator and Killing spinors, Max. Planck Institute für Mathematik Bonn MPI/SFB 85-29 (1986).

|8| N. HITCHIN, Compact four-dimensional Einstein manifolds, J. of Diff. Geom., 9, p 435-441 (1974).

|9| A. LICHNEROWICZ, Spineurs harmoniques, C.R. Acad. Sci. Paris A-B, 257, p. 7-9, (1963).

|10| J.L. KAZDAN - F.W. WARNER, J. of Diff. Geom. 10, p 113-134, (1975).

Some references in Mathematical Physics (concerning the introduction)

|H| L.P. HUGHSTON, P. PENROSE, P. SOMMERS, M. WALLAR, Comm. Math. Phys. 27, p 303, (1982).

|C| E. CREMMER, B. JULIA, J. SCHEREK, Phys. Lett. 76B, p 409, (1978).

|Ch| A.H. CHAMSEDINE, Nucl. Phys. B 185, p 403, (1981).

|Be| E. BERGSHOEFF, M. de ROO, B. de WIT, P. van NIEUWENHUIZEN, Nucl. Phys. B, 195, p 97, (1982).

|D| M.J. DUFF, C.N. POPE in Supergravity 81 (ed. S. Ferrara, J.G. Taylor) Cambridge U.P. London (1982); M.J. DUFF, F.J. TOWES in Unification of the fundamental Interactions II (ed. S. Ferrara, J. Ellis) Plenum NY (1982).

|Ca| P. CANDELAS, G.T. HOROWITZ, A. STROMINGEN, E. WITTEN, Nucl. Phys. B 258, p 46 (1985).

|Bi| B. BIRAN, F. ENGLERT, B. de WIT, H. NICOLAI, Phys. Lett. 124 B, p 45, (1983).

Self-duality Conditions Satisfied by the Spin Connections of Spheres

JH Rawnsley

1. Introduction

The spinor connections in the half-spin bundles over S^4 are $\pm$self-dual solutions of the Yang-Mills equations of smallest nontrivial Pontrjagin number. When expressed in coordinates they give the original BPST instanton on R^4. See Atiyah et al. [1] for details. Trautman [9] considers these connections from the point of view of the Hopf fibration $S^7 \longrightarrow S^4$ and shows that the other Hopf fibrations $S^3 \longrightarrow S^2$ and $S^1 \longrightarrow S^1$ correspond with fundamental solutions of the other field theories in lower dimensions.

In a recent paper, Grossman et al. [4] asked if there was a similar interpretation for the remaining Hopf bundle $S^{15} \longrightarrow S^8$. Working in coordinates they obtain a solution of the Yang-Mills equations on R^4 which satisfies a generalised self-duality condition of the form

$$*F\wedge F \;=\; \pm F\wedge F.$$

This is a conformally invariant equation in 8 dimensions and so holds on S^8; on the other hand the Yang-Mills equations are only conformally invariant in 4 dimensions, so their connection does not necessarily satisfy the Yang-Mills equations when it is transferred to S^8. In fact G Landi [5] shows that the Yang-Mills equations are still satisfied. It is the purpose of this lecture to discuss these questions in a more general setting on S^4n and obtain in a coordinate free way the BPST solution for n = 1 and the GKS solution for n = 2.

Various authors [2,3,6,7,8] have considered higher dimensional analogues of self-duality as these are important in theories of dimensional reduction. The fields we consider are absolute minima of a functional of the kind studied by Tchrakian, and are a special case of his spherically symmetric solutions [8]. Our method differs from his in that we shall do the calculations using Clifford algebras and spinors as befits a lecture in this meeting.

2. Clifford Algebras and Spin Representations

Let V be a real inner product space of dimension n. Its Clifford algebra C(V) is the associative real algebra with 1 generated by V with the relations

$$v^2 = -|v|^2 1, \quad \forall \; v \in V.$$

If $e_1, \ldots, e_n$ is an orthonormal basis of V, then $e_{i_1} e_{i_2} \ldots e_{i_k}$, $i_1 < i_2 < \ldots < i_k$ are a basis for a subspace $C^k(V)$ of dimension $\binom{n}{k}$ and if $C^0(V) = \mathbb{R}\,1$, $C^1(V) = V$,

$$(2.1) \qquad C(V) = C^0(V) + C^1(V) + \ldots + C^n(V)$$

is a direct sum of so $\dim C(V) = 2^n$. $C^k(V)$ is independent of the choice of basis. If V is oriented, we let

$$\varepsilon_n = e_1 \ldots e_n \in C^n(V)$$

for an oriented orthonormal basis. ε_n is independent of the choice of such a basis and satisfies

$$(2.2) \qquad \varepsilon_n^2 = (-1)^{\frac{n(n+1)}{2}} 1, \quad \varepsilon_n v = (-1)^{n-1} v\, \varepsilon_n, \quad v \in V.$$

Multiplication by ε_n sends $C^k(V)$ to $C^{n-k}(V)$, which is analogous to the Hodge * operator on the exterior algebra.

The Clifford algebra is not $\mathbb{Z}$-graded as an algebra, but if we set

$$C_0(V) = \sum_{k \geq 0} C^{2k}(V), \quad C_1(V) = \sum_{k \geq 0} C^{2k+1}(V)$$

then $C(V) = C_0(V) + C_1(V)$ gives C(V) the structure of a $\mathbb{Z}_2$-graded algebra. The Clifford algebra of a direct sum $V \oplus W$ is easily seen to be the $\mathbb{Z}_2$-graded tensor product $C(V) \otimes C(W)$; sometimes we can replace this by the ordinary tensor product. For example, we see from (2.2) that ε_n anticommutes with V if dim V is even. Thus for dim V even, we may define a map $V \oplus W \to C(V) \otimes C(W)$ by

$$(2.3) \qquad (v,w) \mapsto v \otimes 1 + \varepsilon_n \otimes w.$$

Then

$$(v \otimes 1 + \varepsilon_n \otimes w)^2 = v^2 \otimes 1 + v\varepsilon_n \otimes w + \varepsilon_n v \otimes w + \varepsilon_n^2 \otimes w^2$$
$$= -|v|^2 \, 1 \otimes 1 - |w|^2 \varepsilon_n^2 \otimes 1.$$

If $n = 2k$, $\varepsilon_n^2 = (-1)^k$, so if k is also even, (n a multiple of four) then

$$(v \otimes 1 + \varepsilon_n \otimes w)^2 = -(|v|^2 + |w|^2)\, 1 \otimes 1$$

showing $C(V \oplus W) \quad C(V) \otimes C(W)$, the isomorphism being induced by the map (2.3).

Any orthogonal transformation of V generates a canonical automorphism of C(V) and also a canonical antiautomorphism. We denote by $a \mapsto a^*$ the antiautomorphism which satisfies

$$v^* = -v, \qquad v \in V.$$

Then

$$\varepsilon_n^* = (-1)^{\frac{n(n+1)}{2}} \varepsilon_n.$$

Let Rea denote the component of a in $C^0(V)$ in the direct sum (2.1), then

$$\mathrm{Re}(ab) = \mathrm{Re}(ba)$$

and

$$(a,b) = \mathrm{Re}(a^* b)$$

gives C(V) a positive definite inner product for which 1, e_{i_1} ... e_{i_k} , $i_1 < \ldots < i_k$ form an orthonormal basis.

In C(V) let

$$[a,b] = ab - ba$$

then $C^2(V)$ is closed under this bracket, so forming a Lie algebra and $[C^2(V), C^1(V)] \subset C^1(V)$. Then

$$\dot{S}(a)v = [a,v], \qquad a \in C^2(V), \quad v \in C^1(V)$$

gives a homomorphism of Lie algebras

$$\dot{s} : C^2(V) \longrightarrow so(V).$$

The group of units of C(V) generated by

$$\{e^a : a \in C^2(V)\}$$

is the spin group Spin(V) and

$$s(g)v \;=\; g\,v\,g^{-1} \qquad g \in \mathrm{Spin}(V), \quad v \in C^1(V)$$

gives the spin covering map

$$s \;:\; \mathrm{Spin}(V) \longrightarrow SO(V),$$

Clearly $\mathrm{Spin}(V) \subset C_0(V)$.

A Clifford module is a (left) module M for the Clifford algebra. It is known that for dim V even, C(V) has, up to isomorphisms, a unique non-trivial simple Clifford module which we denote by S, called the spin module. For dim V divisible by four, $\varepsilon_n^2 = 1$, and commutes with $C_0(V)$, so under $C_0(V)$, S splits into two submodules $S_+ + S_-$ where

$$\varepsilon_n \;=\; \pm 1 \quad \text{on } S_{\pm}.$$

$S_{\pm}$ are irreducible representations of Spin(V). On the other hand in odd dimensions $C_0(V)$ has a unique non-trivial simple module which is irreducible for Spin(V).

Let W have odd dimension, $e_0 \in W$ a unit vector and $V = e_0^{\perp}$. Any element $a \in C_0(W)$ can be written

$$a \;=\; a_0 + e_0 a_1$$

with $a_0, a_1 \in C(V)$, so we get a map to C(V) via

$$a \mapsto a_0 + a_1 \;:\; C_0(W) \longrightarrow C(V).$$

This is known to be an isomorphism of algebras.

We let C(k) denote $C(\mathbb{R}^k)$, Spin(k) denote $\mathrm{Spin}(\mathbb{R}^k)$ with the standard inner product, then we have $C_0(4k + 1) \simeq C(4k)$, and the spin module S of C(4k) gives the spin representation of Spin(4k + 1) and this splits into two irreducibles $S_{\pm}$ when restricted to Spin(4k) given by the eigenspaces of ε_{4k}.

3. Spin Bundles of Spheres

The kernels of the covering maps $\mathrm{Spin}(k + 1) \longrightarrow SO(k + 1)$

and Spin(k) $\rightarrow$ SO(k) coincide, so

$$\mathrm{Spin}(k+1)/\mathrm{Spin}(k) \;=\; SO(k+1)/SO(k) \;=\; S^k.$$

The fibring

(3.1) $\quad \mathrm{Spin}(k+1) \rightarrow \mathrm{Spin}(k+1)/\mathrm{Spin}(k) \;=\; S^k$

is then spin bundle of S^k. The Maurer-Cartan form $\Theta = g^{-1}dg$ has values in the Lie algebra $C^2(k+1)$. Let e_0 be the base-point in S^k, so (3.1) is given by

(3.2) $\quad g \mapsto g e_0 g^{-1}.$

The isomorphism $C_0(k+1) \simeq C(k)$ leads to

$$C^2(k+1) \simeq C^2(k) + C^1(k)$$

with

$$a + e_0 b \mapsto a + b$$

and so

$$\Theta = \alpha + e_0\beta$$

where α is $C^2(k)$ - valued, which, being the Lie algebra of Spin(k), we conclude that α is the connection form of the spin connection in the principal Spin(k) bundle (3.1). β on the other hand is the soldering form. It has values in $\mathbb{R}^k$, and relative to an orthonormal basis for $\mathbb{R}^k$, its components project to an orthonormal coframe for S^k.

The Maurer-Cartan form Θ satisfies

$$d\Theta + \Theta \wedge \Theta \;=\; 0$$

which implies

$$d\alpha + e_0 d\beta + \alpha\wedge\alpha + \alpha\wedge e_0\beta + e_0\beta\wedge\alpha + e_0\beta \wedge e_0\beta \;=\; 0$$

or

$$d\alpha + \alpha\wedge\alpha + \beta\wedge\beta + e_0(d\beta + [\alpha\wedge\beta]) \;=\; 0.$$

Thus

$$d\alpha + \alpha\wedge\alpha = -\beta\wedge\beta \quad , \qquad d\beta + [\alpha\wedge\beta] = 0.$$

The second of these equations shows that the connection defined by α is, in fact, torsion-free, whilst the first computes its curvature as $F = -\beta\wedge\beta$.

Proposition 1. Suppose K = 4n. If $\underline{S}_\pm$ are the vector bundles associated to the fibration (3.1) by the spin representations $S_\pm$, then the curvatures $F_\pm$ of the connections which α induces in these bundles satisfy

$$*(\overbrace{F_\pm\wedge\cdots\wedge F_\pm}^{n}) = \pm(-1)^n F_\pm\wedge\cdots\wedge F_\pm$$

Proof. The * operator acts on horizontal forms on Spin(4n + 1) and if $e_1,\dots,e_{4n}$ is a fixed oriented orthornormal frame of $\mathbb{R}^{4n}$,

$$\beta = \beta_1 e_1 + \dots + \beta_{4n} e_{4n}$$

defines a coframe $\beta_1,\dots,\beta_{4n}$ for S^{4n}. $\beta_1\wedge\cdots\wedge\beta_{4n}$ is the Riemannian volume form, and so

$$*\beta_{j_1}\wedge\cdots\wedge\beta_{j_p} = \pm\beta_{i_1}\wedge\cdots\wedge\beta_{i_{4n-p}}$$

where $\{j_1,\dots,j_p\}\cup\{i_1,\dots,i_{4n-p}\} = \{1,\dots,4n\}$ and the sign is the sign of the permutation $(j_1,\dots,j_p,i_1,\dots,i_{4n-p})$ of $1,\dots,4n$. It is easy to see we also have

$$e_{j_1}\cdots e_{j_p} e_{i_1}\cdots e_{i_{4n-p}} = \pm\varepsilon_{4n}$$

with the same choice of sign, so

$$*\beta_{j_1}\wedge\cdots\wedge\beta_{j_p} e_{j_1}\cdots e_{j_p} =$$

$$= (-1)^{\frac{p(p-1)}{2}}\beta_{i_1}\wedge\cdots\wedge\beta_{i_{4n-p}} e_{i_1}\cdots e_{i_{4n-p}}\varepsilon_{4n}.$$

Thus if $\beta^p = \overbrace{\beta\wedge\cdots\wedge\beta}^{p}$ then

$$
\begin{aligned}
*\beta^p &= *\sum_{j_1,\dots,j_p} \beta_{j_1}\wedge\dots\wedge\beta_{j_p}\; e_{j_1}\dots e_{j_p} \\
&= *p!\sum_{j_1<\dots<j_p} \beta_{j_1}\wedge\dots\wedge\beta_{j_p}\; e_{j_1}\dots e_{j_p} \\
&= p!(-1)^{\frac{p(p-1)}{2}} \sum_{i_1<\dots<i_{4n-p}} \beta_{i_1}\wedge\dots\wedge\beta_{i_{4n-p}}\, e_{i_1}\dots e_{i_{4n-p}}\, \varepsilon_{4n} \\
&= \frac{p!}{(4n-p)!}(-1)^{\frac{p(p-1)}{2}} \sum_{i_1\dots i_{4n-p}} \beta_{i_1}\wedge\dots\wedge\beta_{i_{4n-p}}\, e_{i_1}\dots e_{i_{4n-p}}\, \varepsilon_{4n} \\
&= \frac{p!}{(4n-p)!}(-1)^{\frac{p(p-1)}{2}} \beta^{4n-p}\, \varepsilon_{4n}.
\end{aligned}
$$

Hence

$$*\beta^{2n} = (-1)^n \beta^{2n} \varepsilon_{4n}$$

and

$$*F^n = *(-1)^n \beta^{2n} = \beta^{2n}\varepsilon_{4n} = (-1)^n F^n \varepsilon_{4n}\,.$$

Since $\varepsilon_{4n} = \pm 1$ on $S_\pm$, we get the result.

For $n = 1$, we get $*F_- = F_-$ giving the well known result that the bundle $\underline{S}_-$ has a self-dual connection.

For $n = 2$ the identity $*F_\pm \wedge F_\pm = \pm F_\pm \wedge F_\pm$ was established for $\mathbb{R}^8$ in [4] using octonions. The above gives an alternative proof since the equation is conformally invariant.

Remark 2. On S^{4n}, $F_\pm$ minimize the functional

$$L(A) = \int_{S^{4n}} |\overbrace{F\wedge\dots\wedge F}^{n}|^2 \text{ vol}$$

for, if $(.)^\pm$ denote the self-dual and antiself dual components of a 2n-form, then

$$L(A) = \int |(\wedge^n F)^+|^2 + |(\wedge^n F)^-|^2 \text{ vol}$$

whilst

$$\int_{S^{4n}} \mathrm{Tr}(\wedge^{2n} F) = \int_{S^{4n}} |(\wedge^n F)^+|^2 - |(\wedge^n F)^-|^2 \text{ vol}$$

and this latter is a topological invariant.

<u>Remark 3.</u> One can show by essentially the same calculations as in Proposition 1 that on S^{2n} the spin curvature satisfies (as a Clifford bundle valued 2-form)

$$*F^p = (-1)^p 2p!/(2n - 2p)!\ F^{n-p}\varepsilon_{2n}.$$

From this it follows , by taking p = 1, that $D*F = 0$ giving Landi's result (5).

To conclude this section we work out the expression for the connection α in suitable coordinates. This allows us to compare it with that of Grossman et al. in dimension 8. We do the calculation in the Clifford algebra; it will then be valid in any Clifford module, and in particular in the spin representations.

We write vectors on S^m in the form $x_o e_o + x$, with $x_o \in \mathbb{R}$, $x \in \mathbb{R}^m$ (identified with $e_o^{\perp}$). The open subset U of S^m where $x_o \neq -1$ can be identified with $\mathbb{R}^m$ by stereographic projection:

$$z = x/(x_o + 1),$$

with inverse map

$$x_o e_o + x = ((1 - z^2)e_o + 2z)/(1 + z^2).$$

Over U the bundle $\mathrm{Spin}(m + 1) \to S^m$ can be trivialized by a section $g : \mathbb{R}^m \to \mathrm{Spin}(m + 1)$, which therefore satisfies

$$g(z)e_o g(z)^{-1} = ((1 - z^2)e_o + 2z)/(1 + z^2).$$

One choice of such a section is

$$g(z) = (1 + e_o z)/\sqrt{(1 + |z|^2)}$$

as is easily verified. Now $\Theta = g^{-1}dg$, so α is given by the part of

$$(1-e_o z)/\sqrt{1+|z|^2}\ .\left\{e_o dz/\sqrt{1+|z|^2} - \tfrac{1}{2}(1+e_o z)d|z|^2/(1+|z|^2)^{\frac{3}{2}}\right\}$$

not involving e_o. This is

$$= -\tfrac{1}{2}d|z|^2/(1+|z|^2)^2 - zdz/(1+|z|^2) + \tfrac{1}{2}|z|^2 d|z|^2/(1+|z|^2)^2.$$

Remembering that $z^2 = -|z|^2$, so $-d|z|^2 = zdz + dz.z$ this simplifies to

$$\alpha = \frac{[dz,z]}{2(1+|z|^2)}\ .$$

If we expand $z = z^{\mu}e_{\mu}$, and put $\Sigma_{\mu\nu} = \frac{1}{4}[e_{\mu}, e_{\nu}]$, this becomes

$$= \frac{4z^{\nu}dz^{\mu}\Sigma_{\mu\nu}}{2(1+|z|^2)} .$$

so

$$(3.3) \qquad \alpha_{\mu} = \frac{2\Sigma_{\mu\nu}z^{\nu}}{(1+|z|^2)} .$$

This is essentially the formula of Grossman et al., taking into account the mathematical convention of using skew hermitian operators for elements of Lie algebras : $\alpha_{\mu} = i A_{\mu}$, where A_{μ} is self-adjoint. Thus (3.3) defines a connection on S^m for any m, and when m = 4n the curvature satisfies the generalized (anti) self-duality condition of Proposition 1 in any spin representation.

References

1. Atiyah, MF; Hitchin, NJ; Singer, IM Self-duality in four-dimensional Riemannian geometry. Proc. Roy. Soc. Lond. A362 (1978) 425-461.

2. Bais, FA: Batenburg, P Yang-Mills duality in higher dimensions. Nucl. Phys. B269 (1986) 363-388.

3. Fairlie, DB; Nuyts, J Integration conditions for first-order differential equations in higher-dimensional gauge theories. J. Math. Phys. 25 (1984) 2025-2027.

4. Grossman, B; Kephart, TW; Stasheff, JD Solutions to Yang-mills equations in eight dimensions and the last Hopf m . Commun. Math. Phys. 96 (1984) 431-437.

5. Landi, G The natural spinor connection on S^8 is a gauge field. Let. Math. Phys. 11 (1986) 171-175.

6. Tchrakian, DH Self-dual gauge field equations on N-dimensional manifolds. Quantum Theory, Fields and Particles. Ed. AO Barut. Reidel (1983), pp 161-177.

7. Tchrakian, DH N-dimensional instantons and monopoles. J. Math. Phys. 21 (1980) 166-169.

8. Tchrakian, DH Spherically symmetric gauge field configurations with finite action in 4p-dimensions (p=integer). Phys. Lett. 150B (1985) 360-362.

9. Trautman, A Solutions of Maxwell and Yang-Mills equations associated with Hopf fibrings. Int. J. Theor. Phys. 16 (1977) 561-565.

Maslov Index and half-forms

M. Cahen
Université Libre de Bruxelles

In this paper we show that the intertwining operators between irreducible unitary representations of the Heisenberg group associated to the same central character and corresponding to transverse lagrangian subspaces, as obtained by the methods of geometric quantization with half-forms, have a transitive composition law.

1. Let $(\mathbb{R}^{2n} (= \mathbb{R}^n \times \mathbb{R}^n), \omega)$ be the standard symplectic vector space; if (x,y), $(x',y') \in \mathbb{R}^{2n}$:

$$\omega((x,y),(x',y')) = x.y' - y.x'$$

where $u.v$ denotes the usual scalar product in $\mathbb{R}^n$.

The Heisenberg Lie algebra $\mathcal{H} = \mathbb{R}^{2n} \oplus \mathbb{R} E$ has multiplication given by :

$$[X + sE, Y + tE] = \omega(X,Y)E \qquad X,Y \in \mathbb{R}^{2n}, \quad s,t \in \mathbb{R}$$

so that E generates the center of $\mathcal{H}$.

The Heisenberg group $H = \mathbb{R}^{2n} \times \mathbb{R}$ has group multiplication given by :

$$(a,s)(a',s') = (a+a', s+s' + \tfrac{1}{2}\omega(a,a')) \qquad a, a' \in \mathbb{R}^{2n}, \quad s, s' \in \mathbb{R}.$$

The exponential map $\mathcal{H} \to H$ is the identity map :

$$\exp(a + sE) = (a, s)$$

Let ℓ be a Lagrangian subspace of $(\mathbb{R}^{2n}, \omega)$ (i.e. ℓ has dimension n and $\omega(X,Y) = 0$, $\forall X,Y \in \ell$) ; the subspace $\ell + \mathbb{R}E$ is an abelian subalgebra of $\mathcal{H}$ and we shall denote by $L = \exp(\ell + \mathbb{R}E)$ the corresponding abelian subgroup of H.

Let χ_c be a character of the center of H :

$$\chi_c(\exp tE) = e^{2\pi ict}$$

We shall assume from now on that $c \in \mathbb{R}^*$. This character extends to a character of L :

$$\chi_c(\exp(X + tE)) = e^{2\pi ict} \qquad X \in \ell$$

Let ρ_ℓ^c denote the unitary representation of H induced by this character of L ; it is realised on the Hilbert space $\mathcal{H}_\ell$ which is the completion (with respect to the norm) of the space $\tilde{\mathcal{H}}_\ell$:

$$\tilde{\mathcal{H}}_\ell = \left\{ \varphi : H \to \mathbb{C} \;\middle|\; \begin{array}{l} \text{(i) } \varphi \text{ is } C^\infty \\ \text{(ii) } \varphi(hk) = \chi_c(k^{-1})\varphi(h) \quad \forall h \in H, \forall k \in L \\ \text{(iii) } \displaystyle\int_{H/L} |\varphi(\dot{n})|^2 \, d\dot{n} < \infty \end{array} \right\}$$

Observe that condition (iii) makes sense because of condition (ii) ; we have denoted $d\dot{n}$ an H invariant measure on H/L. The representation ρ_ℓ^c reads :

$$(\rho_\ell^c(g)\varphi)(h) = \varphi(g^{-1}h) \qquad \forall g, h \; H .$$

<u>Theorem</u> (Stone Von Neuman) . The induced representation ρ_ℓ^c <u>is an irreducible unitary representation of</u> H . <u>Furthermore every irreducible unitary representation</u> R <u>of</u> H <u>such that</u>

$$R(\exp sE) = e^{2\pi ics} I$$

<u>is unitarily equivalent to</u> ρ_ℓ^c.

In particular, given $c \neq 0$, given two lagrangian subspaces ℓ_1 and ℓ_2, the representations $\rho^c_{\ell_i}$ $(i \leq 2)$ of H on the Hilbert space $\mathcal{H}_{\ell_i}$ $(i \leq 2)$ are equivalent ; hence there exists an isometry :

$$J_{21} : \mathcal{H}_{\ell_1} \to \mathcal{H}_{\ell_2}$$

which is an intertwining operator :

$$J_{21} \circ \rho^c_{\ell_1}(g) = \rho^c_{\ell_2}(g) \circ J_{21} \qquad \forall g \in H$$

The operator J_{21} is determined up to a scalar of module 1.

A classical construction [1] is given by :

$$(J_{21}\varphi)(n) = \int_{L_2/L_1 \cap L_2} \varphi(n\, h_2)\chi^c(h_2)\, d\dot{h}_2 \qquad (*)$$

where $d\dot{h}_2$ denotes a positive, L_2 - invariant measure on $L_2/L_1 \cap L_2$, which is such that J_{21} is an isometry.

A choice of invariant measure is equivalent to the choice of a density of weight 1 at a point and can be realised as follows.

If E is a k-dimensional real vector space, the space $\Lambda^k E$ of k-vectors on is 1-dimensional. Recall that an element $w \in \Lambda^k E^*$ defines a density of weight α $(\alpha \in \mathbb{R})$, denoted $|w|^\alpha$ by :

$$|w|^\alpha(v) = |\langle w, v\rangle|^\alpha \qquad v \in \Lambda^k E.$$

If ℓ is a Lagrangian subspaces of $\mathbb{R}^{2n}$, one can identify ℓ with $(\mathbb{R}^{2n}/\ell)^*$, using the symplectic form ω. An element $e \in \Lambda^n \ell$ defines a volume form, or equivalently a density of weight 1, denoted $|e|$ on $\mathbb{R}^{2n}/\ell$.

If ℓ_1, ℓ_2 are two lagrangian planes, $\ell_1 \cap \ell_2$ is the radical (for ω) of $\ell_1 + \ell_2$ and thus $(\ell_1 + \ell_2)/\ell_1 \cap \ell_2$ has a natural symplectic structure, and a 1-density $|\omega'|$. Given $e_1 \in \Lambda^n \ell_1$ and $e_2 \in \Lambda^n \ell_2$, there exists a unique volume form δ on $\ell_2/\ell_1 \cap \ell_2$ such that :

$$|e_1|^{1/2} \otimes |e_2|^{1/2} \otimes |\omega'|^{1/2} = |e_2| \otimes \delta$$

With the notations above, if $d\dot{n}_1$ is the measure on H/L_1 (defining $\mathcal{H}_{\ell_1}$) corresponding to e_1 ($\in \Lambda^n \ell_1$) and $d\dot{n}_2$ is the measure on H/L_2 corresponding to $e_2(\in \Lambda^n \ell_2)$ we choose the operator J_{21}, defined by (*) with $d\dot{h}_2$ corresponding to δ. Then one has :

<u>Proposition</u>. (Lions) a) J_{21} <u>is an isometry and an interwining operator such that</u> $J_{12} = (J_{21})^{-1}$

b) If ℓ_1, ℓ_2, ℓ_3 <u>are three lagrangian subspaces then</u> :

$$J_{13} \circ J_{32} \circ J_{21} = e^{-\frac{i\pi}{4} \tau(\ell_1, \ell_2, \ell_3)} \, Id|_{\mathcal{H}_{\ell_1}}$$

<u>where</u> $\tau(\ell_1, \ell_2, \ell_3)$ <u>is the Maslov index of the three lagrangian subspaces</u>.

Recall that the Maslov index $\tau(\ell_1, \ell_2, \ell_3)$ is the signature of the quadratic form Q on $\ell_1 \oplus \ell_2 \oplus \ell_3$ defined by :

$$Q(X_1 + X_2 + X_3) = \omega(X_1, X_2) + \omega(X_2, X_3) + \omega(X_3, X_1)$$

Geometric quantization on the space of orbits of H in the dual $\mathcal{H}^*$ of $\mathcal{H}$ provides a realisation of the induced representations ρ^c_ℓ ; the construction of the intertwining operator J_{21} corresponds to the "half densities" setting. We prove in the next paragraph that geometric quantization, in the "half forms" setting gives intertwining operators F_{21} which "avoid the Maslov cocycle". More precisely for transverse lagrangian subspaces ℓ_1, ℓ_2, ℓ_3,

$$F_{13} \circ F_{32} \circ F_{21} = Id|_{\mathcal{H}_{\ell_1}}$$

2. Let $\mathcal{H}^* = \{ (u,v,w) \; ; \; u,v \in \mathbb{R}^n , \;\; w \in \mathbb{R} \}$ be the dual of $\mathcal{H}$. The coadjoint action of the element (x,y,s) of H on $\mathcal{H}^*$ is :

$$(x,y,s) \cdot (u,v,w) = (u + wy, \; v - wx, \; w)$$

Thus an orbit of H in $\mathcal{H}^*$, of maximal dimension is

$$\mathcal{O}_{w_o} = \{ (u,v,w_o) \mid w_o \text{ a constant} \neq 0 \} \sim \mathbb{R}^{2n}$$

The canonical symplectic form ω on $\mathcal{O}_{w_o}$ is given by

$$\omega(X^*, Y^*) = \langle \xi , [Y , X] \rangle \qquad \xi \in \mathcal{O}_{w_o} \; ; \; X , Y \in \mathcal{H}$$

where :

$$X^*_\xi = \frac{d}{dt} (\exp - tX . \xi) \Big|_{t = 0}$$

Observe that, as homogeneous space $\mathcal{O}_{w_o} = H/_{\exp \mathbb{R}\varepsilon}$; we shall denote by $\pi : H \to \mathcal{O}_{w_o}$ the corresponding projection.

If we introduce coordinates

$$p = - \frac{v}{w_o} \qquad\qquad q = \frac{u}{w_o}$$

on $\mathcal{O}_{w_o}$ one has :

$$(u, v, w_o) = (p, q, o) . (o, o, w_o)$$

$$\omega = -w_o \; dp \quad dq$$

In what follows we shall choose $w_o = -1$, i.e consider only this particular orbit. Remark that the choice of coordinates (p,q) on $\mathcal{O} (= \mathcal{O}_{-1})$ corresponds to a choice of section σ of H over $\mathcal{O}$:

$$\sigma \; : \mathcal{O} \to H : (u, v, w_o) = (p, q, o) . (o, o, w_o) \longrightarrow (p, q, o)$$

If one applies the methods of geometric quantization to the symplectic manifold $(\mathcal{O} (= \mathbb{R}^{2n}) \; , \; \omega = dp \wedge dq)$ one constructs :

a) a complex line bundle on θ ; this line bundle is trivial and we realise it as :

$$B^c = H \times_{\chi_c} \mathbb{C} \sim \mathbb{R}^{2n} \times \mathbb{C} \xrightarrow{\pi} \mathbb{R}^{2n}$$

A section ψ : $\mathbb{R}^{2n} \longrightarrow B^c$, can be viewed as a function $\hat{\psi}$ on $\mathbb{R}^{2n}$ with complex values or as a function $\tilde{\psi}$ on H , with the obvious equivariance property :

$$\tilde{\psi}(gk) = \chi_c(k^{-1})\,\tilde{\psi}(g) \qquad\qquad k \in \exp \mathbb{R}E$$

The link between these various viewpoints is given by :

$$\psi(\pi g) = [g, \tilde{\psi}(g)] \quad \text{and} \quad \hat{\psi}(x) = \tilde{\psi}(\sigma(x))$$

where $[g,v]$ denotes the equivalence class of (g,v) in $H \times_{\chi_c} \mathbb{C}$.

b) a covariant derivative on B^c , with curvature $\frac{\omega}{i\hbar}$. The connection form α is chosen to be :

$$\alpha = \frac{1}{2i\hbar}(pdq - qdp)$$

where $\hbar = \frac{h}{2\pi}$ = Planck constant $(2\pi)^{-1}$.

The covariant derivative of a section ψ in the direction X , tangent to the orbit θ , reads :

$$\widehat{\nabla_X \psi} = X\hat{\psi} + \alpha(X)\,\hat{\psi}$$

Hence :

$$\widetilde{\nabla_{A\partial_p + B\partial_q} \psi} = R^{*}_{(A,B,o)}\,\tilde{\psi} \qquad \text{if} \quad c = h^{-1}$$

where $R^{*}_{(A,B,C)}$ is the vector field on H , having for value at g :

$$R^{*}_{(A,B,C)}(g) = \left.\frac{d}{dt}\right|_0 g\, \text{expt}(A,B,C)$$

and (A,B,C) denotes an element of $\mathcal{H} = \mathbb{R}^n + \mathbb{R}^n + \mathbb{R}$.

c) an invariant hermitian structure h on B^c. Here it reads, up to multiplication by a constant real factor,

$$h(\psi, \psi')(\xi) = \hat{\psi}(\xi)\ \overline{\hat{\psi}'(\xi)} \qquad \xi \in \theta$$

Indeed, as α is purely imaginary :

$$X(h(\psi, \psi')) = h(\nabla_X \psi, \psi' + h(\psi, \nabla_X \psi')$$

for any vector field tangent to the orbit.

d) an invariant real polarization P, i.e a rank n sublundle of $T\mathbb{R}^{2n}$, invariant by translation and such that $\omega|_{P\times P} = 0$. Such a polarization P corresponds to a lagrangian subspace ℓ of $\mathbb{R}^{2n}$ ($\ell = P_{(o,\, o,\, w_o)}$).

Now if L_g denotes the left translation by g in H :

$$P_{\pi(g)} = \pi_* \circ L_g \ (\ell + \mathbb{R}E)$$

The space $\Gamma^\infty_P(B^c)$ of smooth polarised sections of B^c, i.e the space of sections ψ such that $\nabla_X \psi = 0$, for all $X \in P$, corresponds to the space of functions $\tilde{\psi}$ on H such that :

$$R^*_{(A,B,O)} \tilde{\psi} = 0 \qquad \forall\ (A,B) \in \ell$$

$$R^*_{(o,o,D)} \tilde{\psi} = -2\pi i c D\ \tilde{\psi} \qquad \forall\ D \in \mathbb{R}$$

Such a function $\tilde{\psi} : H \longrightarrow \mathbb{C}$ has the equivariance property of the functions of $\tilde{\mathcal{H}}_\ell$, namely :

$$\tilde{\psi}(hk) = \chi_c(k^{-1})\, \tilde{\psi}(h) \qquad \forall\, h \in H,\ \forall\, k \in L$$

e) a representation of H on $\Gamma^\infty_P(B^c)$ by integrating the representation of $\mathcal{H}$ given by :

$$\frac{1}{i\hbar}\ Q(A, B, D)\,\psi$$

Here $(A, B, D) \in \mathcal{H}$ is considered as a function on $\theta\ (\subset \mathcal{H}^*)$ and :

$$Q(f)\psi = i\hbar\ \nabla_{X_f} \psi + f\psi$$

for a function f such that its hamiltanian vector field $X_f(\, i(X_\xi)\omega = df)$ preserves P ($[X_\xi, P] \subset P$).

Observe that :

$$\widehat{\frac{1}{i\hbar} Q\,(A,\, B,\, D)\psi}\;(p,q) = (-\, A\partial_p - B\partial_q - \frac{1}{2i\hbar}\,(2D + Aq - Bp))\,\hat{\psi}\,(p,q)$$

Thus if $L^*_{(A,\, B,\, D)}$ is the vector field on H given by :

$$L^*_{(A,\, B,\, D)}\,(g) = \frac{d}{dt}\;\exp -t\,(A,\, B,\, D).\, g\,\Big|_{t\,=\,0}$$

one has :

$$\widetilde{\frac{1}{i\hbar}\; Q\,(A,\, B,\, D)\psi} = L^*_{(A,\, B,\, D)}\,\widetilde{\psi} \qquad \text{if } c = h^{-1}$$

This implies that the representation of H on $\Gamma^{\infty}_{P}(B^c)$ coïncides with the representation ρ^c_ℓ on $\widetilde{\mathcal{H}}'_\ell$ ($=\widetilde{\mathcal{H}}_\ell$ without the condition (iii))

f) a bundle of half forms Q ; this is a complex line bundle together with an isomorphism

$$i = Q \otimes Q \longrightarrow \Lambda^n F_o$$

Here F is an invariant positive definite polarization; i.e F is a complex n dimensional subspace of $(\mathbb{R}^{2n}_\xi)^{\mathbb{C}}$ such that $\omega_\xi\big|_{F_\xi \times F_\xi} = 0$ and $i\,\omega_\xi(X,\bar{X}) > 0$ for all $0 \neq X \in F$.

We denote by F_o the bundle of 1-forms vanishing on F and by $\Lambda^n F_o$ its top exterior power.

Using this bundle we can define a Hilbert structure on the space of polarized section of B^c. If P is (as above) an invariant real polarization, let us choose a frame of P_ξ , $\{v_1 \ldots v_n\}$, invariant by translations and such that

$$v_j = u_j + \bar{u}_j \qquad j \leqslant n$$

where $\{u_1 \ldots u_n\}$ is a unitary frameof F :

$$i\omega(u_j, \bar{u}_k) = \delta_{jk}$$

Define the invariant volume form α on $H/_L \sim \mathbb{R}^{2n}/P$ such that :

$$\alpha(w_1, \ldots w_n) = \left|\frac{\omega^n}{n!}(w_1, \ldots, w_n, v_1, \ldots v_n)\right|$$

where $\{w_1, \ldots w_n\}$ is any n-tuple of vectors tangent to $\mathbb{R}^{2n}$.

These various points show that geometric quantization constructs $\mathcal{H}_\ell$, ρ_ℓ^c and a given choice of invariant measure on $H/_L$.

We can now also define a pairing between $\mathcal{H}_{\ell_1}$ and $\mathcal{H}_{\ell_2}$ for two transverse polarizations P_1 and P_2 or equivalently two transverse lagrangian planes ℓ_1 and ℓ_2 [2]. We have considered a unitary frame $\{u_1 \ldots u_n\}$ of the positive polarization F. Let then $\tilde{v}_1^{(i)} \ldots \tilde{v}_n^{(i)}$ $(i \leqslant 2)$ be a frame of P_i $(i \leqslant 2)$ such that :

$$\tilde{v}_k^{(i)} = u_k + \sum_{l=1}^{1} \bar{u}_l \; Z_{(i)k} \qquad k \leqslant n$$

or equivalently $P_F \; \tilde{v}_k^{(i)} = u_k$ where P_F is the projection on F parallely to $\bar{F}$. Then a section $\overset{(1)}{\psi}$ of $\mathcal{H}_{\ell_1}$ is paired with a section $\overset{(2)}{\psi}{}'$ of $\mathcal{H}_{\ell_2}$ by :

$$\langle \overset{(1)}{\psi}, \overset{(2)}{\psi}{}' \rangle_{\mathcal{H}_{\ell_1} \mathcal{H}_{\ell_2}} = \int_{\mathbb{R}^{2n}} \widehat{\overset{(1)}{\psi}} \; \overline{\widehat{\overset{(2)}{\psi}{}'}} \; \det{}^{1/2} (1 - Z_1 \bar{Z}_2) \left|\frac{\omega^n}{n!}\right|$$

The important facor is the square root of the determinant which is uniquely defined if one imposes $\det^{1/2} I = 1$; indeed $I - Z_1 \bar{Z}_2$ belongs to a convex set of $GL(n, \mathbb{C})$. It can be proven that this pairing does not depend on the auxiliary polarization F nor of the unitary frame $\{u_1 \ldots u_n\}$.

This pairing induces an intertwining operator :

$$F_{21} : \mathcal{H}_{\ell_1} \longrightarrow \mathcal{H}_{\ell_2}$$

by :

$$\langle \overset{(1)}{\psi}, \overset{(2)}{\psi'} \rangle_{\mathcal{H}_{\ell_1} \mathcal{H}_{\ell_2}} = \langle F_{21} \overset{(1)}{\psi}, \overset{(2)}{\psi'} \rangle_{\mathcal{H}_{\ell_2}}$$

In the next paragraph we compute the composition of those intertwining operators for three mutually transverse polarizations.

3. Consider three real polarisations P_1, P_2, P_3 on $\mathbb{R}^{2n}$ invariant by tranlations and mutually transverse and the corresponding lagrangian subspaces ℓ_1, ℓ_2, ℓ_3.

<u>Lemma</u> [I] : <u>There exists a symplectic basis</u> $\{ e_j, f_j ; j \leqslant n \}$ of $\mathbb{R}^{2n}$ <u>and an integer</u> k $(0 \leqslant k \leqslant n)$ <u>so that if</u> $x = \sum_j (p^j e_j + q^j f_j)$ for $x \in \mathbb{R}^{2n}$ <u>the polarizations</u> P <u>are spanned by</u> :

$$P_1 = > \partial_{p^j} \qquad j \leqslant n <$$

$$P_2 = > \partial_{q^j} \qquad j \leqslant n <$$

$$P_3 = > \partial_{p^j} + \epsilon_j \partial_{q^j} \qquad j \leqslant n <$$

<u>where</u> $\epsilon_j = 1$ <u>if</u> $j \leqslant k$ <u>and</u> $\epsilon_j = -1$ <u>if</u> $j > k$.

Consider an auxiliary positive definite polarization F spanned by $F = > \partial_{\bar{z}^j} < = > 1/2 \, (\partial_{p^j} + i \partial_{q^j}) <$ where $z^j = p^j + iq^j$.

Consider as before $\alpha = \frac{1}{2i\hbar}$ (pdq - qdp)

The polarised sections $\Gamma^{\infty}_{P_a}(B^c)$ are given (as functions on $\mathbb{R}^{2n}$) by :

$$\Gamma^{\infty}_{P_1}(B^c) = \{ \overset{(1)}{\psi} \in C^{\infty}(\mathbb{R}^{2n}) \mid X\overset{(1)}{\psi} + \alpha(X)\overset{(1)}{\psi} = 0 \quad \forall X \in P_1 \}$$

$$= \{ \overset{(1)}{\psi} = e^{-\frac{i\pi pq}{\hbar}} \psi_1(q) \}$$

The frame $v_j^{(1)}$ of P_1 is given by $v_j^{(1)} = \sqrt{2}\, \partial_{z_j} + 2\, \partial_{\bar{z}_j} = 2\partial_{p_j}$

so that the volume form α' on $\mathbb{R}^{2n}/P_1$ is given by

$$\alpha'(\partial_{q^1}, \ldots, \partial_{q^n}) = \left| \frac{\omega^n}{n!}(\partial_{q^1}, \ldots, \partial_{q^n}, \sqrt{2}\partial_{p^1}, \ldots \sqrt{2}\partial_{p^n}) \right| = 2^{n/2}$$

and

$$\langle \overset{(1)}{\psi}, \overset{(1)}{\psi'} \rangle_{\mathcal{H}_1} = 2^{n/2} \int_{\mathbb{R}^n} \psi_1(q)\, \overline{\psi_1'(q)}\, dq^1 \cdots dq^n$$

Similarly

$$\Gamma^{\infty}_{P_2}(B^c) = \{ \overset{(2)}{\psi} = e^{\frac{i\pi pq}{\hbar}} \psi_2(p) \}$$

$$v_j^{(2)} = \sqrt{2}\, i\, \partial_{\bar{z}_j} - \sqrt{2}\, i\, \partial_{z_j} = -\sqrt{2}\, \partial_{q_j}$$

$$\alpha^2(\partial_{p^1}, \cdots, \partial_{p^n}) = \left| \frac{\omega^n}{n!}(\partial_{p^1}, \ldots, \partial_{p^n}, -\sqrt{2}\partial_{q^1}, \ldots, -\sqrt{2}\partial_{q^n}) \right| = 2^{n/2}$$

and

$$\langle \overset{(2)}{\psi}, \overset{(2)}{\psi'} \rangle_{\mathcal{H}_2} = 2^{n/2} \int_{\mathbb{R}^n} \psi_2(p)\, \overline{\psi_2'(p)}\, dp^1 \cdots dp^n$$

In the same way

$$\Gamma^{\infty}_{P_3}(B^c) = \{ \overset{(3)}{\psi} = e^{\frac{i\pi\epsilon}{2\hbar}(p^2 - q^2)} \psi_3(v) \quad , v^j = p^j - \epsilon_j q^j \}$$

$$v_j^{(3)} = (1 - i\epsilon_j)\partial_{\bar{z}_j} + (1 + i\epsilon_j)\partial_{z_j} = \partial_{p_j} + \epsilon_j \partial_{q_j}$$

$$\alpha^3(\partial_{v^1}, \ldots, \partial_{v^n}) = 1$$

and

$$\langle \overset{(3)}{\psi}, \overset{(3)}{\psi'} \rangle_{\mathcal{H}_3} = \int_{\mathbb{R}^n} \psi_3(v)\, \overline{\psi_3'(v)}\, dv^1 \cdots dv^n$$

The pairings between $\mathcal{H}_1, \mathcal{H}_2, \mathcal{H}_3$ are determined as follows

Consider the unitary frame of F $\{u_j = \sqrt{2}\,\partial_{\bar z_j}\}$ which is the image under the projection P^F on F of :

$$\{\tilde v_j^{(1)} = \sqrt{2}\,\partial_{p_j} = u_j + \bar u_j\} \subset P_1 \qquad \text{so that} \quad Z_1 = \mathrm{Id}$$

$$\{\tilde v_j^{(2)} = \sqrt{2}\,i\,\partial_{q_j} = u_j - \bar u_j\} \subset P_2 \qquad \text{so that} \quad Z_2 = -\mathrm{Id}$$

$$\{\tilde v_j^{(3)} = \frac{\sqrt{2}}{1-i\epsilon_j}\left(\partial_{p_j} + \epsilon_j \partial_{q_j}\right) = u_j + \frac{1+i\epsilon_j}{1-i\epsilon_j}\bar u_j\} \subset P_3 \quad \text{so that } Z_3 = \mathrm{diag}(i\epsilon_j)$$

Then the pairing between $\mathcal{H}_1$ and $\mathcal{H}_2$ is

$$\langle \overset{(1)}{\psi}, \overset{(2)}{\psi}{}' \rangle_{\mathcal{H}_1 \mathcal{H}_2} = \int_{\mathbb{R}^{2n}} e^{-\frac{2i\pi pq}{h}} \psi_1(q)\, \overline{\psi_2'(p)}\, 2^{n/2}\, dp^1 \cdots dp^n dq^1 \cdots dq^n$$

and induces an intertwining operator F_{21} of $\mathcal{H}_1$ in $\mathcal{H}_2$:

$$F_{21} \overset{(1)}{\psi} = e^{\frac{i\pi pq}{h}} \int_{\mathbb{R}^n} e^{-\frac{2i\pi pq}{h}} \psi_1(q)\, dq^1 \cdots dq^n$$

It is unitary if h = 1 , which we assume in what follows

Similarly

$$\langle \overset{(2)}{\psi}{}', \overset{(3)}{\tilde\psi} \rangle_{\mathcal{H}_2 \mathcal{H}_3} = \int_{\mathbb{R}^{2n}} e^{\frac{i\pi\epsilon}{2h}(2p^2 - 4pv + v^2)} \psi_2'(p)\, \tilde\psi_3(v)\, 2^{n/4} e^{\frac{i\pi}{8}(n-2k)} dp^1 dp^n dv^1 dv^n$$

as $\det^{1/2}(1 - Z_2 \bar Z_3) = \prod_{j=1}^n (1 - i\epsilon_j)^{1/2} = \prod_{j=1}^n \left(2^{1/2} e^{-\frac{i\epsilon_j \pi}{4}}\right)^{1/2} = 2^{n/4} e^{-\frac{i\pi}{8}\left(\sum_{j=1}^n \epsilon_j\right)}$

So that

$$F_{32} \overset{(2)}{\psi}{}' = e^{\frac{i\pi\epsilon(p^2 - Q^2)}{2h}} \int_{\mathbb{R}^n} e^{\frac{i\pi\epsilon}{h}(p^2 - 2pv + \frac{1}{2}v^2)} \psi_2'(p)\, 2^{n/4} e^{\frac{i\pi}{8}(n-2k)} dp^1 \cdots dp^n$$

Which is unitary for h = 1

One also has

$$\langle \overset{(3)}{\tilde{\psi}}, \overset{(1)}{\psi} \rangle_{\mathcal{H}_3 \mathcal{H}_1} = \int_{\mathbb{R}^{2n}} e^{\frac{i\pi}{h}(\epsilon q^2 + \frac{\epsilon}{2} v^2 + 2vq)} \tilde{\psi}_3(v) \overline{\psi_1(q)}\, 2^{n/4} e^{\frac{i\pi}{8}(n-2k)} dv^1 \cdots dv^n dq^1 \cdots dq^n$$

$$F_{13} \overset{(3)}{\tilde{\psi}} = e^{-\frac{i\pi pq}{h}} \int_{\mathbb{R}^n} 2^{-n/4} e^{\frac{i\pi}{8}(n-2k)} e^{\frac{i\pi}{h}(\epsilon q^2 + \frac{\epsilon}{2} v^2 + 2vq)_2} \tilde{\psi}_3(v) dv^1 \cdots dv^n$$

which is also unitary for $h = 1$.

Calculating the composition of these intertwining operators one gets :

$$(F_{13} \circ F_{32} \circ F_{21}) \overset{(1)}{\psi} = e^{-\frac{i\pi pq}{h}} e^{\frac{i\pi}{4}(n-2k)} \int_{\mathbb{R}^{3n}} e^{\frac{i\pi \epsilon v'^2}{h}} e^{\frac{2\pi i}{h} p(q-q')} \psi_1(q') dq' dp\, dv'$$

where $v'^j = -p^j + v^j + \epsilon_j q^j$

As $\int_{\mathbb{R}} e^{2\pi i \epsilon \frac{v^2}{2}} dv = \begin{cases} e^{\frac{i\pi}{4}} & \text{if } \epsilon = 1 \\ e^{-\frac{i\pi}{4}} & \text{if } \epsilon = -1 \end{cases}$ one has, for $h = 1$:

$$(F_{13} \circ F_{32} \circ F_{21}) \overset{(1)}{\psi} = e^{-\frac{i\pi pq}{h}} \psi_1(q)\, e^{\frac{i\pi}{4}(n-2k)} e^{\frac{i\pi}{4}(2k-n)} = \overset{(1)}{\psi}$$

So that the composition is a transitive law.

Theorem. The intertwining operators induced by half form pairings between transverse polarizations have a transitive law of composition.

Remarks 1) The introduction of half forms induces in the pairings some phase factors; without them, one gets back the operators J_{ji} ; the above computation shows that $J_{13} \circ J_{32} \circ J_{21} = e^{-\frac{i\pi}{4}(n-2k)}$ Id ; remark that $n-2k$ is precisely the Maslov index of the three lagrangian planes corresponding to P_1, P_2, P_3

2) This theorem is of importance when one uses moving polarisations which is what one does when one quantizes functions which do not preserve the chosen polarisation.

Acknoledgments This work has been done in collaboration with S. Gutt; it is a pleasure to thank her.
J. Rawnsley also pointed to us that this transitiveness of the intertwining operators has been studied by Guillemin and Sternberg.

Bibliography

1. J. Lions, M. Vergne "Weyl representation, Maslov index and theta Series" Birkhauser (1980). Prog. Math 6.

2. J. Rawnsley : Com. Math. Phys. 58, p. 1 (1978) see also notes "Half forms" from 1980 C.N.R.S. Luminy.

SPIN-3/2 FIELDS ON BLACK HOLE SPACETIMES

Peter C. Aichelburg

Institut für Theoretische Physik, Universität Wien
Boltzmanngasse 5, A-1090 Vienna
AUSTRIA

ABSTRACT

A review is given of what is known about stationary black hole configurations with non-zero supercharge in N = 1 and N = 2 supergravity.

The study of spin-3/2 fields on stationary black holes is motivated by the fundamental role that these fields play in supergravity theories. The invariance of the theory under local supersymmetry transformations give rise to conserved spinor charges, which can be expressed as surface integrals at spatial infinity.

It is therefore of some interest to look for "classical" particle-like field configurations which carry supercharge. In general relativity there exist vacuum solutions which are time-independent, asymptotically flat, regular outside an event horizon and classically stable. However, the fact that these black holes are highly selective with respect to their allowed 'charges' (no-hair theorems), has led to the conjecture that black holes with supercharge do not exist[1].

Here I shall give a brief summary of what is known about stationary black hole configurations in N = 1 and N = 2 supergravity.

Within the Einstein-Maxwell theory a generic stationary black hole is completely characterized by its mass (M), angular momentum (a), electric (e), and magnetic (q) charge.

This theory can be embedded in N = 2 supergravity whose field content is: The gravitational vierbein field $e^a = e^a_\mu dx^\mu$, the electromagnetic potential one-form $A = A_\mu dx^\mu$ and two Majorana spinor-valued one-forms ψ^j, combined to a complex (Dirac) field $\psi = \psi^1 + i\psi^2 = \psi_\mu dx^\mu$ (Rarita-Schwinger field). All fields are Grassmann-valued, the bosonic fields (e,A) being even elements while ψ is odd (anticommutating).

The theory is invariant under local supergauge transformations generated by a complex spinor field $\varepsilon = \varepsilon^1 + i\varepsilon^2$ (ε^j Majorana). The infinitesimal form reads:

$$\delta e^a = -\frac{ik}{2}(\bar{\varepsilon}\gamma^a\psi - \bar{\psi}\gamma^a\varepsilon) \tag{1a}$$

$$\delta A = \frac{i}{2}(\bar{\varepsilon}\psi - \bar{\psi}\varepsilon) \tag{1b}$$

$$\delta\psi = \frac{1}{k}\hat{D}\varepsilon\ . \tag{1c}$$

(The explicit form of the field equations, notations and conventions may be found in Ref. 2, especially $k^2 = 4\pi G$, signature (+ - - - -),

$$\hat{D} = D - \frac{k}{2}\hat{F}^{ab}\sigma_{ab}\gamma\ ,\quad D = d + \frac{1}{2}\omega^{ab}\sigma_{ab},\quad \gamma = \gamma_a e^a\ .$$

The global conserved quantity associated with this symmetry is the (spinorial) supercharge[3,4])

$$S = -\frac{i}{k}\oint_{S^2_\infty}\gamma_5\gamma\wedge\psi \tag{2}$$

which also acts as generator for the (asymptotic) global supersymmetry transformations. We consider only asymptotically flat configurations, and ψ has to fall off like $O(r^{-2})$ in order to render S finite.

A possible way to study the question whether black holes can carry supercharge is to consider perturbations of the spin-3/2 field on the black hole background. Retaining from the N = 2 supergravity field equations only linear terms in ψ yield:

$$-\frac{1}{4k^2}R^{bc}\wedge {}^*e_{abc} = {}^*t_a(F) \tag{3a}$$

$$d^*F = 0 \tag{3b}$$

$$\gamma \wedge \hat{D} \wedge \psi = 0 \,. \tag{3c}$$

The first two equations is just the Einstein-Maxwell system, while the last equation is the (linear) Rarita-Schwinger equation on the given background. As a solution of the system (3a,3b) one takes the Kerr-Newman field and tries to solve Eq. (3c) imposing the following conditions:

i) stationarity

ii) fall off at spatial infinity

iii) regularity (on and outside the horizon).

Because of the local gauge freedom (1a - c) any field of the form

$$\psi = \frac{1}{k} \hat{D} \varepsilon \tag{4}$$

is automatically a solution to (3c). Moreover, the supercharge (2) is only invariant under gauge transformations for which the gauge spinor ε tends to zero at spatial infinity. Therefore one has to distinguish between gauge-generated and non-gauge configurations (see Fig. 1).

NON-GAUGE FIELDS

Let us first consider non-gauge configurations and impose the additional condition:

iv) $\psi \neq \frac{1}{k} \hat{D} \varepsilon$ (non-gauge).

A detailed analysis of the spin-3/2 modes on a Kerr-Newman black hole has shown[5] that there are no solutions of Eq. (3c) that satisfy the conditions i) - iv) unless the background parameters are such that

$$k^2 M^2 = e^2 + q^2 \tag{5}$$

i.e. the charge of the hole equals its mass (in units where $4\pi G = 1$).

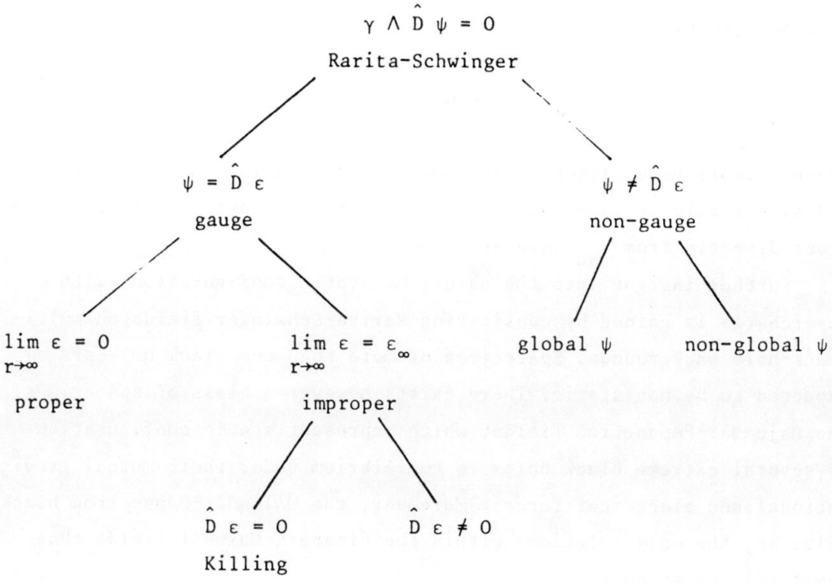

Fig. 1: The classification of solutions to the Rarita-Schwinger equation are shown. Of interest are the non-gauge global and non-global (monopole-like) as well as the improper gauge generated (superpartners) configurations.

This singles out the extreme Reissner-Nordstrøm metric, since a has to vanish in order that a horizon exists. It was shown[6,7] that the Reissner-Nordstrøm field indeed admits spin-3/2 perturbations satisfying the conditions i) - iv). These perturbations are characterized by two complex constants and give rise to a non-zero supercharge. The linear solution was later generalized (by rather tedious calculations) to an exact one, taking into account the full back-reaction of the Rarita-Schwinger field onto the Einstein-Maxwell fields[8]. Embacher[9] has shown that this solution is the most general one.

The uniqueness of the extreme Reissner-Nordstrøm metric in support ing a spin-3/2 supercharge has a geometrical explanation: Among the Kerr-Newman black holes the extreme Reissner-Nordstrøm metric is single out by the existence of Killing (supercovariantly constant) spinors

χ_{sc}, satisfying

$$\hat{D}\,\chi_{sc} = 0 . \tag{6}$$

These spinors give rise to additional gauge invariant Rarita-Schwinger modes. Actually it is possible to construct the linear superhair solutions directly from χ_{sc} (see Ref. 10).

Further insight into the nature of static configurations with supercharge is gained by considering Rarita-Schwinger fields on multi-black-hole backgrounds. Spacetimes of more than one black hole are expected to be non-static. There exists however a class of spacetimes, the Majumdar-Papapetrou fields, which represent static configurations of several extreme black holes in equilibrium under their mutual gravitational and electrical forces. Moreover, the Majumdar-Papapetrou black holes are the only solutions within the Einstein-Maxwell fields that admit Killing spinors.

The Majumdar-Papapetrou metric for n black holes with the mass parameters M_J equal to the individual electric charges Q_J is given in an orthonormal basis by:

$$e^0 = V^{-1}\,dt , \qquad e^i = V\,dx^i , \qquad i = 1,2,3 ,$$

$$V = 1 + \sum_{J=1}^{n} \frac{M_J}{|\vec{x} - \vec{x}_J|} , \qquad \vec{x}_J \neq \vec{x}_K \quad \text{for} \quad J \neq K , \tag{7}$$

together with the Maxwell field

$$F = - V^{-2}\, V_{,i}\; dt \wedge dx^i . \tag{8}$$

On this background consider a complex vector-spinor $\psi = \psi_a\, e^a$. Writing ψ as

$$\psi_a = \frac{1}{2}\begin{pmatrix} \beta_a + \alpha_a \\ \beta_a - \alpha_a \end{pmatrix} \tag{9}$$

and imposing (3c) plus the conditions i) - iv) one arrives at the following results[2)]:

a) there is a supergauge in which $\beta_o = \alpha_i = 0$,

b) α_o has the form

$$V^{1/2}\alpha_o = \sum_{J=1}^{n} M \frac{(x_J - x)^i}{|\vec{x}_J - \vec{x}|^3}\sigma_i c_J , \tag{10}$$

where the c_J are complex constant two-spinors,

c) the remaining Rarita-Schwinger equation for β_j reads:

$$\varepsilon_{mij}[V^{3/2}(\sigma_j\alpha_o + \beta_j)]_{,i} = 2iV_{,i}\,\sigma_i\,\sigma_m V^{1/2}\alpha_o . \tag{11}$$

Although the local integrability conditions for these equations are trivially satisfied, there are global conditions coming from the non-trivial topology of the manifold. Integrating the l.h.s. of Eq.(11) over a closed two-surface enclosing any of the $\vec{x}_J$ gives zero, while the r.h.s. leads to

$$0 = \sum_{\substack{K=1\\(K\neq J)}}^{n} M_J M_K \frac{(x_J - x_K)^i}{|\vec{x}_J - \vec{x}_K|^3}\sigma_i(c_J + c_K) , \quad J = 1,\ldots,n . \tag{12}$$

These n conditions on the spinors c_J leave just one of them unconstrained. Therefore, the number of arbitrary parameters describing the Rarita-Schwinger field on Majumdar-Papapetrou backgrounds is independent of the number of holes. For the solutions on Majumdar-Papapetrou backgrounds, the supercharge may be expressed directly by the constants c_J:

$$S = \begin{pmatrix} q \\ q \end{pmatrix} , \qquad q = \sum_{J=1}^{n} M_J c_J . \tag{13}$$

Thus, the constraints (12) imply for a system of two black holes with equal masses that the supercharge is zero. (The explicit solution for n = 2 is given in Ref. 11.) A possible explanation of these results is to interpret the constraints as equilibrium conditions for static configuration. This would imply that there is a new kind of interaction between black holes due to the existence of supercharge. A step towards such an interpretation was made in Ref. 12, by relating the supercharge to changes in the mass and charge parameters of the background.

MONOPOLE-LIKE SOLUTIONS

There is another possibility to look at the constraints: Since the conditions (12) follow from the global integrability conditions, violating them implies the non-existence of global solution for ψ on the given background. However, ψ being a gauge field, it is conceivable to consider only its supercovariant field strength $\hat{D} \wedge \psi$ as the fundamental object and to give up global existence of ψ. If one requires regularity for $\hat{D} \wedge \psi$, then ψ may have line-singularities which may be shifted by gauge transformations of the form (1). This is analogous to the Dirac monopole in electrodynamics. In a recent paper we have studied "monopole-like solutions" on Majumdar-Papapetrou backgrounds[13]. The existence of "strings" is associated with non-zero global quantities:

$$P_J = \frac{1}{4\pi i} \oint_{S_J^2} V^{1/2} \, \hat{D} \wedge \psi \tag{14}$$

which we call "spinorial magnetic-type charges", where the integral is to be taken around the points $\vec{x}_J$. Assuming that there is no line-singularity going off to infinity, the total spinorial magnetic-type charge vanishes, i.e.

$$P = \sum_{J=1}^{n} P_J = \frac{1}{4\pi i} \oint_{S_\infty^2} \hat{D} \wedge \psi = 0 \, , \tag{15}$$

but the individual P_J are only zero if the constraints (12) are satisfied. At present it is not clear whether these magnetic-type charges are of topological origin.

GAUGE-GENERATED CONFIGURATIONS

Consider now ψ-configurations of the type (4) for which the gauge spinor ε tends to a constant ε_∞ at spatial infinity. The gauge-induced supercharge may then be written as:

$$S = -\frac{i}{k^2}\oint_{S^2_\infty} \gamma_5\gamma \wedge \hat{D}\,\varepsilon = (-\,i\gamma^a P_a^{ADM} + \frac{i}{k}\,e + \frac{i}{k}\,\gamma_5 q)\varepsilon_\infty \qquad (16)$$

where

$$P_a^{ADM} = \frac{1}{4k^2}\oint_{S^2_\infty} \omega^{bc} \wedge {}^*e_{abc} \qquad (17)$$

is the usual ADM four-momentum of the configuration, while e and q is the total electric and magnetic charge, respectively:

$$e = \oint_{S^2_\infty} {}^*F\,, \qquad\qquad q = \oint_{S^2_\infty} F\,. \qquad (18)$$

(16) expresses the global N = 2 supersymmetry algebra with central charges e, q.

We wish to identify all configurations which can be obtained by "short range" gauge transformations ($\varepsilon \to 0$) from one another thereby dividing the solutions into equivalence classes.

If the background is extreme (5), the matrix on the l.h.s. of (16) is singular and in the rest frame (7) reads

$$S = iM(1 - \gamma_0)\varepsilon_\infty\,. \qquad (19)$$

As a consequence there are ε_∞ leaving S invariant.

A suitable way to construct one representant of each equivalence class explicitly is to impose the gauge condition

$$\gamma^\mu\,\psi_\mu = 0 \qquad (20)$$

at the linearized level. This implies that ε satisfies the supercovariant Dirac equation

$$\gamma^{\mu} \hat{D}_{\mu} \varepsilon = 0 . \tag{21}$$

As pointed out by Cibbons[14)], these conditions plus the requirement of ε being regular at the horizon and time-independent project out all the short range gauge transformations. Thus any time-independent regular solution of (21) is uniquely determined by its asymptotic value ε_{∞}. The solutions are spanned by:

$$\varepsilon = V^{-1/2} \begin{pmatrix} c \\ -c \end{pmatrix} = - i\gamma_5 \chi_{sc} \tag{22}$$

and

$$\chi_{sc} = V^{-1/2} \begin{pmatrix} c \\ c \end{pmatrix} . \tag{23}$$

Clearly, only the first class gives rise to a non-zero supercharge.

Hajicek[15)] has argued that the extreme Reissner-Nordstrøm black holes may be considered as solitons because they are stable against quantum radiation. In N = 2 supergravity these configurations are partially supersymmetric. Gauge transformations with χ_{sc} leave the fields invariant. On the other hand the extreme Reissner-Nordstrøm solitons are degenerated in the sense that there exist gauge transformed configurations induced by (22). These "superpartners" carry a supercharge which is given by (19).

Applying the same reasoning to the Schwarzschild black hole in N = 1 supergravity, Gibbons has argued that, because there are no regular time-independent solutions to the Dirac equation, no superpartners exist. This means that regular static gauge configurations $\psi = k^{-1} D \varepsilon$, which carry supercharge, are projected out by the gauge condition. However, unless a gauge condition of the type (21) is dictated by the quantum theory, classically these configurations are acceptable.

Let me end by mentioning that recently we were able to iterate the gauge transformations (1a - c) with the gauge spinor (22) to all orders[16]. This leads to the exact superpartners of the Majumdar-Papapetrou spacetimes, satisfying the full non-linear N = 2 field equations. The supercharge gives rise to an intrinsic angular momentum in the metric and to the corresponding quantum mechanical magnetic moment.

REFERENCES

1) Cordero, P. and Teitelboim, C., Phys. Lett. 78B, 80 (1978).
2) Embacher, F., Gen. Rel. Grav. 16, 909 (1985).
3) Deser, S. and Teitelboim C., Phys. Rev. Lett. 39, 249 (1977).
4) Teitelboim, C., Phys. Lett. 69B, 240 (1977).
5) Aichelburg, P.C. and Güven, R., Phys. Rev. D24, 2066 (1981).
6) Güven, R., Phys. Rev. D25, 3117 (1982).
7) Aichelburg, P.C. and Güven, R., Phys. Rev. D27, 456 (1983).
8) Aichelburg, P.C. and Güven, R., Phys. Rev. Lett. 51, 1613 (1983).
9) Embacher, F., Class. Quantum Grav. 2, 323 (1985).
10) Aichelburg, P.C. and Güven, R., Phys. Lett. 135B, 291 (1984).
11) Embacher, F. and Aichelburg, P.C., Phys. Rev. D30, 2457 (1985).
12) Aichelburg, P.C. and Embacher, F., Class. Quantum Grav. 2, 65 (1985).
13) Aichelburg, P.C. and Embacher, F., Gen. Rel. Grav. 18, 705 (1986).
14) Gibbons, G.W., "Black Holes, Solitons in Ungauged Extended Supergravity" in Proceedings of the XVth GIFT Seminar, eds. F. Aguila et al. (World Scientific Publ. 1985).
15) Hajicek, P., Phys. Rev. D26, 3384 (1982).
16) Aichelburg, P.C. and Embacher, F., Phys. Rev. D34, 3006 (1986).

INDECOMPOSABLE CONFORMAL SPINORS AND OPERATOR PRODUCT EXPANSIONS IN A MASSLESS QED MODEL *)

Yassen S. Stanev[1] and Ivan T. Todorov[2]

---

*) Talk presented by I.T. Todorov at the Conference on Spinors in Physics and Geometry, Trieste, 11-13 September 1986.

1. Department of Physics, University of Sofia, Sofia, Bulgaria.
2. International School for Advanced Studies, Trieste, Italy and Institute for Nuclear Research and Nuclear Energy, Sofia, Bulgaria.

ABSTRACT

A 4-dimensional conformal model is constructed which shares the following properties with quantum electrodynamics (QED): (a) standard Maxwell equations for $F_{\mu\nu}$ (in a local covariant gauge); (b) standard Ward identities; (c) non-vanishing conformally invariant current-current Wightman function proportional to the second order perturbation theory expression (for a massless electron field). Two versions of the model are presented: one with anomalous dimension of the charged spin ½ field ($d_\psi = 3/2 + \lambda$), and a second one with $\lambda = 0$, but with an indecomposable (8-component) conformal spinor field. In both cases $F_{\mu\nu}$ is part of an indecomposable decaplet and has a logarithmic 2-point function.

1. INTRODUCTION

It has been customary to believe that unitary representations of a symmetry group - which are always fully reducible - are the only relevant ones in quantum physics, where unitarity is dictated by conservation of probability. Paradoxically, the basic massless field equations in relativistic physics appear in the context of (non-unitary) indecomposable representations. The partial differential operators in the Maxwell (and the massless Dirac or Weyl) equations play the role of intertwining maps whose kernels (the set of solutions of the corresponding homogeneous equations) are invariant subspaces for distinguished indecomposable representations of the conformal group (also appearing as a natural playground for gauge transformations). Keeping in sight the topic of our Conference, we shall fix our attention on indecomposable conformal spinors.

A 4-component Dirac spinor field ψ of canonical dimension, $d_{\psi} = 3/2$, already belongs to an indecomposable representation of the conformal group G = SU(2,2): the subset of solutions of the massless Dirac equation $\not{\partial}\psi = 0$ forms a G-invariant subspace with no G-invariant complement. This representation is, however, induced by a fully reducible representation of an 11-parameter parabolic subgroup H, the stability subgroup of a point, say x=0, in Minkowski space (more precisely, by the direct sum of two conjugate 2-dimensional representations of H; it can be viewed as an irreducible representation, if we extend both G and H by space reflection). We call the representations of G induced by irreducible representations of H, elementary (for a review of elementary local field representations of G and their applications - see [T1]). Here we shall deal with more general conformal spinor representations that are induced by indecomposable representations of H.

Most of us have first encountered the 8-component conformal spinors in Dirac's manifestly covariant formalism [D3] (Dirac himself gives credit

– in [D3(a)] – to Veblen, whose ideas, continuing the tradition of Hermann Weyl were making part of the Princeton folklore in the 1930's*) and have only partly been reflected in publications like [V1]). We have in mind 8-component spinor fields $\Psi(\xi)$ that are homogenous functions of degree -2 on the (O(4,2)-invariant) cone $\underset{\sim}{\xi}^2+\xi_5^2=\xi_0^2+\xi_6^2 \ (\neq 0)$ and transform under the bispinor (or "bitwistor") representation of SU(2,2) (for a recent review, see [T2]). Segal et al. [P2] are introducing them as tensor products of the (infinite dimensional) scalar field representation of G of weight 2 with the direct sum of the defining 4-dimensional ("twistor") representation of SU(2,2) and of its dual.

In Minkowski space an 8-component spinor appears as a pair of Dirac fields $\psi(x)$ and $\psi_+(x)$ of dimension 3/2 and 5/2 respectively. Two types of such indecomposable 8-spinors have been studied. We are working with spinors that have standard covariance properties under translations, $U(a)\psi_{(+)}(x)U^{-1}(a)=\psi_{(+)}(x+a)$, but obey an indecomposable infinitesimal special conformal transformation law**)

$$2[\psi(x), C_\lambda] = \{x_\lambda(x\partial+\tfrac{3}{2})-\tfrac{1}{2}x^2\partial_\lambda+\tfrac{1}{4}[\gamma_\lambda,\not{x}]\}\psi(x), \tag{1.1a}$$

$$2[\psi_+(x), C_\lambda] = \{x_\lambda(x\partial+\tfrac{5}{2})-\tfrac{1}{2}x^2\partial_\lambda+\tfrac{1}{4}[\gamma_\lambda,\not{x}]\}\psi_+(x)+\gamma_\lambda\psi(x). \tag{1.1b}$$

Paneitz, Segal and Vogan [P2] define a <u>spanor</u> as an 8-component (Lorentz) spinor that transforms in an indecomposable way under translations (rather than under special conformal transformations). From a representation theoretic point of view the difference is inessential: in the first case one

---

*) I would like to thank Irving E. Segal, who pointed out to me the role of Oswald Veblen (and his Princeton Seminar) in the early applications of conformal invariance to relativistic physics.

**) We are using a slightly different convention from ref. [F5] where an i factor was present in the last term of (1.1b).

is inducing by an indecomposable representation of the stability subgroup of x=0; in the second, by the stability subgroup of a point at infinity. From the point of view of compactified Minkowski space the two points are interchangeable; the two subgroups are conjugate in G. However, the local properties of the fileds in the two cases differ. (Indecomposable spinors with non-trivial transformation properties with respect to translations have been also studied by Castell [C1].)

For a long time, theorists regarded the extra components coming from the manifestly covariant formalism as a nuisance. One was getting rid of them by imposing various types of subsidiary conditions (see e.g., [M1,2], [F2]). Paolo Budinich was, for quite some time, the only one around who took the 8-component spinors seriously and tried to give them a direct physical meaning - see [B3,4] . Later, also the 6-vector manifestly covariant representation was used without subsidiary conditions for a conformal gauge formulation of massless QED - see [B1,2] [F5] [P4] [S3,4] [T3] [Z1] .

The most ambitious programme we are aware of, that aims to give a realistic meaning of indecomposable conformal spinors, belongs to Segal et al. ([P2] [S2] and references therein) who propose to use the different irreducible factors of the spanor representation for the classification of basic leptons (the electron, the muon and their neutrinos). The present attempt is much more modest in scope. We start with a conformal QED model based on the (linear) Maxwell equations and on covariant operator product expansions (OPEs) [F2] [D4] [C2] . It involves a charged Dirac ("electron") field of -typically-anomalous dimension ($3/2+\lambda$, $\lambda>0$). Recalling that the dimension of the electron field is a gauge dependent parameter ([A2]) we study the limit $\lambda\to 0$. It turns out that in this (generalized) Landau gauge the electron field appears necessarily as an indecomposable conformal spinor, the ψ_+ part corresponding in some sense to the limit of $\frac{1}{\lambda}\not\partial\psi(x,\lambda)$ for $\lambda\to 0$.

The main part of the paper (Secs. 2,3) is devoted to our QED model: in order to give weight to the final outcome - the necessity of dealing with an indecomposable 8-component conformal spinor in the limit of zero anomalous dimension - a convincing argument is needed to demonstrate that the model has indeed a sound basis. We shall see that it follows in an essentially unique way from three main assumptions (i) standard Maxwell equations, (ii) conformal invariance of (Wightman) correlation functions (under a yet to be specified representation of G) and (iii) a non-vanishing 2-point function of the electromagentic current.

2. THE MODEL. BASIC ASSUMPTIONS

2A. The Bose sector. Maxwell equations

We attempt to construct a (non-free) conformal QED model with a minimal number of fields: a conserved (observable) electromagnetic current J^{μ} - that will ultimately be identified, via an OPE, as a composite of a charged (electron) field ψ and its conjugate; the Maxwell stress tensor $F_{\mu\nu}$ related to J^{μ} by a weak version of Maxwell's equations and an auxiliary vector field h_{μ} (which cannot be identified with an electromagentic potential) whose presence will be dictated by the indecomposable conformal transformation law for $F_{\mu\nu}$, necessary in order to give a conformal invariant meaning of its 2-point function.

Our starting assumption - so basic, that we do not give it a number - is the one of conformal invariance. Its exact technical meaning - specifying fields' transformation laws under which the theory is invariant - will be gradually revealed in the course of our study. (We know of no *a priori* principle that tells us what is the precise conformal law for various interacting quantum fields; our objective is to derive, as much as possible, such laws starting from some general principles). An important part of this assumption is the requirement that the vacuum is invariant - under the whatever conformal law will emerge - so that field covariance properties and invariance of the equations of motion are directly translated into invariance properties of vacuum expectation values.

Our first postulate is a standard requirement in any local formulation of QED.

I. We assume a weak form of the first Maxwell equation

$$\langle \partial_{\nu_1} F^{\mu_1\nu_1}(x_1)\, \partial_{\nu_2} F^{\mu_2\nu_2}(x_2)\rangle_c = \langle J^{\mu_1}(x_1)\, J^{\mu_2}(x_2)\rangle_0 = \langle \partial_{\nu_1} F^{\mu_1\nu_1}(x_1)\, J^{\mu_2}(x_2)\rangle_0 , \qquad (2.1)$$

and a (strong) operator form of the second one:

$$\partial_\mu F_{\nu\rho}(x) + \partial_\nu F_{\rho\mu}(x) + \partial_\rho F_{\mu\nu}(x) = 0 , \qquad \left(\partial_\mu \equiv \frac{\partial}{\partial x^\mu} \right). \qquad (2.2)$$

Current conservation, $\partial_\mu J^\mu(x) = 0$, is also assumed to hold as a conformally invariant operator equation.

Comments. It has been proven ([F3] [S6]) that $J^\mu - \partial_\nu F^{\mu\nu}$ cannot vanish as an operator if there exists a local charged field. The Gauss law $J^0 = \partial_i F^{0i}$ (= div $\underset{\sim}{E}$) should only be assumed to hold for matrix elements between physical states (that is, between vectors Φ of a Poincaré invariant subspace $\mathcal{H}'$ of the indefinite metric space $\mathcal{H}$, for which $\langle \Phi | \Phi \rangle \geqslant 0$). We shall adopt here without further mentioning the general framework of [S6] , which allows us to interpret Eq.(2.1) as saying that the vacuum $|0\rangle$ and the gauge invariant state $J^\mu|0\rangle$ belong to $\mathcal{H}'$. We note that although Eq.(2.2) guarantees the existence of an electromagentic 4-potential A_μ , it does not ensure its locality. Although we shall be able to write down a (singular) Lagrangian density in terms of A_μ (in Sec. 3C below) that reproduces the equations of motion for our model, we shall not attempt to derive OPEs involving A_μ's, which according to past experience lead to ambiguities and difficulties.

Before formulating our second postulate, we recall a fact: the most general conformally invariant expression for the currents' 2-point Wightman function is proportional to the result of second order perturbation theory when the mass of the charged field is set equal to zero:

$$\langle J^{\mu}_{(x)} J_{\nu(0)} \rangle_c = \frac{a e^2}{\pi^4} \frac{r^{\mu}{}_{\nu(x)}}{x_+^6} \tag{2.3a}$$

$$= \frac{a e^2}{6\pi} \int \theta_+(p)(p^{\mu} p_{\nu} - p^2 \delta^{\mu}{}_{\nu}) e^{ipx} d_4 p \,, \tag{2.3b}$$

where $d_4 p = (2\pi)^{-4} d^4 p$, $\theta_+(p) = \theta(p^0)\theta(-p^2)$ $(p^2 \equiv \underset{\sim}{p}^2 - p_0^2)$,

$$r^{\mu}{}_{\nu} = \delta^{\mu}{}_{\nu} - 2\frac{x^{\mu} x_{\nu}}{x_+^2} \,, \quad x_+^2 = x^2 + i0x^0. \tag{2.4}$$

(Note: no ambiguity is encountered in the 1-loop calculation of Wightman 2-point functions; only the time ordered Green function involves a divergence at the 1-loop level.) Second order perturbation theory for spinor QED gives (2.3) with a=1.

II. <u>The 2-point function of the electromagnetic current is given by (2.3) with a > 0 (in accord with Wightman positivity).</u>

<u>Comment:</u> the current-current Wightman function is directly related to an observable quantity, the total cross-section of electron-positron annihilation. The qualitative agreement of (2.3) with experiments on high energy e^+e^--annihilation into hadrons was instrumental - in the early 70's - for determining the total number of quarks contributing to the annihilation cross section, and for making conformal invariance popular. It is ironic that in the first serious attempt to implement conformal invariance in QED (see the paper of Adler [A1] which was a culmination point of earlier work, started by Johnson and Baker [J1] , as well as subsequent efforst [T1,3,4] [F4,5] [P4] [S3,4]) this property was lost: Maxwell equations and standard assumptions about the conformal properties of the electromagentic field lead to a vanishing current 2-point function. A way out, as we shall see, is to consider a more general indecomposable conformal (and dilation) law for $F_{\mu\nu}$.

Proposition 2.1 The general form of the vanishing at infinity 2-point Wightman function for the Maxwell stress tensor consistent with (2.1) (2.2) (2.3) (and going to the standard free expression for $e \to 0$) is

$$\langle F_{\mu_1\nu_1}(x_1)F_{\mu_2\nu_2}(x_2)\rangle_0 = \frac{1}{2\pi^2 x_{12+}^4}\Big\{\big(r_{\mu_1\mu_2}(x_{12})r_{\nu_1\nu_2}(x_{12}) - r_{\mu_1\nu_2}(x_{12})r_{\mu_2\nu_1}(x_{12})\big)\Big(1+\frac{ae^2}{6\pi^2}\ln\frac{\ell^2}{x_{12+}^2}\Big) - \frac{ae^2}{12\pi^2}\big(\eta_{\mu_1\mu_2}\eta_{\nu_1\nu_2} - \eta_{\mu_1\nu_2}\eta_{\mu_2\nu_1}\big)\Big\}, \tag{2.5}$$

where $r_{\mu\nu}$ is given by (2.2) and ℓ is an arbitrary (positive, e-independent) scale.

Proof. The general Poincaré invariant expression for the 2-point function of $F_{\mu\nu}$ is

$$\langle F_{\mu_1\nu_1}(x)F_{\mu_2\nu_2}(0)\rangle_0 = f(t)\big(\eta_{\mu_1\nu_2}\partial_{\nu_1}\partial_{\mu_2} + \eta_{\nu_1\mu_2}\partial_{\mu_1}\partial_{\nu_2} - \eta_{\mu_1\mu_2}\partial_{\nu_1}\partial_{\nu_2} - \eta_{\nu_1\nu_2}\partial_{\mu_1}\partial_{\mu_2}\big)\frac{1}{4\pi^2 x_+^2} - g(t)\big(\eta_{\mu_1\mu_2}\eta_{\nu_1\nu_2} - \eta_{\mu_1\nu_2}\eta_{\mu_2\nu_1}\big)\frac{1}{4\pi^2 x_+^4}, \quad t = \ln\frac{\ell^2}{x_+^2} \tag{2.6}$$

$$(\eta_{\mu\nu} = \mathrm{diag}(-+++))$$

(the choice x_+^2 for the Lorentz interval - see (2.4) - reflects the x-space analyticity of Wightman functions derived from energy positivity). Eq. (2.2) gives $f' = g + \frac{1}{2}g'$;

it then follows from (2.6) that

$$\langle \partial_{\nu_1}F^{\mu_1\nu_1}(x)F_{\mu_2\nu_2}(0)\rangle_0 = 2f'\,\frac{x_{\nu_2}\delta^{\mu_1}_{\mu_2} - x_{\mu_2}\delta^{\mu_1}_{\nu_2}}{\pi^2 x_+^6}.$$

Taking finally (2.1) and (2.3) into account, we find

$$-\partial_{\mu_1}\partial^{\nu_2}\langle F^{\mu_1\nu_1}(x)\, F_{\mu_2\nu_2}(0)\rangle_0 = 6f'\,\frac{\Gamma(x)^{\mu_1}{}_{\mu_2}}{\pi^2 x_+^6} + 4f''\,\frac{x^2\delta^{\mu_1}{}_{\mu_2} - x^{\mu_1}x_{\mu_2}}{\pi^2 x_+^8} = a\,\frac{e^2}{\pi^4}\,\frac{\Gamma(x)^{\mu_1}{}_{\mu_2}}{x_+^6}\;;$$

thus $f' = \frac{ae^2}{6\pi^2}\ (f''=0),\ g = \frac{ae^2}{6\pi^2} + Ce^{-2t}$. To recover (2.5) we fix the integration constant in $f(=1+\frac{ae^2}{6\pi^2}t)$ to fit the free propagator limit and use the vanishing of the 2-point function at infinity (which reflects the uniqueness of the vacuum) to set C=0.

The expression (2.5) is not even dilation (let alone conformal) invariant under the standard homogeneous law $U(\rho)F_{\mu\nu}(x)U^{-1}(\rho) = \rho^2 F_{\mu\nu}(\rho x)$, $U(\rho)|0\rangle = |0\rangle$, $\rho > 0$. (It appears though, as the kernel of an invariant intertwining map acting on a restricted space of test functions, which are annihilated upon integration with the free 2-point Wightman function of the Maxwell field [P3] . This does not seem, however, relevant for our purposes, since restricting the test function space is not justified in a local quantum field theoretical framework.) Happily, a more general indecomposable, dilation law, which admits a conformal extension [F1] - is also available (see [S1] [D1]) allowing us to interpret (2.5) as (a piece of) a conformal invariant 2-point function of an indecomposable multiplet. The most economic way to make (2.5) consistent with dilation invariance is to assume the existence of a skewsymmetric companion $G_{\mu\nu}$ of $F_{\mu\nu}$ such that

$$U(\rho)F_{\mu\nu}(x)U^{-1}(\rho) = \rho^2\left\{F_{\mu\nu}(\rho x) + \frac{ae^2}{12\pi^2}G_{\mu\nu}(\rho x)\ln\rho\right\}, \tag{2.7a}$$

$$U(\rho)G_{\mu\nu}(x)U^{-1}(\rho) = \rho^2 G_{\mu\nu}(\rho x) \qquad (\rho>0)\,. \tag{2.7b}$$

With this law (2.5) is reproduced and is dilation invariant if

$$\langle F_{\kappa\lambda}(x_1) G_{\mu\nu}(x_2)\rangle_0 = \frac{1}{\pi^2 x_{12+}^4}\left(r_{\kappa\mu}(x_{12}) r_{\lambda\nu}(x_{12}) - r_{\kappa\nu}(x_{12}) r_{\lambda\mu}(x_{12})\right) = \langle G_{\kappa\lambda}(x_1) F_{\mu\nu}(x_2)\rangle_0 ,$$

$$\langle G_{\kappa\lambda}(x_1) G_{\mu\nu}(x_2)\rangle_0 = 0 . \tag{2.8}$$

Dilation invariance of Eq.(2.2) implies that $G_{\mu\nu}$ should satisfy the same equation so that it can be represented as a curl of a 4-potential h_μ :

$$G_{\mu\nu} = \partial_\mu h_\nu - \partial_\nu h_\mu . \tag{2.9}$$

If we demand that h_μ (and its derivative) is the only field accompanying $F_{\mu\nu}$ in its indecomposable conformal law, then conformal invariance of Eq.(2.2) and of the 2-point function (2.5) imply the following infinitesimal special conformal transformations:

$$2[F_{\mu\nu}(x), C_\lambda] = \{x_\lambda(2 + x\partial) - \tfrac{1}{2}x^2\partial_\lambda\} F_{\mu\nu}(x) + x_{[\mu} F_{\nu]\lambda}(x) - x^\rho \eta_{\lambda[\mu} F_{\nu]\rho}(x) +$$

$$+ \frac{ae^2}{12\pi^2}\left(x_\lambda G_{\mu\nu}(x) + \eta_{\lambda[\mu} h_{\nu]}(x)\right) \qquad (x_{[\mu} F_{\nu]\lambda} = x_\mu F_{\nu\lambda} - x_\nu F_{\mu\lambda} \ etc.); \tag{2.10a}$$

we fix the remaining freedom by demanding that the conformal law for h_μ is homogeneous:

$$2[h_\mu(x), C_\lambda] = \{x_\lambda(1 + x\partial) - \tfrac{1}{2}x^2\partial_\lambda\} h_\mu(x) + \eta_{\mu\lambda}\, x.h(x) - x_\mu h_\lambda(x). \tag{2.10b}$$

Eqs. (2.5) and (2.10) also determine the 2-point functions of h:

$$\langle F_{\mu\nu}(x)\, h_\rho(0)\rangle_0 = \frac{\eta_{\rho[\mu}\, x_{\nu]}}{2\pi^2 x_+^4}\,, \tag{2.11a}$$

$$\langle h^\mu(x)\, h_\nu(0)\rangle_0 = \frac{3\, r^\mu{}_\nu(x)}{4 e^2 x_+^2}\,. \tag{2.11b}$$

Remark Since h_μ is an auxiliary field, which does not have a counterpart in the conventional formulation of QED, its normalization is arbitrary. Our choice, which allows to write down the 2-point function (2.11a) as the curl of a free vector-field propagator, $\langle F_{\mu\nu}(x)\, h_\rho(0)\rangle_0 = (\partial_\mu \eta_{\nu\rho} - \partial_\nu \eta_{\mu\rho})\,\frac{x_+^{-2}}{4\pi^2}$ will be truly justified in Sec. 3, where it will be demonstrated to yield equality among three charges: one defined by a standard Ward identity, another appearing as coupling constant of a Dirac-like equation (involving the field h_μ in the place of the vector potential), and a third, the coefficient in the exponential in the definition (3.14) of the current as a composite operator. Conversely, the normalization of $h^\mu(x)$ is uniquely determined by the requirement that these three potentially different charges actually coincide. We also note that an indecomposable law of the type (2.10) together with the corresponding invariant Maxwell equation (2.13) below has been considered in the first reference [S3] . General indecomposable conformal laws were studied recently in [F6].

The dilation invariance of the first Maxwell equation (2.3) indicates that $\partial_\nu G^{\mu\nu}(x) = 0$ at least in the weak sense. Thus, h_μ behaves as a free vector field with a longitudinal propagator (although it is not purely longitudinal, since its curl, $G_{\mu\nu}$ has a non-vanishing 2-point function (2.8) with $F_{\kappa\lambda}$). The following assumption will essentially fix the remaining freedom in the Bose sector of our model, and will serve, at the same time, as a characteristic of the physical subspace.

III. The field h_μ has zero matrix elements between physical states; in particular, it is orthogonal to the current:

$$\langle h_\mu(x_1)\, J_\nu(x_2)\rangle_0 = 0. \tag{2.12}$$

To complete our discussion of the Bose sector, there remain three more questions. We should like to find: (i) the(possibly indecomposable) conformal law for J_μ, consistent with current conservation; (ii) the equations of motion for h_μ; (iii) a conformal invariant first Maxwell equation, consistent with (2.1). The three questions appear closely intertwined. We are unable to settle anyone of them without taking the two others. We shall therefore give the answer to all three and then exhibit their interdependence and consistency.

<u>Proposition 2.2.</u> The operator Maxwell equation

$$\partial_\nu F^{\mu\nu}(x) - \frac{a e^2}{24\pi^2}\partial^\mu \partial h(x) = J^\mu(x) \tag{2.13}$$

is invariant under the infinitesimal law (2.10) provided that the current obeys the indecomposable conformal law

$$2[J^\mu(x), C_\lambda] = \left\{x_\lambda(3 + x\partial) - \tfrac{1}{2}x^2\partial_\lambda\right\} J^\mu(x) + \delta^\mu_\lambda\, x.J(x) - x^\mu J_\lambda(x) + \\ + \frac{a e^2}{6\pi^2} G^\mu{}_\lambda(x) + \frac{a e^2}{12\pi^2} x_\lambda \partial_\nu G^{\mu\nu}(x). \tag{2.14}$$

Current conservation requires - and is ensured by - the conformal gauge condition

$$\Box \partial h = 0. \tag{2.15}$$

Its conformal invariance makes necessary the operator equation

$$\partial_\nu G^{\mu\nu} = 0 \qquad \text{or} \quad \Box h^\mu = \partial^\mu \partial h \,. \tag{2.16}$$

Eqs. (2.13-16) are uniquely determined by the requirements that they should be linear equations involving $F^{\mu\nu}$, h_μ and J^μ, provided our three postulates are satisfied.

The proof of the first (direct) part of the proposition is straightforward. The uniqueness claim can be be verified as follows. The most general indecomposable conformal law for J^μ could also involve h_μ directly (i.e. not only through $G_{\mu\nu}$). However, such a contribution would contradict the conformal invariance of (2.12) and should therefore be dropped. Given the laws (2.10) and (2.14), it is straightforward to verify that Eq.(2.13) is the only linear in F, h and J equation invariant under these laws and reducing to the physical space Maxwell equation (2.3). Eqs. (2.15) and (2.16) then follow from current conservation (and its invariance) as indicated.

We can thus conclude our study of the Bose sector of the model by demanding

IV. There is a conformally invariant (linear) form of the Maxwell equation involving only F, h and J and therefore coinciding (due to Proposition 2.2) with Eq. (2.13).

Eqs. (2.15) and (2.16), indicating that h^μ behaves as a (generalized) free field, then follow as a corollary from Proposition 2.2.

2B. A charged spinor field. Ward identities

Following the general pattern of the discussion of the Bose sector we shall again start with a model independent assumption, valid beyond the conformal limit. Let $\psi(x)$ be the "electron" field and $\bar{\psi}$, its Dirac conjugate. We should like to make sure that $J^0(x)$ is the density of the charge operator that generates phase transformations for ψ and $\bar{\psi}$.

V. The operator Ward identities (WI's) for time ordered product

$$\frac{\partial}{\partial x^\mu}(T J^\mu(x) \psi(y)) = \delta(x^0-y^0)[J^0(x), \psi(y)] = -e\,\delta(x-y)\,\psi(y)$$
$$\frac{\partial}{\partial x^\mu}(T J^\mu(x) \bar{\psi}(y)) = e\,\delta(x-y)\,\bar{\psi}(y)\,, \qquad \delta(z) \equiv \delta^4(z) \tag{2.17}$$

are satisfied. They will serve as a definition of the electric charge e.

Corollary. Eq.(2.17) implies the standard relations between 2- and 3-point functions. Combined with the Maxwell equation (2.13) it gives

$$-\frac{ae}{24\pi^2}\frac{\partial}{\partial x_3^\mu}\langle T\psi(x_1)\bar{\psi}(x_2)\Box h^\mu(x_3)\rangle_0 = \langle T\psi(x_1)\bar{\psi}(x_2)\rangle_0\left[\delta(x_{23})-\delta(x_{13})\right]. \tag{2.18}$$

The analysis [S5] of the OPE of $h^\mu(x+z)\psi(x)$ (summarized in Sec. 3A below), conformal invariance (with the elementary conformal law (2.10b) for h_μ) and the (free) equations of motion for h_μ together with (2.18) allow to find the 3-point function of the charged field with h_μ :

$$ae\,\langle\psi(x_1)\bar{\psi}(x_2)h^\mu(x_3)\rangle_0 = 3i\,\langle\psi(x_1)\bar{\psi}(x_2)\rangle_0\left(\frac{x_{13}^\mu}{x_{13+}^2} - \frac{x_{23}^\mu}{x_{23+}^2}\right), \tag{2.19}$$

so that

$$\langle\psi(x_1)\bar{\psi}(x_2)\,G_{\mu\nu}(x_3)\rangle_0 = 0\,. \tag{2.20}$$

We now proceed to formulate the conformal law for ψ implicit in (2.19).

VI. The Dirac field $\psi(x)$ of dimension $\frac{3}{2}+\lambda$ satisfies the homogeneous special conformal transformation law

$$2[\psi(x), C_\mu] = \left\{x_\mu\left(\tfrac{3}{2}+\lambda+x\partial\right) - \tfrac{1}{2}x^2\partial_\mu + \tfrac{1}{4}[\gamma_\mu, \not{x}]\right\}\psi(x). \tag{2.21}$$

The theory is also γ_5 - invariant i.e. invariant under the involutive transformation

$$\psi(x)\otimes\bar{\psi}(y)\to-\gamma_5\psi(x)\otimes\bar{\psi}(y)\gamma_5 \quad (\gamma_5^2=1),\; J_\mu\to J_\mu\,,\; h_\mu\to h_\mu\,. \tag{2.22}$$

Corollary: The genenral conformal and γ_5-invariant 2-point function of ψ is

$$\langle\psi(x)\bar{\psi}(0)\rangle_0 = \frac{N_\lambda}{2\pi^2}\,\frac{\not{x}}{(x_+^2)^{2+\lambda}} = \tag{2.23a}$$

$$= -\frac{2\pi N_\lambda}{4^\lambda\Gamma(\lambda)\Gamma(2+\lambda)}\int\theta(p^0)\,i\not{p}\,(-p^2)_+^{\lambda-1}\,e^{ipx}\,d_4p\,; \tag{2.23b}$$

a standard choice for the normalization constant N_λ is

$$N_\lambda = 4^\lambda\Gamma(2+\lambda) \qquad (N_0=1)\,; \tag{2.24}$$

it guarantees Wightman positivity for $\lambda\geq 0$ and fits the free (massless) Dirac propagator for $\lambda = 0$.

We end our list of general requirements by the following assumption concerning the fields entering the OPE of the product of ψ with $\bar{\psi}\gamma_\mu$ and h_μ. (An additional restriction on these fields will be needed in the derivation of the "Dirac quation" in Sec. 3A below.)

VII. There exist a conformal OPE for $\bar{\psi}(x+z)\gamma_\mu\psi(x)$ only involving the fields h_μ, $F_{\mu\nu}$, their derivatives and their normal products. The lowest dimensional field entering the OPE of $h_\mu(x+z)\psi(x)$ is ψ. All fields entering these OPEs that are not linear in ψ (and ψ_+, for $\lambda=0$), h,F (and their derivatives) are othogonal to $\bar{\psi}$, h_μ, $F_{\mu\nu}$ respectively.

Remarks: (1) The last property (orthogonality) is automatic for disjoint representations since 2-point functions play the role of intertwining maps.

(2) Whenever studied, the convergence of OPEs has only been established when both sides are applied to a finite energy state (like the vacuum). Combining these results with locality we can also use OPEs within correlation functions, if restricted to a space-time region that is space-like separated from the arguments of the "spectator fields" in the vacuum expectation value.

3. OPEs AND NONLINEAR EQUATIONS OF MOTION

3A. The $h_\mu \psi$ expansion. A substitute for the Dirac equation

Using the postulates of Sec.2 for the contribution of the field ψ and its derivatives to the OPE of $h_\mu \psi$ one derives (for more details see [S5]):

$$\frac{iea}{3} h_\mu(x+z)\psi(x) = \frac{z_\mu}{z_+^2}\psi(x) + \Gamma(\lambda)\sum_{n=0}^{\infty}\frac{1}{n!}\int_0^1 du\left(-\frac{u(1-u)}{4}z^2\Box\right)^n\left\{\frac{(1-u)^{1+\lambda}}{2\Gamma(n+\lambda)}\partial_\mu + \right.$$

$$\left. + \frac{(1-u)^{1+\lambda}}{4\Gamma(n+1+\lambda)}\gamma_\mu \not\partial + \frac{u(1-u)^{1+\lambda}}{4\Gamma(n+1+\lambda)}\not z \not\partial \partial_\mu - \frac{z_\mu (1-u)^{2+\lambda}}{4\Gamma(n+1+\lambda)}\Box - \frac{z_\mu \not z\, u(1-u)^{2+\lambda}}{8\Gamma(n+2+\lambda)}\Box\not\partial\right\}\psi(x+uz) + \tag{3.1a}$$

$$+ R_\mu(x,z) =$$

$$= \left\{\frac{z_\mu}{z_+^2} + \frac{1}{4(2+\lambda)}\left(2\partial_\mu + \frac{1}{\lambda}\gamma_\mu\not\partial\right)\right\}\psi(x) + O(z), \tag{3.1b}$$

where ∂ stands for $\partial/\partial x$, $R_\mu(x,z)$ has a vanishing 2-point function with $\bar\psi(y)$ (according to postulate VII). It is easy to verify that the approximate expression (3.1b) indeed reproduces the singular and the finite (nonvanishing) terms of the small z expansion of the 3-point function (2.19):

$$ae\langle h_\mu(x+z)\psi(x)\bar\psi(0)\rangle_0 = 3i\langle\psi(x)\bar\psi(0)\rangle_0\left\{-\frac{z_\mu}{z_+^2} + \frac{x_\mu}{x_+^2} + O(z)\right\}. \tag{3.2}$$

To this end it is sufficient to use (2.22) to derive the relation

$$-\frac{1}{2(2+\lambda)}\left(\partial_\mu + \frac{1}{2\lambda}\gamma_\mu \not{\partial}\right)\langle \psi(x)\bar{\psi}(0)\rangle_0 = \frac{x_\mu}{x_+^2}\langle \psi(x)\bar{\psi}(0)\rangle_0 . \tag{3.3}$$

The missing terms in (3.1b) can only be smaller (for $z\to 0$) than those kept in the right hand side, if we assume that the remainder $R_\mu(x,z)$ tends to zero for $z\to 0$; that would be the case if there were no conformal fields of dimension $5/2+\lambda$ contributing to R_μ. In order to obtain a Dirac-like equation we only need the following weaker assumption (in addition to the postulates of Sec. 2).

VIII. The remainder $R_\mu(x,z)$ in Eq.(3.1a) satisfies in the limit $z\to 0$ a Rarita-Schwinger type subsidiary condition

$$\gamma^\mu R_\mu(x,0) = 0 . \tag{3.4}$$

(A similar requirement has been used by M. Palchik [P1] in a conformal OPE approach to the Thirring model.)

<u>Corollary.</u> If we define the renormalized normal product

$$N(\not{h}\psi)(x) = \lim_{z\to 0} \frac{\not{h}(x+z)+\not{h}(x-z)}{2}\psi(x) \tag{3.5a}$$

$$= \lim_{z\to 0}\left\{\left(\not{h}(x+z) + \frac{3i}{ae}\frac{\not{z}}{z_+^2}\right)\psi(x)\right\} \tag{3.5b}$$

then Eqs.(3.1) and (3.4) imply the 'Dirac equation'

$$\not{\partial}\psi(x) \quad \left(= i\,\frac{2a\lambda e}{3} N(\not{h}\psi)(x)\right) = ie\,N(\not{h}\psi)(x) \tag{3.6}$$

for anomalous dimension λ related to the current normalization "a" by

$$\lambda = \frac{3}{2a} . \tag{3.7}$$

We note that the similarity between Eq.(3.6) and the massless Dirac equation is somewhat deceiving: the field h_μ differs in an essential way from the electromagnetic potential A_μ , its propagator (2.11b) being purely longitudinal. One should be, therefore, ready to explore a violation of Eq.(3.7) studying, in particular, the limit $\lambda \to 0$ (for $e \neq 0$) – see Sect. 4 below. For $\lambda \neq 0$, however, we shall indeed adopt (3.7).

Eq.(3.6) is the first non-linear relation in our model – obtained from the OPE (3.1) and the assumption (3.4). To complete our analysis of dynamical equations we only need to express the current J^μ as a composite of the charged fields ψ and $\bar{\psi}$. Before doing that we shall exhibit the gauge freedom inherent to Eqs.(2.4) (2.13) and (3.6), since it will give us an important criterion for writing down higher order terms in the OPE for $\bar{\psi}\gamma_\mu\psi$.

Let s(x) be a dimensionless conformal field (of an elementary – i.e. homogeneous – conformal law) satisfying the invariant 4th order equation

$$\Box^2 s = 0. \tag{3.8}$$

Then the Bose sector equations of the model are invariant under gauge transformations of the type

$$h_\mu(x) \to h_\mu(x) + \partial_\mu s(x), \tag{3.9}$$

$$F_{\mu\nu}(x) \to F_{\mu\nu}(x) + \tilde{F}_{\mu\nu}(x) \qquad (J_\mu(x) \to J_\mu(x)) \tag{3.10a}$$

where $\tilde{F}_{\mu\nu}$ is a solution of the free equation [F5]

$$\partial_\nu \tilde{F}^{\mu\nu}(x) = \frac{a e^2}{24\pi^2} \partial^\mu \Box s(x). \tag{3.10b}$$

Eq.(3.6) with λ given by (3.7) also remains invariant under these transformations – at least at the classical level – provided

$$\psi(x) \longrightarrow e^{ies(x)}\psi(x). \tag{3.11}$$

It can also be given meaning at the quantum level defining the normal ordered exponential for the generalized free field s(x).

3B. The OPE for $\bar{\psi}\gamma^\mu\psi$. A gauge invariant definition of the current.

A central result of [S5] is the derivation of the following generalized conformal OPE

$$\begin{aligned}\bar{\psi}(x+z)\gamma^\mu\psi(x) = \mathrm{tr}\Big\{\gamma^\mu\langle\bar{\psi}(x+z)\psi(x)\rangle_0 : \exp\Big\{-ie\int_0^1 du\Big[z_\kappa h^\kappa(x+uz) + \\ + \frac{B+1}{16}\left(z^2\gamma_\kappa\gamma_\nu - 4(1-u)\not{z}\, z_\kappa\gamma_\nu\right)G^{\kappa\nu}(x+uz) - \frac{3i\pi^2}{ae}\sum_{n=0}^{\infty}\frac{1}{(n!)^2}\int_0^1 du\, u(1-u)\times \\ \times\left(-\frac{u(1-u)}{4}z^2\Box\right)^n\Big[\frac{z^2 z_\nu}{n+1} + \frac{B+1}{4}\Big(\frac{z^4\not{\partial}\gamma_\nu}{2(n+1)} + \frac{1-u}{n+1}z^2\not{z}(\not{\partial}z_\nu - z\partial\gamma_\nu) + \\ + 2z^2(\not{z}\gamma_\nu - z_\nu)\Big)\Big]J^\nu(x+uz)\Big\}:\Big\} + \ldots\end{aligned} \tag{3.12}$$

where J^ν should be regarded as a shorthand for the combination of Bose fields in the left hand side of the Maxwell equation (2.13). This expansion reproduces the 3-point functions (2.19),

$$\begin{aligned}\langle\bar{\psi}(x+z)\gamma^\mu\psi(x)J_\nu(0)\rangle_0 = \frac{ieN_\lambda}{2\pi^4 z_+^{2\lambda}}\Big\{(B+1)\frac{2x^\mu x_\nu + x^\mu z_\nu + z^\mu x_\nu - \delta^\mu_\nu(x^2+xz)}{(x+z)_+^4\, x_+^4} - \\ -(B-1)\frac{z^\mu}{z_+^2 x_+^2 (x+z)_+^2}\Big(\frac{x_\nu}{x_+^2} - \frac{x_\nu+z_\nu}{(x+z)_+^2}\Big)\Big\},\end{aligned} \tag{3.13}$$

as well as the one with $F_{\mu\nu}$; it is conformal and gauge invariant. Additional contributions to the right hand side (if any) should be orthogonal to the basic Bose fields and gauge covariant by themselves (they are also expected to vanish for $z \to 0$). Eq.(3.12) (unlike (3.1)) does not define a standard OPE, though it is not written as a series in a set of conformal fields and their derivatives. Instead, it involves a normal exponential of a combination of Bose fields (reminiscent to the one encountered in the operator solution of the massless Thirring model - see [D2]).

Eq.(3.12) allows, in particular, to define the current J_μ as a composite field by

$$J_\mu(x) = i a e \lim_{z\to 0} \frac{\partial}{\partial z^\mu}\left[\frac{z^{2\lambda}}{N_\lambda}\bar{\psi}(x+z)\not{z}\psi(x) - \frac{2}{\pi^2 z^2}:\exp\left(-ie\int_0^1 du\, z.h(x+uz)\right):\right] =$$

$$= i a e \lim_{z\to 0} \frac{\partial}{\partial z^\mu}\left[\frac{z^{2\lambda}}{N_\lambda}:\bar{\psi}(x+z)\not{z}\exp\left(ie\int_0^1 du\, z.h(x+uz)\right)\psi(x):\right] \tag{3.14a}$$

where

$$h_\mu(x) = \frac{\pi^2}{2N_\lambda}\lim_{z\to 0}\frac{\partial}{\partial z^\mu}\left[(z^2)^{1+\lambda}\bar{\psi}(x+z)\not{z}\psi(x)\right]. \tag{3.14b}$$

It is the identification of the factor in the exponential with the charge e defined by (2.17) that has dictated the normalization of h (cf. Remark after Eq. (2.11)). Its equality with the coupling constant in the Dirac equation (3.6) - achieved for the special choice (3.7) of the anomalous dimension - is instrumental in deducing a gauge invariant "OPE" of type (3.12) by just exponentiating the term linear in the Bose field (a term which, in turn, is deduced from conformal invariance and comparison between 2- and 3-point functions.

3C. Remark: a singular lagrangian for the model?

The formal consistency of our QED model is confirmed by the possibility to write down a lagrangian density

$$L = \frac{1}{2} F^{\mu\nu}(G_{\mu\nu} - \partial_\mu h_\nu + \partial_\nu h_\mu) - \frac{1}{2}(\partial_\mu A_\nu - \partial_\nu A_\mu) G^{\mu\nu} -$$

$$- \frac{a e^2}{48\pi^2}(\partial h)^2 + ie\bar{\Psi}\not{h}\Psi - \frac{1}{2}\bar{\Psi}\overleftrightarrow{\not{\partial}}\Psi \tag{3.15a}$$

$$= -\frac{1}{2}(\partial^\mu A^\nu - \partial^\nu A^\mu)(\partial_\mu h_\nu - \partial_\nu h_\mu) - \frac{a e^2}{48\pi^2}(\partial h)^2 - \bar{\Psi}(\not{\partial} - ie\not{h})\Psi \tag{3.15b}$$

which yields all equations of motion at the classical level. (The form (3.15b) is obtained from (3.15a) by excluding $F_{\mu\nu}$ and $G_{\mu\nu}$ from the constraints obtained by varying (3.15a) with repsect to these two fields.) Eq.(2.16) is obtained by varying with respect to A_μ; the Maxwell equation (2.13) follows by variation with respect to h_μ. Eq.(2.2) is automatic for $F_{\mu\nu} = \partial_\mu A_\nu - \partial_\nu A_\mu$ (which is obtained from (3.15a) by variation in $G^{\mu\nu}$). Current conservation – for $J^\mu = ie\bar{\Psi}\gamma^\mu\Psi$ – is a consequence of the "Dirac" equation $(\not{\partial} - ie\not{h})\Psi = 0$ and its conjugate; it then yeilds Eq.(2.15) as before.

We shall now show how does one find the h-field propagator form such a singular lagrangian.

The Euclidean h-propagator should appear, in momentum space, as the "inverse" of the singular matrix $\frac{a e^2}{12\pi^2} p^\mu p_\nu$. To define it properly we first invert the regularized matrix $\frac{a e^2}{24\pi^2} p^\mu p_\nu + \varepsilon \Pi^\mu{}_\nu p^2$ with the result

$$\Delta^\mu{}_\nu(p,\varepsilon) = \frac{1}{p^2}\left(\frac{1}{\varepsilon}\Pi^\mu{}_\nu + \frac{24\pi^2}{a e^2}\frac{p^\mu p_\nu}{p^2}\right), \quad \left(\Pi^\mu{}_\nu = \delta^\mu_\nu - \frac{p^\mu p_\nu}{p^2}\right) \tag{3.16}$$

The h-field p-space Schwinger function is now defined as the limit

$$\Delta^\mu{}_\nu(p) = \lim_{\varepsilon\to 0}\frac{d}{d\varepsilon}\left\{\varepsilon\Delta^\mu{}_\nu(p,\varepsilon)\right\} = \frac{12\pi^2}{a e^2}\frac{p^\mu p_\nu}{p^4}, \tag{3.17a}$$

so that

$$\langle h^{\mu}(x) h_{\nu}(0) \rangle_E = \frac{24\pi^2}{ae^2} \int \frac{p^{\mu} p_{\nu}}{p^4} e^{ipx} d_4 p = \frac{3}{ae^2} \frac{r^{\mu}{}_{\nu}(x)}{x^2}, \tag{3.17b}$$

in accord with (2.11b). A similar argument allows to recover the 2-point function (2.11a). By contrast, the Lagrangian (3.15) does not lead to the $F_{\mu\nu}$ propagator (2.5), which could, in principle, be recovered if we add to (3.15) a non-local quadratic ("kinetic") term in A_{μ}. That, however would change the equations of motion for h_{μ}, and would (according to our preliminary analysis) only be consistent with a weak form of current conservation. We conjecture that such an extended (non-local) Lagrangian model does in fact exist and differs from ours by a wider unphysical sector. (The two models should lead to the same correlation functions for the fields h_{μ}, $F_{\mu\nu}$, ψ and $\bar{\psi}$.) If this is correct, then the OPE approach will appear to provide a more economic (i.e. using a smaller set of field operators and states) method for constructing a QED model with the same physical content.

4. Indecomposable conformal spinors for zero anomalous dimesnion

There is no smooth limit for $\lambda \to 0$ in the OPE (3.1) consistent with a non-vanishing electric charge. We shall show, however, that there exists a model for $\lambda = 0$ satisfying our basic postulates (at the same level of 2- and 3-point functions, as the model of Sec. 3 with $\lambda = \frac{3}{2a}$) provided we introduce an indecomposable 8-component conformal spinor (ψ, ψ_+) (obeying (1.1)) and adopt a modified form of the "Dirac" equation.

Proposition 4.1 The counterpart of the OPE (3.1) in a theory satisfying I-VII but involving an indecomposable pair (ψ, ψ_+) of weight $(\frac{3}{2}, \frac{5}{2})$ (instead of a Dirac field ψ of dimension $3/2 + \lambda$, $\lambda = \frac{3}{2a} > 0$) is

$$ie\,h_\mu(x+z)\Psi(x) = \sum_{n=0}^{\infty} \frac{1}{(n!)^2} \int_0^1 du\,(1-u)\, e^{uz\partial}\left(-z^2 \frac{u(1-u)}{4}\Box\right)^n \Big\{\Big(z_\mu (1-u)\not\partial +$$

$$+ \frac{1}{2} z_\mu \not z \frac{u(1-u)}{n+1}\Box - \gamma_\mu + \frac{1}{2} z^2 \frac{u(1-u)}{n+1} \not\partial \partial_\mu - \not z\, u \partial_\mu\Big)\Big(\frac{3}{4a}\Psi_+(x) - b\not\partial\Psi(x)\Big)$$

$$- \frac{3}{4a}\Big(z_\mu \frac{1-u}{n+1}\Box - 2\partial_\mu\Big)\Psi(x)\Big\} + \frac{3}{a}\frac{z_\mu}{z_+^2}\Psi(x) + R_\mu(x,z)$$

$$(\langle R_\mu \bar\Psi\rangle_0 = 0 = \langle R_\mu \bar\Psi_+\rangle) \tag{4.1}$$

where ∂_μ and $\Box$ are differentiations with respect to x (as in (3.1)) Eqs.(4.1) and (3.4) imply the generalized Dirac equation

$$\not\partial\Psi(x) - \frac{3}{2a}\Psi_+(x) = ie\,N(\not h\Psi)(x) \qquad \text{for } b = \frac{1}{2}\left(1 + \frac{3}{4a}\right) \tag{4.2}$$

where the normal product $N(\not h\Psi)$ is defined as in (3.5).

<u>Sketch of proof</u>: In verifying the infinitesimal conformal invariance of (4.1) we set x=0 and use

$$(z\partial)^m \Box^n (x_\nu(x\partial + d) - \frac{1}{2}x^2\partial_\nu)\,\phi(x)\Big|_{x=0} =$$

$$= \{m(2n+m+d-1)\, z_\nu (z\partial)^{m-1}\Box^n - \frac{1}{2} m(m-1) z^2 (z\partial)^{m-2}\Box^n \partial_\nu +$$

$$+ 2n(n+d-2)(z\partial)^m \Box^{n-1}\partial_\nu\}\,\phi(0) \qquad \text{for } \phi = \Psi, \not\partial\Psi, \Psi_+ .$$

To prove that (4.1) reproduces the 3-point function (2.19) we first note that the term with coefficient b does not contribute to this matrix element since $\langle \not\partial\Psi(x)\bar\Psi(0)\rangle_0 = 0$. Secondly, we use the indecomposable law (1.1) to derive

$$\langle \psi_+(x)\bar{\psi}(0)\rangle_0 = -\frac{2\not{x}}{x_+^2}\langle \psi(x)\bar{\psi}(0)\rangle_0 \quad \left(= \frac{-1}{\pi^2 x_+^4}\right) \tag{4.3}$$

(the last equation being valid for the canonical normalization of the free electron propagator). To verify that (4.1) does indeed yield (2.19) we multiply both sides of (4.1) by $\bar{\psi}(0)$, take the vacuum expectation value, perform the differentiations and the summation, and finally, do the integral in u using repeatedly the identity

$$\frac{1}{(x+uz)^2\{(x+uz)^2+u(1-u)z^2\}} = \frac{1}{u(1-u)z^2}\left\{\frac{1}{(x+uz)^2} - \frac{1}{(x+uz)^2+u(1-u)z^2}\right\}$$

along with the integration formula

$$\int_0^1 \frac{u^{\alpha-1}(1-u)^{\beta-1}du}{(au+b)^{\alpha+\beta+n}} = \frac{1}{(a+b)^\alpha b^\beta}\sum_{k=0}^{n}\binom{n}{k}\frac{B(\alpha+k,\beta+n-k)}{(a+b)^k\, b^{n-k}} \tag{4.4}$$

where $B(\mu,\nu) = \dfrac{\Gamma(\mu)\Gamma(\nu)}{\Gamma(\mu+\nu)} = \displaystyle\int_0^1 u^{\mu-1}(1-u)^{\nu-1}du$.

The proof that these two properties actually determine the contribution to the OPE (4.1) from ψ and ψ_+ follows the pattern of ref. [S5].

In the above argument we have only verified the consistency of the expansion (4.1) with the 3-point function (2.19). The conformally invariant time ordered Green function

$$ae\langle T h_\mu(x_3)\psi(x_1)\bar{\psi}_+(x_2)\rangle_0 = \frac{-3i}{4\pi^2(x_{12}^2+i0)^2}\partial_{3\mu}\frac{\not{x}_{31}\not{x}_{32}}{(x_{32}^2+i0)} = \frac{3i\not{x}_{12}}{4\pi^2(x_{12}^2+i0)}\partial_{3\mu}\frac{\not{x}_{32}}{x_{32}^2+i0}$$

satisfies

$$-\frac{ae}{24\pi^2}\Box_3\partial_3^\mu\langle T h_\mu(x_3)\psi(x_1)\bar{\psi}_+(x_2)\rangle_0 =$$

$$=\langle T\psi(x_1)\bar{\psi}_+(x_2)\rangle_0\left(1+\frac{1}{4}\not{x}_{31}\not{\partial}_3\right)\delta(x_{32})$$

$$=\frac{1}{2}\langle T\psi(x_1)\bar{\psi}(x_2)\rangle_0\,\not{\partial}_3\,\delta(x_{32})$$

instead of (the counterpart of) (2.18) and thus violates the Ward identity *) involving ψ_+ Another possible contribution to the above 3-point function, proportional to $\frac{\not{x}_{31}\gamma_\mu}{x^2_{12+}x^2_{31+}x^2_{32+}}$ has a non-trivial transverse part (i.e., a non-vanishing curl) and hence violates the equations of motion for h_μ, combined with Eq.(2.20).

The admissible conformal structure for $\langle h_\mu\psi\bar{\psi}_+\rangle_0$ which does agree with both equations of motion and the Ward identity has a form similar to (2.19):

$$ae\langle h^\mu(x+z)\psi(x)\bar{\psi}_+(0)\rangle = 3i\left(\frac{x^\mu+z^\mu}{(x+z)^2_+}-\frac{z^\mu}{z^2_+}\right)\langle\psi(x)\bar{\psi}_+(0)\rangle_0 \tag{4.5}$$

where $\langle\psi(x)\bar{\psi}_+(0)\rangle_0(=\langle\psi_+(x)\bar{\psi}(0)\rangle_0)$ is given by (4.3). On the other hand, the conformal invariant 2-point function $\langle\psi_+(x)\bar{\psi}_+(0)\rangle_0$ is proportional to $\langle\not{\partial}\psi(x)\bar{\psi}_+(0)\rangle_0$:

$$\langle\psi_+(x)\bar{\psi}_+(0)\rangle_0 = N_+\,\frac{4\not{x}}{\pi^2 x^6_+}\;. \tag{4.6}$$

It is straightforward to verify that the OPE (4.1), for b given by (4.2), only agrees with Eq.(4.5) if

$$\left\langle\left\{\frac{3}{4a}\left(\psi_+(x)-\not{\partial}\psi(x)\right)-\not{\partial}\psi(x)\right\}\bar{\psi}_+(0)\right\rangle_0 = 0 \tag{4.7}$$

---

*) It agrees, however, with the neutrino interpretation of the projection of ψ that satisfies the free massless Dirac equation discussed in the concluding remark below.

or

$$\mathcal{N}_+ = 1 + \frac{4}{9} a. \tag{4.8}$$

Remark The conformal law for $\bar{\Psi}_+$ (needed to derive (4.5) and (4.6)) is obtained from (1.1b) by conjugation; in particular $[\bar{\Psi}_+(0), C_\nu] = -\bar{\Psi}\gamma_\nu$. The γ_5 law (2.22) extends to Ψ_+ as follows:

$$\Psi_+(x) \otimes \bar{\Psi}(y) \to \gamma_5 \Psi_+(x) \otimes \bar{\Psi}(y)\gamma_5, \; \Psi_+(x) \otimes \bar{\Psi}_+(y) \to -\gamma_5 \Psi_+(x) \otimes \bar{\Psi}_+(y)\gamma_5. \tag{4.9}$$

To summarize our discussion we conclude with the following remarks.

Some very natural postulates - conventional Maxwell equations and Ward identities in a conformal OPE framework with a non-vanishing current propagator - force us to describe the charged spinor field for zero anomalous dimension by an indecomposable pair (Ψ, Ψ_+) of weight $(\frac{3}{2}, \frac{5}{2})$ satisfying the modified Dirac type equation (4.2). The model is characterized by three constants: the electric charge e (with its three equivalent definitions discussed in the remark after Eq.(2.11)), the coefficient a in the current propagator (2.3) and the factor B in front of the transverse part of the current-field 3-point function (3.13). There is a distinguished value of B, B=-1, for which the first term in the braces in (3.13) vanishes. For such a B the charge density J^0 appears to commute with the 0-mass projection of the field Ψ. This would allow to interpret the invariant subspace of the one-spin ½-particle space, spanned by solutions of the massless Dirac equation as describing neutrinos - as proposed in [P2].

We do not know whether constructing 4-point (and higher) correlation function consistent with the conformal OPEs and with the equations of motion will further restrict the remaining parameters. It is conceivable that the model does exist for a continuous range of values of the electric charge e (as it is the case with the massless Thirring model).

Acknowledgements

It is a pleasure to thank Galen Sotkov for numerous discussions. One of us (I.T.) would like to thank Professor P. Budinich for his hospitality at the International School for Advanced Studies in Trieste.

REFERENCES

[A1] S. Adler, Short distance behaviour of quantum electrodynamics and an eigenvalue condition for α , Phys. Rev. D5(1972) 3021-3047, ibid D7 (1973) 1948

[A2] S. Adler, W. Bardeen , Quantum electrodynamics without photon self-energy parts: an application of Callan-Symanzik scaling equations, Phys.Rev. D4 (1971) 3045-3054, ibid D6 (1972) 734.

[B1] F. Bayen, M. Flato, C. Fronsdal, A. Haidari, Conformal invariance and gauge fixing in QED, Phys. Rev. D32 (1985) 2673-2682.

[B2] B. Binegar, C. Fronsdal, W. Heidenreich, Conformal QED, J. Math. Phys. 24 (1983) 2828-2846.

[B3] P. Budinich, Quarks as conformal semispinors, ICTP, Trieste preprint IC/79/88.

[B4] P. Budinich, P. Furlan, On a "conformal spinor wave equation", Phys. Lett. 107B (1981) 434-436.

[C1] L. Castell, Analysis of spacetime structure in elementary particle physics, Nuovo Cim. 46A (1966) 1-38; The relativistic position operator at subatomic level, Nuovo Cim. A49 (1967) 285-290; The physical aspects of the conformal group $SO(4,2)_o$, Nucl. Phys. B4 (1968) 343-352.

[C2] N.S. Craigie; V.K. Dobrev, I.T. Todorov, Conformal covariant composite operators in quantum chromodynamics, Ann. Phys. (N.Y.) 159 (1985) 411-444.

[D1] G.F. Dell'Antonio, On dilation invariance and the Wilson expansion, Nuovo Cim. 12A (1972) 756-762.

[D2] G.F. Dell'Antonio, Y. Frishman, D. Zwanziger, Thirring model in terms of currents: solution and lightcone expansion, Phys. Rev. D6 (1972) 988-1007.

[D3] P.A.M. Dirac, (a) Wave equations in conformal space, Ann. Math. 37 (1936) 429-442; (b) Relativistic wave equations, Proc. Roy. Soc. London Ser. A 155 (1936) 447-459.

[D4] V. Dobrev, G. Mack, V.B. Petkova, S.G. Petrova, I.T. Todorov, Harmonic Analysis, Lecture Notes in Physics 63 (Springer, Berlin 1977).

[F1] S. Ferrara, R. Gatto, A. Grillo, Logarithmic scaling and spontaneous breaking, Phys. Lett. 42B (1972) 264-266.

[F2] S. Ferrara, R. Gatto, A. Grillo, Conformal Algebra in Space-Time and Operator Product Expansion, Springer Tracts in Modern Physics 67 (Springer, Berlin 1973).

[F3] R. Ferrari, L. Picasso, F. Strocchi, Some remarks on local operators in quantum electrodynamics, Commun. Math. Phys. 35 (1974)25-38.

[F4] E.S. Fradkin, A.A. Kozhevnikov, M. Ya. Palchik, A.A. Pomeransky, Maxwell equations in conformal invariant electro dynamics, Comm. Math. Phys. 91 (1983) 529-541.

[F5] P. Furlan, V.B. Petkova, G.M. Sotkov, I.T. Todorov, Conformal quantum electrodynamics and nondecomposable representations, Riv. Nuovo Cim. 8, n. 3 (1985) 1-50.

[F6] P. Furlan, V.B. Petkova, G.M. Sotkov, Indecomposable representations of the conformal group: a nonsingular photon-Weyl graviton system, ISAS preprint 7/86/E.P. and J. Math. Phys. (1987).

[J1] K. Johnson, R. Willey, M. Baker, Vacuum polarization in quantum electrodynamics, Phys. Rev. 163 (1967), 1699-1715; Zh. Eksp. Teor. Fiz. 52 (1967) 318 (transl.: Soviet Phys. -JEPT 25 (1967) 205); for later developments see M. Baker, K. Johnson, Applications of conformal symmetry in quantum electrodynamics, Physica 96A (1979) 120-130.

[M1] G. Mack, Abdus Salam, Finite component field representations of the conformal group, Ann. Phys. (N.Y.) 53 (1969) 174-202.

[M2] G. Mack, I.T. Todorov, Conformal invariant Green functions without ultraviolet divergences, Phys. Rev. D8 (1973) 1764-1787.

[P1] M. Pal'chik, Dynamical problems of conformal invariant field theory (in Russian) in New Developments in Quantum Field Theory, Proceedings of the III. Primorsko School (October 1977) Sofia, pp. 240-258.

[P2] S.M. Paneitz, I.E. Segal, D.A. Vogan, Jr., Analysis in spacetime bundles IV. Natural bundles deforming into and composed of the same invariant factors as the spin and form bundles, MIT preprint May 1986 (to be published in Journ. Funct. Anal).

[P3] V.B. Petkova, G.M. Sotkov, Exceptional representations of the conformal group and applications. I,II. Bulg.J. Phys. 10 (1983) 144-171, 256-277.

[P4] V.B. Petkova, G.M. Sotkov, I.T. Todorov, Conformal gauges and renormalized equations of motion in massless quantum electrodynamics, Commun. Math. Phys. 97 (1985) 227-256.

[S1] Abdus Salam, J. Strathdee, Nonlinear realizations II. Conformal symmetry, Phys. Rev. 184 (1969) 1760-1768.

[S2] I.E. Segal, C^*-quantization of induced bundles, MIT preprint (1986); I.E. Segal et al., Restoration of invariance under space inversion, MIT(1986).

[S3] G.M. Sotkov, Models of conformal quantum electrodynamics, Dissertation (Sofia 1983, in Bulgarian), partly contained in G.Sotkov,D. Stoyanov, Conformal quantization of electrodynamics, J.Phys. A16 (1983)2817-1826.

[S4] G.M. Sotkov, Ya. S. Stanev, I.T. Todorov, Conformal invariant model of massless QED, Bulg. J. Phys. 12 (1985) 535-551.

[S5] Ya. S. Stanev, I.T. Todorov, A conformal invariant QED model in terms of operator product expansions, Bulg. J. Phys. 14 (1987).

[S6] F. Strocchi, A.S. Wightman, Proof of the charge superselection rule in local relativistic quantum field theory, J. Math. Phys. 15 (1974) 2198-2224, ibid 17 (1976) 1930-1931.

[T1] I.T. Todorov, Local field representations of the conformal group and their applications, in: Mathematics + Physics, Lectures on Recent Results, Vol. 1, Ed. L. Streit (World Scientific, Singapore (1985)), pp. 195-338.

[T2] I.T. Todorov, Conformal Description of Spinning Particles, ISAS Trieste preprint, 1/81/E.P. Trieste Notes in Physics (Springer, Berlin 1986).

[T3] I.T. Todorov, Non-singular conformal invariant formulation of massless quantum electrodynamics, ISAS Trieste preprint 4/85/EP in: Quantum Field Theory and Quantum Statistics, Vol. 1, pp. 315-331.

[T4] I.T. Todorov, M.C. Mintchev, V.B. Petkova, Conformal Invariance in Quantum Field Theory (Scuola Normale Superiore, Pisa 1978).

[V1] O. Veblen, Projective Relativitätstheorie, Ergebnisse der Mathematik und ihrer Grenzgebiete 2, no. 1 (1933).

[Z1] R.P. Zaikov, On conformal invariance in gauge theories: quantum electrodynamics, Teor. Mat. Fiz. 65(1985) 70-78 (transl. Theor. Math. Phys. 65 (1986) 1016-1022 .

NONLINEAR SPINOR REPRESENTATIONS

R. Rączka *
International School for Advanced Studies (SISSA)
Trieste, Italy

and

Institute for Nuclear Studies, Warsaw, Poland

*) Permanent address: Institute for Nuclear Studies,
00-681 Warsaw, Hoża 69, Poland.

1. Introduction

The field and string theories in higher dimensional space-times play now important role in understanding the physical four-dimensional space-time quantum field theories |1|. The many attractive features of higher-dimensional field theories are diminished by the fact that in these models the number of Dirac fields - when the theory is finally restricted to the physical four-dimensional space-time - is very high.

It was pointed out by Budinich |2| that pure spinors may provide a framework for fermion theory in which the number of independent spinor components - due to Cartan constraints - is considerably diminished. In this work we wish to call attention to the fact that pure spinors exhibit a special case of nonlinear irreducible representation of the rotation group. Consequently we first develope in Sec.2 a general theory of nonlinear representations of arbitrary topological group. Next in Sec.3 we apply this construction for complex and real rotation groups. We show a rather interesting fact that the covariant Cartan constraints - which reduce an ordinary spinor to a pure spinor - are directly connected with the stability subgroup H_m of the highest weight vector Ψ_m of the linear irreducible representation determined by the highest weight m.

We give also two Tables which illustrate the reduction of a dimension of a nonlinear spinor representation with respect to a linear one. It is evident then that nonlinear spinors are the most elementary building blocks for a physical theory.

In Sec.4 we discuss some application and generalizations of the considered theory.

2. Nonlinear group representations.

Let G be a topological group and let N be a nonlinear topological space. We say that the map $g \to T_g$ is a nonlinear representation of G in N if the following conditions are satisfied:

1^0. With each $g \in G$ there is associated a transformation $T_g : n \to T_g n$ of N into N.

2^0. If e is the identity element e of G then $T_e = I$.

3^0. The mapping $(g, n) \to T_g n$ of $G \times N$ into N is continuous.

4^0. For $g_1, g_2 \in G$ and $n \in N$ we have

$$T_{g_1 g_2} n = T_{g_1} (T_{g_2} n)$$

The novelty of nonlinear representations consists in the condition that the carrier space N for T_g operators is nonlinear. Hence the analysis and a classification of nonequivalent nonlinear representations is very difficult. For instance all linear irreducible representations of Poincaré group $G = T^4 \rtimes SO(3,1)$ are classified |3|: on the other hand a classification of nonlinear representations of Poincaré group would correspond to a classification of all nonlinear relativistic wave equations, which is clearly, for a time being, unsolve problem.

We now give a general construction of nonlinear group representations for arbitrary simple Lie group. Let $m = (m_1, \ldots, m_n)$ be a highest weight of a linear irreducible representation of a simple Lie group G. Let L^m be a linear carrier space of a linear irreducible representation T^m of G in L^m. Let ψ_m be the highest vector in L^m: it follows from the Cartan Weyl theory |3| that the Cartan subalgebra $\{X_i\}$, $i=1,\ldots,n$ diagonalizes on ψ_m i.e.

$$X_i \psi_m = m_i \psi_m . \tag{2.1}$$

Let $H_m \subset G$ be the stability subgroup of ψ_m i.e. for $h \in H_m$ we have

$$T_h \psi_m = \psi_m .$$

It follows from the fundamental Mackey Decomposition Theorem |3| that there exists a Borel set $C_m \subset G$ such that for every $g \in G$ we have

$$g = ch , \quad c \in C_m , \quad h \in H_m . \tag{2.2}$$

Using (2.2), for every $g \in G$ we obtain

$$T_g \psi_m = T_{ch} \psi_m = T_c T_h \psi_m = T_c \psi_m \equiv \psi_m(c) . \tag{2.3}$$

It follows from (2.3) that the set

$$N^m = \{ \psi_m(c) \; ; \; c \in C_m \} \tag{2.4}$$

forms a nonlinear carrier space for T_g: in fact we have

$$T_{g_0} \psi_m(c) = T_{g_0} T_c \psi_m = T_{c_{g_0 c}} T_{h_{g_0 c}} \psi_m = \psi_m(c_{g_0 c}) \tag{2.5}$$

where $c_{g_0 c}$ is defined by the Mackey decomposition (2.2) i.e.

$$g_0 c = c_{g_0 c} h_{g_0 c} .$$

It follows from the above construction that

$$\dim N^m = \dim C_m = \dim ({}^G/H_m)$$

Clearly by (2.5) N^m is isomorphic to the homogeneous space $C_m = {}^G/H_m$.

We see therefore that with every linear irreducible representation T^m acting in L^m we can associate a nonlinear representation T^{N^m} given by (2.5) in the nonlinear space N^m. The dimension of T^{N^m} is in general much lower than dim of T^m - especially for higher dimensional L^m spaces.

From the above construction it follows a general method of construction of nonlinear group representation:

1° Take a linear irreducible representation T^m of G in L^m characterized by the highest weight $m = (m_1, \ldots, m_n)$

2° Find the stability subroup H_m of the highest weight vector ψ_m.

3° Find the Mackey set $C_m = {}^G/H_m \simeq N^m$.

Then the nonlinear representation T^{N^m} of G in N^m is given by the formula

$$T_{g_0} \psi_m(c) = \psi_m(c_{g_0 c}), \qquad g_0 c = c_{g_0 c} h_{g_0 c} .$$

The obtained representation is irreducible.

It is evident fromthe above construction that the key role in the construction of nonlinear representation is played by the Mackey Decomposition Theorem |3|. Since this theorem holds for any topological group one can easily extend the above construction of nonlinear representation to all topological groups including infinite dimensional groups like e.g. diffeomorphism groups of R^n or Virasoro groups.

3. Nonlinear irreducible spinor representations.

We now give an explicit construction of nonlinear spinor representations of complex and real rotation groups |4|.

A. Consider first $SO(2\nu, \mathbb{C})$ complex rotation groups defined in $\mathbb{C}^{2\nu}$ complex space of dimension 2ν. There are two highest weights corresponding to fundamental spinor representations |3|

(3.1) $$m_{\pm} = (\tfrac{1}{2}, \tfrac{1}{2}, \ldots \tfrac{1}{2}, \pm \tfrac{1}{2})$$

Let L^{m_+} be the carrier space of irreducible spinor representation associated with m_+ - highest weight and let Ψ_{m_+} be the highest weight vector in L^{m_+}. Then we have |4|:

Theorem 3.1. The stability group H_{m_+} of Ψ_{m_+} is the semi-direct product group

(3.2) $$H_{m_+} = T^{\binom{\nu}{2}} \otimes SL(\nu, \mathbb{C})$$

where $T^{\binom{\nu}{2}}$ is the abelian $\binom{\nu}{2}$- dimensional subgroup of $SO(2\nu, C)$. ▼

It is noteworthy that in the present case the Mackey set $C_{m_+} = SO(2\nu, C)/H_{m_+}$ can be also represented as a subgroup of $So(2\nu, C)$. In fact we have |4|:

Theorem 3.2. The Mackey set C_{m_+} may be represented as the connected solvable subgroup

$$C_{m_+} = T^{\binom{\nu}{2}} \circledS T^1 \tag{3.3}$$

with dim $C_{m_+} = 1 + \binom{\nu}{2}$. ▼

It follows from the general theory presented in Sec.2 that the action of T_g in N^m is given by the formula

$$T_g \psi_{m_+}(c) = \psi_{m_+}(c_{gc})$$

where c_{gc} is the element of C_{m_+} given by the Mackey decomposition $gc = c_{gc} h_{gc}$.

Since the dimension of the linear spinor representation of L^m is $d^m = 2^{\nu-1}$ and the dimension of N^m is $1+\binom{\nu}{2}$ there must exist

$$2^{\nu-1} - [1 + \binom{\nu}{2}]$$

constraints which restrict the carrier space L^m to N^m. It is noteworthy that these constraints can be expressed in the covariant manner in terms of original spinors. In fact let Γ_a $a=1,\ldots 2\nu$ be the generalized Dirac matrices satisfying the anticommutation relations

$$\{\Gamma_a, \Gamma_b\} = 2 g_{ab} I \quad , \quad a, b = 1, \ldots, 2\nu. \tag{3.4}$$

Let C be the matrix satisfying the relations

$$C \Gamma_a = (-1)^\nu \Gamma_a^T C \, , \quad C C^T = 1 \, , \quad C^2 = (-1)^{\nu(\nu+1)/2}$$

Let $\tilde{\psi} = \psi^T C$ and let $\Gamma_{a_1, a_2, \ldots, a_k}$ be the polyvectors formed from Γ_a . Then set

$$(3.5)\quad \overline{\Psi}\,\Gamma_{a_1,\ldots,a_k}\Psi = 0 \ , \ k = 0, 1, \ldots, \nu\text{-}1 .$$

One readily calculates that there is $2^{\nu-1} - [1 + \binom{\nu}{2}]$ independent equations following from (3.5) |4|. Consequenlty the reduction of a linear carrier space L^m to the nonlinear N^m is carried out by the imposion of $2^{\nu-1} - [1 + \binom{\nu}{2}]$ bilinear covariant conditions (3.5) on spinors ψ in L^m .

Clearly the imposition of suplementary conditions (3.5) on spinors considerably restricts the dimension of the carrier space N^m of nonlinear representation. This fact is illustrated by the following table which for various SO(2ν,$\mathbb{C}$) groups gives the dimension of linear and nonlinear spinor representation and the number of constraints;

Table I

ν	2	3	4	5	6	10
$2^{\nu-1}$	2	4	8	16	32	512
$1+\binom{\nu}{2}$	2	4	7	11	16	46
$2^{\nu-1}-1-\binom{\nu}{2}$	0	0	1	5	16	466

B. Consider now SO($2\nu + 1$, $\mathbb{C}$) complex rotation groups. In this case we have only one fundamental spinor representation characterized by the highest weight $m = (\frac{1}{2}, \ldots, \frac{1}{2})$. The carrier space L^m has dimension 2^{ν} . Let ψ_m be the highest weight vector in L^m. Then we have |4|:

Theorem 3.3. The stability group H_m of ψ_m is the semi-direct product group

$$(3.6)\qquad H_m = R \circledS SL(\nu, \mathbb{C})$$

where R is the solvable group whose Lie algebra r has the following structure

$$r = t^{\binom{\nu}{2}} \dot{+} e^{\nu}$$

with $t^{\binom{\nu}{2}}$ - a $\binom{\nu}{2}$- dimensional abelian subalgebra and e^{ν} a ν- dimensional vector space in SO(2ν +1, $\mathbb{C}$) Lie algebra. ▼

In the present case C_m may be also represented as a subgroup of SO(2ν +1, $\mathbb{C}$): in fact we have:

Theorem 3.4. The carrier space N^m may be represented as a solvable Lie group C whose Lie algebra c has the following structure

$$c = (t^{\binom{\nu}{2}} \dot{+} f^{\nu}) \oplus t^{1}$$

with $t^{\binom{\nu}{2}}$ a $\binom{\nu}{2}$ -dimensional abelian algebra and f^{ν} a ν-dimensional vector space in SO(2ν +1, $\mathbb{C}$) Lie algebra and

$$\dim N^m = 1 + \nu + \binom{\nu}{2} . \tag{3.7}$$ ▼

As previously the reduction of L^m to N^m may be carried out by imposition of 2 - 1 - ν - $\binom{\nu}{2}$ independent covariant constraints on spinor ψ of the form

$$\tilde{\psi}\, \Gamma_{a_1,\ldots,a_k} \psi = 0 \quad , \quad k < \nu \tag{3.8}$$

C. Consider now real pseudo-orthogonal groups SO(2ν - h,h), h=0,1,..., ν. In this case we have again two fundamental spinor representations defined by m_+ and m_- highest weights respectively. The structure of the stability subgroup H_m of the highest weight vector is now much more complicated than in complex case. In fact we have |4|:

Theorem 3.5.

$$H_{m_+} = R \otimes [SU(\nu - h) \times SL(h, \mathbb{R})] \tag{3.9}$$

where R is a solvable subgroup of SO(2ν - h,h) whose Lie algebra r has the form

$$r = t^{\binom{h}{2}} \dotplus d^{2h(\nu-h)}$$

where $t^{\binom{h}{2}}$ is the abelian subalgebra and $d^{2h(\nu-h)}$ a $2h(\nu-h)$ -dimensional vector space in the SO(2ν -h h) Lie algebra. The N^m carrier space has the dimension

$$\dim N^{m_+} = \nu^2 - \nu + 2 - \frac{h}{2}(h-1). \tag{3.10}$$ ▼

Notethat in compact case h=0 corresponding to SO(2ν) rotation group we have

$$H_{m_+} = SU(\nu) \quad \text{and} \quad N^{m_+} = SO(2\nu)/SU(\nu)$$

Hence N^m is a symmetric Cartan space |3|. In turn for h=ν we have |4|

$$H_{m_+} = T^{\binom{\nu}{2}} \otimes SL(\nu, (\mathbb{R}) \quad \text{and} \quad N^{m_+} = T^{\binom{\nu}{2}} \otimes T^1$$

The table below illustrates the reduction of the dimension for the irreducible spinor representations

Table II

ν	2	3	4	5	6	10
2^{ν}	4	8	16	32	64	1024
$1+\quad+(_{2})$	4	7	11	16	22	56
$2^{\nu}-1-\nu-\binom{\nu}{2}$	0	1	5	16	42	968

We see that the number of constraints sharply increases when the dimension of the linear spinor representation increases.

D. Consider now odd-dimensional real pseudo-orthogonal groups $SO(2\nu+1-h\ \ h)$, $h = 0,1,\ldots,\nu$. In this case we have only one fundamental spinor representation L^m. Let Ψ_m be the highest weight vector in L^m. Then we have |4|:

Theorem 6.

$$H_{m_+} = R \circledS [SU(\nu-h)\times SL(h,\mathbb{R})] \tag{3.11}$$

where R is a solvable group whose Lie algebra r has the following structure

$$r = t^{\binom{\nu}{2}} \dot{+} d^{2h(\nu-h)+h}$$

where $t^{\binom{h}{2}}$ is a $\binom{h}{2}$-dimensional abelian Lie algebra and $d^{2h(\nu-h)+h}$ is a $2h(\nu-h)+h$ - dimensional vector space in $SO(2\nu+1-h,h)$ Lie algebra with

$$\dim N^m = 2+\nu+\nu^2-\frac{h}{2}(h+1) \,. \tag{3.12}$$

4. Discussion.

The presented general formalism allows for the explicit construction of nonlinear representation of arbitrary topological group. If G is a symmetry group of a given physical system then our formalism allows to construct the irreducible representation with absolutely minimal dimension which - as follows from the presented Tables - is much smaller than the corresponding dimension of a linear irreducible representation. Since Nature likes simplest objects one might hope that the presented formalism will find applications in physical problems.

The nonlinear spinors present an interesting object for the second quantized theory . Even if we take a free spinor field theory in n-dimensional space time R^n with the action integral $S(\Psi)$ then due to quadratic constraints $(\Psi \equiv \{\psi_\alpha\})$

$$\overset{i}{\gamma}_{\rho\sigma} \psi_\rho \psi_\sigma = 0 , \quad i = 1, \ldots, \dim H_m$$

the propagator will have the form

$$\langle 0 | T \psi_\alpha^*(x) \psi_\beta(y) | 0 \rangle = Z^{-1} \int e^{iS(\Psi)} \times$$

$$\times \psi_\alpha^*(x) \psi_\beta(y) \prod_{i=1}^{\dim H_m} \delta(\overset{i}{\gamma}_{\rho\sigma} \psi_\rho \psi_\sigma) D\Psi$$

Consequently this propagator will be drastically different from the free propagator in R^n.

We see therefore that a quantum field theory for interacting nonlinear spinors may be entirely different from that for linear spinors.

We note finally that the construction of nonlinear group representations presented in Sec.2 may be also applied for a construction of arbitrary nonlinear spinor or tensor representation of arbitrary semisimple (or in general topological) group.

Acknowledgment

The author thanks Professor P.Budinich and Professor P. Furlan for inspiring discussions.He also gratefully acknowledges the warm hospitality extended to him by Professor P.Budinich and the administrative staff of the ISAS during his stay in Trieste.

References

1. See e.g. M.J.Duff, "Not standard superstring review", preprint CERN-TH.4749/87.
2. P.Budinich, "On the possible role of pure conformal spinors in physics", ISAS preprint 36/83/EP, Trieste.
3. A.O.Barut and R.Rączka, "Theory of Group Representations and Applications (PWN, Warsaw, 1980).
4. P.Furlan and R.Rączka, J.Math.Phys.26,3021(1985).

NONLINEAR WAVE EQUATIONS FOR INTRINSIC SPINOR COORDINATES

P. Furlan

Dipartimento di Fisica Teorica, Università di Trieste, Italy

International School for Advanced Studies, Trieste, Italy

Istituto Nazionale di Fisica Nucleare (INFN), Sezione di Trieste

I. INTRODUCTION

The aim of this talk is to show the application of the theory of nonlinear spinor representations[1], just exposed to you by R. Raczka[2], to the study of Dirac-like wave equations. We want to show how a <u>linear</u> massless spinor wave equation in $\mathbb{R}^{p,q}$ of the form

$$\Gamma^a \frac{\partial}{\partial x^a} \psi(x) = 0 \quad , \qquad x \in \mathbb{R}^{p,q} , \tag{1.1}$$

once we impose that $\psi(x)$ carries out a nonlinear spinor representation of SO(p,q), produces a system of <u>nonlinear</u> and <u>nonpolynomial</u> equations for the independent spinor components. In turn, when we express these spinor components in terms of intrinsic spinor coordinates, the nonpolynomiality disappears and we remain with a system of nonlinear equations with quadratic nonlinearities. We can consider the resulting nonlinear field theory as a σ -type model spinor field theory. In fact the spinor ψ of, e.g. SO(ν,ν) satisfies a set of $2^{\nu-1} - 1 - \binom{\nu}{2}$ quadratic constraints

$$\tilde{\psi}\, \Gamma_{a_1 \dots a_k} \psi = 0 \qquad \text{for } k = 0, 1, \dots, \nu-1 , \tag{1.2}$$

where $\Gamma_{a_1 \dots a_k}$ are polyvectors generated by Γ_a .

In what follows we shall freely use all information already given to you by R. Rączka[2], and we shall adopt the same notations and conventions. We shall use many times the term "pure spinor", to indicate any spinor obtained by the action of the SO(p,q) group on the standard spinor ψ_m (chosen to be the highest weight eigenspinor), even if this name, due to E. Cartan[3], strictly applies to the complex SO(2ν, $\mathbb{C}$) and SO($2\nu+1$, $\mathbb{C}$) cases only.

In the following we shall take into account the groups SO($\nu+1$, ν) and SO(ν,ν) only. In fact, from Rączka's talk[2] we see that these ones (together with SO($\nu+2,\nu-1$) and SO($\nu+1$, $\nu-1$)) are the only (real) cases in which the carrier space of the nonlinear spinor representations may be represented - up to a set of Haar measure zero - not simply as an homogeneous space SO(p,q)/H (H = stability subgroup of ψ_m) but as a group space. This carrier space is obtained by the action of SO(p,q) on the standard spinor ψ_m .

If we denote by h the Lie algebra of H and by c the set of generators complementary to h in so(p,q), we see that in the above mentioned cases c is also a Lie algebra, and then the elements of the pure spinor spaces can be parametrized by the group elements of exp(c).

If we denote the generic element of the Lie algebra c by

$$\tilde{X} := \sum_{i=1}^{\dim c} a^i X_i \quad , \tag{1.3}$$

where $\{X_i\}$ are the algebra's generators, then the pure spinors are given by

$$\psi = e^{\tilde{X}} \psi_m := \psi(\{a^i\}) \, , \tag{1.4}$$

and $\{a^i\}$ appear to be the intrinsic spinor coordinates.

Unfortunately the explicit dependence of ψ from $\{a^i\}$ is not available straightforwardly. We have first to be able to express $e^{\tilde{X}}$ as a product of one-parameter subgroups, i.e.

$$e^{\tilde{X}} = \prod_{j=1}^{\dim c} e^{b^j X_j} \quad , \tag{1.5a}$$

where

$$b^j = b^j(a^i) \, , \tag{1.5b}$$

getting now

$$\psi = \psi(\{b^j\}) . \qquad (1.5c)$$

It is only at this stage that we can find explicitly the non-linear Cartan relations expressing the pure spinor components in terms of the independent ones. This in turn allows us to reduce drastically the number of equations deduced from the starting Dirac-like one (1.1) and finally it allows to express them in terms of intrinsic coordinates only.

Let us see in detail the SO($\nu+1,\nu$) and SO(ν,ν) cases.

II. EXPONENTIATION OF THE COMPLEMENTARY SUBALGEBRA c IN THE FORM OF ONE-PARAMETER SUBGROUPS

IIa. THE SO($\nu+1,\nu$) CASE

The complementary subalgebra c is given in terms of[1]

$$\begin{array}{lll} \binom{\nu}{2} \text{ generators} & \tilde{Q}^{k\ell} = -\tilde{Q}^{\ell k} & , \; k \neq \ell = 1,\ldots,\nu \\ \nu \quad '' & \tilde{F}^{k} & , \; k = 1,\ldots,\nu \\ 1 \text{ generator} & \tilde{D} & , \end{array} \qquad (2.1)$$

satisfying the following commutation relations

$$[\tilde{D}, \tilde{Q}^{r\nu}] = -\tilde{Q}^{r\nu} \quad , \quad r = 1,\ldots,\nu-1 \qquad (2.2a)$$

$$[\tilde{D}, \tilde{F}^{\nu}] = -\tilde{F}^{\nu} \qquad (2.2b)$$

$$[\tilde{F}^{k}, \tilde{F}^{\ell}] = 2\tilde{Q}^{k\ell} \quad , \quad k \neq \ell = 1,\ldots,\nu , \qquad (2.2c)$$

all other commutators being zero. We see that it has the following structure:

$$c = r^1 \oplus \left(t^{\binom{\nu}{2}} \dotplus f^{\nu} \right) , \tag{2.3}$$

where r^1 is represented by the generator $\tilde{D}$ and the rest is a solvable Lie algebra with $t^{\binom{\nu}{2}}$ a $\binom{\nu}{2}$-dimensional Abelian Lie algebra and f^{ν} a ν-dimensional vector space in so(ν +1, ν).

Let us define the general element of c by

$$\tilde{X} := \alpha \tilde{D} + \sum_{k<\ell=1}^{\nu} c_{k\ell} \tilde{Q}^{k\ell} + \sum_{k=1}^{\nu} f_k \tilde{F}^k := U + V \tag{2.3a}$$

where

$$U := \alpha \tilde{D} + \sum_{r<s=1}^{\nu-1} c_{rs} \tilde{Q}^{rs} + \sum_{r=1}^{\nu-1} f_r \tilde{F}^r \tag{2.3b}$$

$$V := \sum_{r=1}^{\nu-1} c_{r\nu} \tilde{Q}^{r\nu} + f_\nu \tilde{F}^\nu . \tag{2.3c}$$

Applying the Zassenhaus formula (see Appendix), we get

$$e^{\tilde{X}} = e^{U+V} = e^U e^V e^{C_2(U,V)} e^{C_3(U,V)} \cdots =$$

$$= \left[\prod_{k=1}^{\nu} \exp(g_k \tilde{F}^k) \right] \left[\prod_{\ell<m=1}^{\nu} \exp\left(\tilde{d}_{\ell m} \tilde{Q}^{\ell m} \right) \right] e^{\alpha \tilde{D}} , \tag{2.4}$$

where

$$g_r := f_r \quad , \qquad r = 1, \ldots, \nu-1$$

$$g_\nu := \frac{1-e^{-\alpha}}{\alpha} f_\nu \tag{2.5}$$

$$\tilde{d}_{rs} = -\tilde{d}_{sr} := c^{rs} - \epsilon(s-r) f^r f^s \quad , \qquad r \neq s = 1, \ldots, \nu - 1$$

$$\tilde{d}_{r\nu} = -\tilde{d}_{\nu r} := \frac{1-e^{-\alpha}}{\alpha} c_{r\nu} + 2\,\frac{1-\alpha-e^{-\alpha}}{\alpha^2} f_r f_\nu \,, \quad r = 1, \ldots, \nu - 1 \,. \tag{2.5}$$

These relations are similar to those ones expressing the Euler angles in terms of the standard parameters of the SO(3) group elements written as exponentials of the so(3) Lie algebra.

II b. THE SO(ν,ν) CASE

This case is obtained from the preceding one deleting all $\tilde{F}^k$ generators. Then the exponentiation of the generic element of the complementary subalgebra c is given by $e^{\tilde{y}}$, where

$$\tilde{y} := \alpha \tilde{D} + \sum_{k<l=1}^{\nu} c_{kl} \tilde{Q}^{kl} \,, \tag{2.6}$$

and from Eqs. (2.4) and (2.5) we have

$$e^{\tilde{y}} = \left[\prod_{k<l=1}^{\nu} \exp\left(\tilde{c}_{kl} \tilde{Q}^{kl}\right) \right] e^{\alpha \tilde{D}} \,, \tag{2.7}$$

with

$$\tilde{c}_{rs} = -\tilde{c}_{sr} := c_{rs} \quad , \qquad r \neq s = 1, \ldots, \nu - 1$$

$$\tilde{c}_{r\nu} = -\tilde{c}_{\nu r} := \frac{1-e^{-\alpha}}{\alpha} c_{r\nu} \,, \qquad r = 1, \ldots, \nu - 1 \,. \tag{2.8}$$

III. INTRINSIC SPINOR COMPONENTS FOR PURE SPINORS

III a. THE SO(ν+1,ν) CASE

The generic SO(ν+1, ν) pure spinor is given by

$$\psi := e^{\tilde{X}} \psi_m , \tag{3.1a}$$

where $e^{\tilde{X}}$ is given by Eq. (2.4) and

$$\psi_m = \begin{Vmatrix} 1 \\ 0 \\ 0 \\ \vdots \\ 0 \end{Vmatrix} \tag{3.1b}$$

is the standard (highest weight eigen-) spinor introduced by Cartan[3] in his specific spinor representation (see ref. 1 for more details). We shall use in the following the same convention as Cartan[3], denoting the spinor components by a totally antisymmetrized set of indices, like

$$\psi_{i_1 \dots i_p} , \quad p = 0,1,\dots,\nu ; \qquad i_1,\dots i_p = 1,2,\dots,\nu . \tag{3.2}$$

Thus, inserting Eqs. (2.4) and (2.5) into Eqs. (3.1a,b), and making use of Cartan's explicit representation of $C\ell(\nu+1,\nu)$, we get the explicit form of the SO($\nu+1,\nu$) pure spinor components[1]

$$\psi_{i_1 \dots i_{2p}} = (2p-1)!!\, e^{\alpha/2}\, d_{[i_1 i_2}\, d_{i_3 i_4} \cdots d_{i_{2p-1} i_{2p}]} \tag{3.3a}$$

$$\psi_{i_1 \dots i_{2q+1}} = -(2q+1)!!\, e^{\alpha/2}\, g_{[i_1}\, d_{i_2 i_3} \cdots d_{i_{2q} i_{2q+1}]} ,$$

where

$$d_{k\ell} = - d_{\ell k} := \tilde{d}_{k\ell} + g_k g_\ell\, \epsilon(\ell - k) , \quad k \neq \ell = 1,\dots,\nu \tag{3.3b}$$

(and we use the convention $(-1)!! \equiv 1$ in Eq. (3.3a)).

It is easy to see that the components (3.3a) satisfy the Cartan's quadratic relations for pure spinors[3] or, equivalently, the Caianiello's pfaffians' relations[4].

III b. THE SO(ν, ν) CASE

The generic SO(ν, ν) pure semi-spinor of first kind[3] (or, equivalently, Weyl spinor with positive chirality) is given by

$$\psi := e^{\tilde{y}} \psi_m \quad , \tag{3.4}$$

where $e^{\tilde{y}}$ is given by Eq. (2.7) and ψ_m is given again by Eq. (3.1b). We remark that ψ, being a pure semi-spinor of the first kind, has all components with an odd number of indices equal to zero.

Then, inserting Eqs. (2.7) and (2.8) into Eq. (3.4), and taking into account Cartan's[3] explicit representation of Cl(ν, ν), we obtain the explicit form of the SO(ν, ν) pure spinor components:

$$\psi_{i_1 \dots i_{2p}} = (2p-1)!!\, e^{\alpha/2}\, \tilde{c}_{[i_1 i_2} \cdots \tilde{c}_{i_{2p-1} i_{2p}]} \quad , \tag{3.5}$$

(with the convention (-1)!! ≡ 1 again).

Cartan's[3] or Caianiello's[4] relations for pure spinor components are satisfied as before.

IV. LINEAR WAVE EQUATIONS FOR (NON INDEPENDENT) SPINOR COMPONENTS

IV a. THE SO($\nu+1, \nu$) CASE

We remind the reader that, following Cartan[3], we use the isotropic or Witt basis with coordinates y^0, y^l, $y^{l'}$ $(l = 1,\dots,\nu)$

and the corresponding Clifford algebra $Cl(\nu+1, \nu)$ basis, given by the matrices $H_0, H_\ell, H_{\ell'}$ $(\ell=1,\dots,\nu)$ (see ref. (1) for more details).

Now we can rewrite Eq. (3.3a) as follows:

$$\psi_{i_1\dots i_{2p}} = (2p-1)!!\, \psi_0^{1-p}\, \psi_{[i_1 i_2}\, \psi_{i_3 i_4} \cdots \psi_{i_{2p-1} i_{2p}]} \tag{4.1a}$$

$$\psi_{i_1\dots i_{2q+1}} = (2q+1)!!\, \psi_0^{-q}\, \psi_{[i_1}\, \psi_{i_2 i_3} \cdots \psi_{i_{2q} i_{2q+1}]} ,$$

with

$$\psi_0 = e^{\alpha/2}$$

$$\psi_i = -e^{\alpha/2} g_i , \quad i = 1, \dots, \nu \tag{4.1b}$$

$$\psi_{ij} = -\psi_{ji} = e^{\alpha/2} d_{ij} , \quad i \neq j = 1, \dots, \nu .$$

Inserting the expressions (4.1a,b) into the massless Dirac-like equation (1.1), we get the following system of 2^ν first order linear partial differential equations for the spinor components

$$\partial_0 \psi_{i_1\dots i_r} - 2r\, \partial_{[i_1} \psi_{i_2\dots i_r]} + \sum_{i_{r+1}=1}^{\nu} \partial^{i_{r+1}} \psi_{i_{r+1} i_1 \dots i_r} = 0 , \tag{4.2}$$
$$r = 0, 1, \dots, \nu .$$

Let us write the first $r = 0, 1, 2, 3$ cases:

$$\begin{cases} \partial_0 \psi_0 + \sum_{i_1=1}^{\nu} \partial^{i_1} \psi_{i_1} = 0 \\ \partial_0 \psi_{i_1} - 2\partial_{i_1} \psi_0 - \sum_{i_2=1}^{\nu} \partial^{i_2} \psi_{i_1 i_2} = 0 \\ \partial_0 \psi_{i_1 i_2} - 4\partial_{[i_1} \psi_{i_2]} + \sum_{i_3=1}^{\nu} \partial^{i_3} \psi_{i_1 i_2 i_3} = 0 \\ \partial_0 \psi_{i_1 i_2 i_3} - 6\partial_{[i_1} \psi_{i_2 i_3]} - \sum_{i_4=1}^{\nu} \partial^{i_4} \psi_{i_1 i_2 i_3 i_4} = 0 \end{cases} \tag{4.3}$$

Then we have that the remaining equations are redundant, since the following proposition holds:

<u>Proposition</u>. If the spinor components satisfy Eqs. (4.3), then they <u>automatically</u> satisfy the whole system of equations (4.2).

<u>Proof</u>. In the case $r = q$ <u>odd</u> the system (4.2) can be rewritten as

$$\frac{1}{2}q(3-q)\Big\{(\partial_0\psi_{[i_1})-2(\partial_{[i_1}\psi_0)-\sum_{i_{q+1}=1}^{\nu}(\partial_{i_{q+1}}\psi_{[i_1}{}^{i_{q+1}})\Big\}\psi_{i_2\ldots i_q]}+ \tag{4.4a}$$

$$\cdot\Big\{(\partial_0\psi_{[i_1i_2i_3})-6(\partial_{[i_1}\psi_{i_2i_3})-\sum_{i_{q+1}=1}^{\nu}(\partial_{i_{q+1}}\psi_{[i_1i_2i_3}{}^{i_{q+1}})\Big\}\psi_{i_4\ldots i_q]}=0\,,$$

while in the case $r = p$ <u>even</u> it can be rewritten as

$$\left(1-\frac{p}{2}\right)\Big\{\partial_0\psi_0+\sum_{i_{p+1}=1}^{\nu}(\partial_{i_{p+1}}\psi^{i_{p+1}})\Big\}\psi_{i_1\ldots i_p}-\frac{p(p-2)}{2}\Big\{\partial_0\psi_{[i_1}-2\,\partial_{[i_1}\psi_0-$$

$$-\sum_{i_{p+1}=1}^{\nu}(\partial_{i_{p+1}}\psi_{[i_1}{}^{i_{p+1}})\Big\}\psi_{i_2\ldots i_p]}+\binom{p}{2}\Big\{\partial_0\psi_{[i_1i_2}-4\partial_{[i_1}\psi_{i_2}+ \tag{4.4b}$$

$$+\sum_{i_{p+1}=1}^{\nu}(\partial_{i_{p+1}}\psi_{[i_1i_2}{}^{i_{p+1}})\Big\}\psi_{i_2\ldots i_p]}+\binom{p}{3}\Big\{\partial_0\psi_{[i_1i_2i_3}-$$

$$-6\,\partial_{[i_1}\psi_{i_2i_3}-\sum_{i_{p+1}=1}^{\nu}(\partial_{i_{p+1}}\psi_{[i_1i_2i_3}{}^{i_{p+1}})\Big\}\psi_{i_4\ldots i_p]}=0\,.$$

We see that, if Eqs. (4.3) are satisfied, then the expressions within the braces of Eqs. (4.4a,b) are identically zero, making the system of equations (4.2) identically satisfied. ▼

From Eq. (4.1a) we see that not all spinor components appearing in the system (4.3) are independent; if we express $\psi_{i_1i_2i_3}$ and $\psi_{i_1i_2i_3i_4}$ in terms of the independent components $\psi_0, \psi_{i_1}, \psi_{i_1i_2}$ through Eq. (4.1a), we get a system of <u>nonlinear</u> differential equations with <u>nonpolynomial</u> nonlinearities.

IV b. THE SO(ν,ν) CASE

We can rewrite Eq. (3.5) as follows:

$$\psi_{i_1 \ldots i_{2p}} = (2p-1)!!\, \psi_0^{1-p}\, \psi_{[i_1 i_2} \cdots \psi_{i_{2p-1} i_{2p}]} \tag{4.5a}$$

with

$$\psi_0 = e^{\alpha/2}$$

(4.5b)

$$\psi_{ij} = -\psi_{ji} = e^{\alpha/2}\, \tilde{c}_{ij} \quad , \quad i \neq j = 1, \ldots, \nu .$$

Inserting the expressions (4.5a,b) into Eq. (1.1), we get the following system of $2^{\nu-1}$ first order <u>linear</u> partial differential equations for the first kind semi-spinor components (with only an even number of indices):

$$2q\, \partial_{[i_1} \psi_{i_2 i_3 \ldots i_q]} + \sum_{i_{q+1}=1}^{\nu} \partial^{i_{q+1}} \psi_{i_1 \ldots i_q i_{q+1}} = 0 \quad , \quad q = 1, 3, \ldots, 2\left[\frac{\nu-1}{2}\right]+1 . \tag{4.6}$$

Let us write the first $q = 1,3$ cases:

$$\begin{cases} 2\partial_{i_1} \psi_0 + \sum_{i_2=1}^{\nu} \partial^{i_2} \psi_{i_1 i_2} = 0 \\ 6\, \partial_{[i_1} \psi_{i_2 i_3]} + \sum_{i_4=1}^{\nu} \partial^{i_4} \psi_{i_1 i_2 i_3 i_4} = 0 \end{cases} \tag{4.7}$$

Then we have that the remaining equations are redundant, since the following proposition holds:

<u>Proposition</u>. If the semi-spinor components satisfy Eqs. (4.7), then they <u>automatically</u> satisfy the whole system of equations (4.6).

Proof. For q odd (since we have a first kind semispinor) the system (4.6) can be rewritten as

$$\binom{q}{3}\left\{6\left(\partial_{[i_1}\psi_{i_2 i_3}\right)+\sum_{i_{q+1}=1}^{\nu}\left(\partial_{i_{q+1}}\psi_{i_1 i_2 i_3}{}^{i_{q+1}}\right)\right\}\psi_{i_4\ldots i_q]}+$$
$$+\frac{q(3-q)}{2}\left\{2\left(\partial_{[i_1}\psi_0\right)+\sum_{i_{q+1}=1}^{\nu}\left(\partial_{i_{q+1}}\psi_{[i_1}{}^{i_{q+1}}\right)\right\}\psi_{i_2\ldots i_q]}=0\,. \tag{4.8}$$

We see that, if Eqs. (4.7) are satisfied, then the expressions within the braces of Eq. (4.8) are identically zero, making the system of equations (4.6) identically satisfied.

Also in Eq. (4.7) not all semi-spinor components are independent; if we express

$$\psi_{i_1 i_2 i_3 i_4}=3\,\psi_0^{-1}\,\psi_{[i_1 i_2}\psi_{i_3 i_4]}\,, \tag{4.9}$$

we get a system of <u>nonlinear</u> differential equations with <u>nonpolynomial</u> nonlinearities.

V. NONLINEAR WAVE EQUATIONS FOR (INDEPENDENT) INTRINSIC SPINOR CO-ORDINATES

V a. THE SO(ν+1, ν) CASE

Inserting the expressions (3.3a) of $\psi_{i_1\ldots i_h}$ in terms of <u>intrinsic spinor coordinates</u> α, g_i, d_{jk}, into the system of equations (4.3), we have

$$\begin{cases} \frac{1}{2}\partial_0\alpha - \sum_{i_1=1}^{\nu}\left(\frac{1}{2}g_{i_1}\partial^{i_1}\alpha + \partial^{i_1}g_{i_1}\right) = 0 \\ \frac{1}{2}(\partial_0\alpha)g_{i_1} + \partial_0 g_{i_1} + \partial_{i_1}\alpha + \sum_{i_2=1}^{\nu}\left(\frac{1}{2}d_{i_1i_2}\partial^{i_2}\alpha + \partial^{i_2}d_{i_1i_2}\right) = 0 \\ \frac{1}{2}\partial_0 d_{i_1i_2} + g_{[i_1}\partial_0 g_{i_2]} + 2\partial_{[i_1}g_{i_2]} - \sum_{i_3=1}^{\nu}\left[(\partial^{i_3}g_{[i_1})d_{i_2]i_3} + \frac{1}{2}g_{i_3}\partial^{i_3}d_{i_1i_2}\right] = 0 \\ g_{[i_1}\partial_0 d_{i_2i_3]} + 2\partial_{[i_1}d_{i_2i_3]} + \sum_{i_4=1}^{\nu}(\partial^{i_4}d_{[i_1i_2})d_{i_3]i_4} = 0 \end{cases} \tag{5.1}$$

We see that the starting massless Dirac-like equation (1.1), once applied to a SO(ν+1,ν) pure spinor, expressed in terms of intrinsic coordinates, has produced a system of nonlinear, first order differential equations with only quadratic nonlinearities. We remark the fact that the use of intrinsic coordinates has been crucial in passing from a non-polynomial form to a "renormalizable" one for our wave equations.

V b. THE SO(ν,ν) CASE

Inserting the expression (3.5) of $\psi_{i_1 \dots i_{2p}}$ in terms of intrinsic spinor coordinates α, c_{ij} into the system of equations (4.7), we get

$$\begin{cases} \partial_{i_1}\alpha + \sum_{i_2=1}^{\nu}\left[\partial^{i_2}c_{i_1i_2} + \frac{1}{2}c_{i_1i_2}\partial^{i_2}\alpha\right] = 0 \\ 2\partial_{[i_1}c_{i_2i_3]} + \sum_{i_4=1}^{\nu}(\partial^{i_4}c_{[i_1i_2})c_{i_3]i_4} = 0 \end{cases} \tag{5.2}$$

As in the previous case, applying the massless Dirac-like equation (1.1) to a SO(ν,ν) pure semi-spinor, expressed in terms of intrinsic coordinates, we have obtained a system of nonlinear, first order differential equations with only quadratic nonlinearities.

We remark that the last two equations (5.2) have been obtained independently by a group of mathematicians from the university of Łódź[5], using the Chevalley formalism of algebraic spinors[6].

APPENDIX

We want to give here a brief account of the Zassenhaus formula, inviting the reader to consult ref. (7) for a deeper understanding .

Let R be a free ring with two generators x,y with rational coefficients. Then we have the following theorem due to Zassenhaus:

Theorem. There exist uniquely determined Lie elements C_n (n=2,3,4...) in R which are exactly of degree n in x,y , such that

$$e^{x+y} = e^{x} e^{y} e^{C_2} e^{C_3} \cdots e^{C_n} \cdots \quad . \qquad \blacktriangledown \tag{A.1}$$

Roughly speaking each C_n is given by any linear combination of multiple commutators of x and y , in which the total power of x and y is n. For example we have

$$C_2 = -\frac{1}{2}[x,y]$$

$$C_3 = -\frac{1}{3}[[x,y],y] - \frac{1}{6}[[x,y],x] . \tag{A.2}$$

REFERENCES

1) P. Furlan, R. Rączka, "Nonlinear spinor representations" J.M.P. 26, 3021, (1985);

P. Furlan, R. Rączka, "A pure spinor nonlinear sigma-type model" Phys. Lett. 152B, 75 (1985);

P. Furlan, R. Rączka, "Intrinsic nonlinear spinor wave equations associated with nonlinear spinor respresentations" J.M.P. 27, 1883 (1986);

P. Furlan, "Generalized Euler angles as intrinsic coordinates for nonlinear spinors" ISAS preprint 33/86/E.P. Trieste (to appear on J.M.P.).

2) R. Raczka, "Theory of nonlinear spinor representations", in these proceedings.

3) E. Cartan, The Theory of Spinors (Hermann, Paris, 1966).

4) E.R. Caianiello, A. Giovannini, "Pure Spinors as Pfaffians Connecting Clifford and Grassmann Algebras", Lett. Nuovo Cimento, 34, 301 (1982).

5) S. Giler et al, "On SO(ν,ν) Pure Spinors" IM UŁ preprint 4/85/IM Łódź,

6) C.C. Chevalley, "The Algebraic Theory of Spinors", Columbia Press, N.Y., 1955.

7) W. Magnus, "On the Exponential Solution of Differential Equations for a Linear Operator", Commun. Pure and Appl. Math., 7, 649 (1954).

TWISTORS - "SPINORS" of SU(2,2), THEIR GENERALIZATIONS AND ACHIEVEMENTS *)

J. Niederle **)
International School for Advanced Studies
I-34014 Trieste

Abstract

The basic ideas and results of the twistor and supertwistor theory are briefly described.

*) An invited talk presented in the "Spinors in Physics and Geometry", Trieste, 11-13 September 1986

**) The permanent address:
Institute of Physics, Czech.Acad.Sci., CS-180 40 Prague 8

1. A BIT OF HISTORY

Twistors were introduced by Roger Penrose /1/ at the end of sixties exploiting the fact that points of the four-dimensional space-time can be regarded as (projective) lines of a three-dimensional complex space CP^3.

It seems unusual that Penrose suggested to use complex numbers for a description of a real object - the Minkowski space-time. However, as indicated for instance in Gindikin's paper /2/ this kind of thought was quite natural in the last century. In fact Julia Plücker (1801 - 1868) introduced a space elements of which were straight lines of R^3. It is four-dimensional and it seems to be the first four-dimensional space introduced in science. Surprisingly nobody has compared it with the later appearing Minkowski space till Penrose.

Let us recall briefly Plücker's work. He considered a three-dimensional projective space P^3 and introduced in it coordinates that treat proper and improper points on the same footing. Thus any point in P^3 is parametrized by Plücker's coordinates (now more frequently called homogeneous coordinates): (x_0,x_1,x_2,x_3) with the usual condition $(\lambda x_0, \lambda x_1, \lambda x_2, \lambda x_3) = (x_0,x_1,x_2,x_3)$ with λ real and $\neq 0$. Planes in P^3 are described by equations

$$x_0 \xi_0 + x_1 \xi_1 + x_2 \xi_2 + x_3 \xi_3 = 0$$

so that any plane is associated with $(\xi_0, \xi_1, \xi_2, \xi_3)$. Thus there is an obvious duality between points and planes of P^3. What about a system of straight lines in P^3? As we shall see this consideration leads to a new geometrical object. Julia Plücker described any straight line in P^3 by means of coordinates of its two points, any $x = (x_0, x_1, x_2, x_3)$ and $\tilde{x} = (\tilde{x}_0, \tilde{x}_1, \tilde{x}_2, \tilde{x}_3)$, i.e. in terms of his famous coordinates

$$p_{ij} = x_i \tilde{x}_j - \tilde{x}_i x_j, \ i,j, = 0,1,2,3 \quad (1.1)$$

which are independent on a particular choice of x and $\tilde{x}$. Since $p_{ij} = -p_{ji}$, $p_{ii} = 0$ and $\{\lambda p_{ij}\} = \{p_{ij}\}$, any straight line in P^3 can be regarded as a point in P^5. However, p_{ij} are not yet independent; they satisfy the relation

$$p_{01}p_{23} - p_{02}p_{13} + p_{03}p_{12} = 0 . \quad (1.2)$$

This relation describes a quadric Q in P^5 since by the coordinate transformations

$$\begin{aligned} p_{01} &= u_0 - u_3, \ p_{23} = u_0 + u_3, \ p_{02} = u_4 - u_1, \\ p_{13} &= u_4 - u_1, \ p_{03} = u_2 - u_3, \ p_{12} = u_2 + u_3, \end{aligned} \quad (1.3)$$

it goes to the form

$$u_0^2 + u_1^2 + u_2^2 - u_3^2 - u_4^2 - u_5^2 = 0 \quad (1.4)$$

with signature (3,3). Thus straight lines in P^3 are isomorphic to points on quadric Q in P^5. What about objects associated with points on quadrics with signature (4,2) and (5,1)? The answer was found by Sophus Lie and Felix Klein and can easily be rederived via complexification. Thus, taking a complex projective space CP^3 instead of P^3 any its point can be parametrized by four complex numbers (z_0, z_1, z_2, z_3) and any its line by Plücker's coordinates p_{ij} satisfying (1.4) but with

u_i complex. This complex quadric has several real forms. For instance taking all u_i real we get quadric (1.4), putting $u_3 = iv_3, u_4 = iv_4$ and $u_o, u_1, u_2, u_5 \in R$ we obtain the quadric

$$S : u_o^2 + u_1^2 + u_2^2 + v_3^2 + v_4^2 - u_5^2 = 0 \qquad (1.5)$$

with signature (5,1), i.e. the sphere S, and, finally, taking $u_3 = iv_3$ and $u_o, u_1, u_2, u_4, u_5 \in R$, we get the quadric

$$H : u_o^2 + u_1^2 + u_2^2 + v_3^2 - u_4^2 - u_5^2 = 0 \ , \qquad (1.6)$$

i.e. the hyperboloid H .

Now, by taking into account transformations (1.3) with particularly chosen u_i , it is easy to see that points of S satisfy the relations

$$p_{23} = \bar{p}_{01},\ p_{13} = -\bar{p}_{02}\ ,\ \mathrm{Im}p_{03} = \mathrm{Im}p_{12} = 0. \qquad (1.7)$$

Thus any line in CP^3 containing a point $z = (z_o, z_1, z_2, z_3)$ should contain the point $\bar{z} = (-\bar{z}_3 \bar{z}_2, - \bar{z}_1, \bar{z}_o)$ as well. Consequently points of real sphere S are isomorphic to lines in CP^3 going through point z and $\bar{z}$. Let us note that since there is only one such line going through each particular point z in CP^3, we are getting a fibration of CP^3 .

Plücker's coordinates p_{ij} associated with hyperboloid H can be treated analogously. They satisfy the relations

$$p_{23} = \bar{p}_{01},\ \mathrm{Im}p_{13} = \mathrm{Im}p_{03} = \mathrm{Im}p_{12} = \mathrm{Im}p_{02} = 0\ . \qquad (1.8)$$

However, to find the corresponding lines it is a bit more complicated. Assuming $p_{o3} \neq 0$, i.e. for instance $p_{O3} = 1$, lines satisfying (1.8) go throught point $z = (1,a,c,0)$ and $\tilde{z} = (0,\bar{c},b,1)$, $a,b \in R$. Denoting straight lines connecting z and $\tilde{z}$ by $w = \lambda z + \mu \tilde{z}$

relations (1.8) can be written in the form

$$\mathrm{Im}(w_1\bar{w}_0 + w_2\bar{w}_3) = 0 \ .$$

This equation defines a subspace N in CP^3 so that points of H are associated with lines of five-dimensional subspace $N \subset CP^3$.

The connection between geometrical objects in CP^3 and in various quadrics was developed in all details. For example it was shown that two straight lines in CP^3 , say $p = (\alpha_i, \beta_i)$ and $p' = (\alpha'_i, \beta'_i)$ parametrized by $\alpha_i, \beta_i, \alpha'_i, \beta'_i \in C$, $i = 1,2$, intersect whenever the function $\rho(p,p')$ (the distance) defined by

$$\rho(p,p') = (\alpha_1 - \alpha'_1)(\beta_2 - \beta'_2) - (\alpha_2 - \alpha'_2)(\beta_1 - \beta'_1) \tag{1.10}$$

vanishes, i.e. whenever

$$\rho(p,p') = 0 \ .$$

In this connection let us remark three things:

i) By using function ρ we can introduce an isotropic cone V_p in any point p of the considered quadric Q , namely

$$V_p = \{ p' \in Q \mid \rho(p,p') = 0 \} \ . \tag{1.11}$$

ii) By restricting ρ to sphere S we obtain the usual Euclidean distance on $S \subset E^5$.

iii) Restriction of ρ to hyperboloid H is more refine but by taking a subspace $M \subset H$ determined by points corresponding to lines with $p_{03} = 1$ it yields the usual Minkowski metric. The intersection $V_p \cap M$ is nothing else than a usual light-cone in $p \in M$. In fact points of Minkowski space $M \subset H$ are associated with lines in five-dimensional space $N \subset CP^3$

which do not intersect the line $z_o = z_3 = 0$. Consequently hyperboloid H turns out to be the conformal closure of Minkowski space M .

We may consider projective transformations of CP^3 preserving N(i.e. mapping lines in N to lines in N and intersecting lines to intersecting lines). They induce the corresponding transformations on M and H . They are just the conformal transformations. The Poincaré transformations are those from them which preserve some additional structure, namely the line $z_o = z_3 = 0$.

Summarizing the geometry of Minkowski space M can be obtained from the Plücker geometry of five-dimensional space of lines $N \subset CP^3$ and vice versa. This is remarkable itself but the increasing interest in this correspondence among physicists has arisen mainly from the fact that it leads to twistor theory providing a simpler description of classical spinning massless particles and of free massless fields and new analytic constructions of solutions of the non-linear Einstein and Yang-Mills field equations in the (anti-)self-dual case.

2. TWISTORS AND TWISTOR CORRESPONDENCE

Twistors can be defined by many essentially equivalent ways. Perhaps the most directly physical approach is to regard a twistor $Z = (Z^a) = (Z^o, Z^1, Z^2, Z^3)$ as an element of C^4, i.e. a twistor space T , representing the momentum P^a and angular-momentum M^{ab} of a classical free zero-mass particle /3/*). Alternatively a

*/ For example $P^o = \frac{1}{2}(Z^3\bar{Z}_1 + Z^2\bar{Z}_o)$, $P^1 = \frac{1}{2}(Z^3\bar{Z}_1 - Z^2\bar{Z}_o), \dots,$
$M^{o1} = -M^{1o} = 1/2(Z^o\bar{Z}_o - Z^1\bar{Z}_1 - Z^2\bar{Z}_2 + Z^3\bar{Z}_3), \dots,$

$M^{13} = -M^{31} = 1/2(Z^o\bar{Z}_1 + Z^1\bar{Z}_o + Z^2\bar{Z}_3 + Z^3\bar{Z}_2)$ with $P_aP^a = 0$, $P^o > 0$ and the Pauli Lubanski vector $S_a = 1/2\, e_{abcd}P^bM^{cd}$ satisfying $S_a = sP_a$, where the helicity s is given by $s = 1/2\, Z^a\bar{Z}_a$ and $\bar{Z} = (\bar{Z}_o, \bar{Z}_1, \bar{Z}_2, \bar{Z}_3) = (\bar{Z}^2, \bar{Z}^3, \bar{Z}^o, \bar{Z}^1)$ is the complex conjugate twistor to Z.

---

twistor can be defined via Robinson congruences or as a solution of the conformally invariant field equation or as a helicity-raising operator but for our purpose we define a twistor as an element of space $C^4 \sim T$ in which a (fourfold) covering group SU(2,2) of the identity-connected component of the conformal group of space-time M acts according to the fundamental representation. In a proprietly chosen basis any twistor $Z = (Z^a) = (Z^o, Z^1, Z^2, Z^3) \in T$ can be written as $Z = (\omega^o, \omega^1, \pi_{o'}, \pi_{1'})$, where ω^α, $\pi_{\alpha'}$ are respectively spinors and cospinors of the Lorentz group contained in SU(2,2). Moreover, there exists a hermitian form $(.,.)$ invariant with respect to SU(2,2). For twistors $Z = (Z^a) = (\omega^\alpha, \pi_{\alpha'})$ and $Y = (Y^b) = (\mu^\beta, \lambda_{\beta'})$ it is given by

$$(Z,Y) = \bar{\omega}^{\alpha'}\lambda_{\alpha'} + \bar{\pi}_\alpha \mu^\alpha = \bar{Z}_aY^a \quad , \qquad (2.1)$$

where $\bar{Z} = (\bar{Z}_a) = (\bar{\pi}_\alpha, \bar{\omega}^{\alpha'})$ is the complex conjugate twistor to Z. By using this hermitian form we may distinguish a twistor Z^a (or $\bar{Z}_a$) to be positive, negative or null according as $(Z,Z) = \bar{Z}_aZ^a$ is positive, negative or zero and split T to T^+, T^- and N respectively.

Following Penrose/1,3/ we now set the correspondence between twistors and objects in the Minkowski space. In order to interpret this correspondence geometrically we pass to the projective twistor space $P(T)$ and to the compactified complex Minkowski space CM.

Space P(T) is defined as space CP^3, i.e. a space whose points are the equivalent classes of proportional twistors,

$$P(T) := \{ \mathbf{Z} \mid \mathbf{Z} = \gamma Z, \gamma \in C, \gamma \neq 0, Z \in T, Z \neq 0 \}. \quad (2.2)$$

Space CM is a compact complex 4-space the points of which are composed not only of the points of the complex Minkowski space but in addition of the points of a complex 3-surface - a light cone - defining the "points at infinity" of the complex Minkowski space. Thus denoting by x^a and $x^{\alpha\beta'}$ the vectorial and spinorial components of a point x (not belonging to the light-cone at infinity) in CM, the correspondence between the complex compactified Minkowski space CM and the projective twistor space P(T) is given by the basic relation

$$\omega^\alpha = i x^{\alpha\beta'} \pi_{\beta'} \quad (2.3)$$

which expresses the conditions for $\mathbf{Z} \in P(T)$ given by $\mathbf{Z} = \{ \gamma Z \mid 0 \neq \gamma \in C, Z = (\omega^\alpha, \pi_{\alpha'}) \in T, \pi_{\alpha'} \neq 0 \}$ to be incident with $x \in$ CM . Namely each point $\mathbf{Z}$ of P(T) (i.e. each projective twistor with $\pi_{\alpha'} \neq 0$) can be interpreted as the set of points in CM incident with it, i.e. with the set of complex $x^{\alpha\beta'}$ satisfying (2.3) with ω^α and $\pi_{\beta'}$ fixed. Inversely, the set of solutions of (2.3) in $\mathbf{Z}$ with $x^{\alpha\beta'}$ held fixed may be interpreted as the set of projective twistors $\mathbf{Z}$ in P(T) incident with point x of CM . In more detail a null twistor Z (with $\pi_{\beta'} \neq 0$) can be interpreted as a null-line in CM. A non-null twistor of P(T) as a totally null plane in CM, i.e. as a complex 2-plane in CM all tangent vectors of which are null, mutually orthogonal since they are of the form $\rho^\alpha \pi^{\alpha'}$, where π is fixed and ρ varies. This plane in CM is called an α- plane.[*/] On the other hand

---

*/ Similarly a β-plane in CM is defined as that whose tangent vectors have the form $\rho^{\alpha}\pi^{\alpha'}$, where now it is the unprimed spinor ρ that is fixed and π that varies. The β-plane in CM corresponds to a projective plane in P(T) .

---

the point in CM corresponds to a projective line in P(T) etc.

The correspondence $P(T) \leftrightsquigarrow CM$ mapping points of P(T) onto α-planes in CM and points in CM onto projective lines in P(T) may be described more abstractly by using flag and Grassmannian manifolds /4/.

Let F be the following flag manifold:

$$F = F(1,2,C^4) := \left\{ (L_1,L_2) \mid L_1,L_2 \text{ are subspaces in } C^4 , \quad \dim L_1 = 1,\ \dim L_2 = 2, L_1 \subset L_2 \right\} \tag{2.4}$$

and $G_{2,4}$ the Grassmannian manifold:

$$G_{2,4}(C) := \left\{ L_2 \mid L_2 \text{ are subspaces in } C^4,\ \dim L_2 = 2 \right\}, \tag{2.5}$$

which is isomorphic to the compactified complex Minkowski space CM. Let μ and ν be natural projections defined by

$$\mu : F \to P(T) , \quad F \ni (L_1,L_2) \mapsto L_1 \in P(T) ,$$

$$\nu : F \to G_{2,4} \sim CM, \quad F \ni (L_1,L_2) \mapsto L_2 \in G_{2,4} \sim CM .$$

Then the correspondence between CM and P(T) is given by

$$\varphi : P(T) \rightsquigarrow CM , \quad \varphi = \nu \circ \mu^{-1},$$

$$\psi : CM \rightsquigarrow P(T) , \quad \psi = \mu \circ \nu^{-1}, \tag{2.6}$$

$$P(T) \xleftarrow{\mu} F \xrightarrow{\nu} CM, \qquad P(T) \underset{\psi}{\overset{\varphi}{\rightleftarrows}} CM$$

This twistor correspondence is essential for finding solutions of massless field equations and (anti-) self-dual Yang-Mills field equations.

3. SUPERTWISTORS AND SUPERTWISTOR CORRESPONDENCE

Major proponents of both twistors /5/ and supersymmetry /6/ have suggested that these two theories are related. This relation was first clarified by Ferber /7/ who introduced supertwistors analogously to twistors, i.e. as quantities describing free massless particles, but in a superspace. Here we shall use a more abstract definition of supertwistors /4/ which provides to introduce various superspaces containing Minkowski space CM. It is based on the notion of supermanifolds with an arbitrary commuting Banach superalgebra Λ playing the role of the field of numbers (for details see /4/).

Let us denote by Σ^N a space of supertwistors for N-extended supersymmetry; Σ^N is a vector space isomorphic to a Λ-superspace $C^{4/N}$ of dimension 4/N on which the fundamental representation of conformal supergroup SU(2,2/N) acts. The coordinates of an element X of Σ^N, i.e. of the supertwistor X, will be denoted by

$$(X^A) = (X^a, X^i), a = 0,1,2,3, \; i=1,\ldots,N \text{ and } A = 0,\ldots 3+N ,$$

where $X^a(X^i)$ belongs to the even (odd) part of a commutative Banach superalgebra Λ over C . In a suitable basis of Σ^N the coordinates of the supertwistor X are $(X^A) = (\omega^\alpha, \pi_{\alpha'}, \vartheta^i)$, where ω , π and ϑ transform under $SL(2,C) \subset SU(2,2\, N)$ as spinors, cospinors and scalars respectively and ϑ^i's in addition form the fundamental representation of the internal symmetry group $SU(N) \subset SU(2,2/N)$. using complex conjugate dual supertwistors $(\bar{X}_A) = (\bar{\pi}_\alpha, \bar{\omega}^{\alpha'}, -\bar{\vartheta}_i)$ to (X^A) we can introduce the hermitian form on $\Sigma^N \times \Sigma^N$:

$$({}_{(1)}X, {}_{(2)}X) = {}_{(1)}\bar{X}_A\ {}_{(2)}X^A = {}_{(1)}\bar{\pi}_\alpha\ {}_{(2)}\omega^\alpha + {}_{(1)}\bar{\omega}^{\alpha'}\ {}_{(2)}\pi_{\alpha'} - {}_{(1)}\bar{\vartheta}_i\ {}_{(2)}\vartheta^i, \tag{3.1}$$

which is invariant under transformations from SU(2,2/N). The infinitesinal transformations from SU(2,2/N) are explicitly given in /4/.

In twistor theory space CM is isomorphic to Grassmannian manifold $G_{2,4}(C)$ given by (2.5), i.e. to the set of 2-dimensional subspaces in $T \sim C^4$. A similar situation happens for supertwistors. However, since in this case some extra odd dimensions appear, we have more possibilities.

Let us consider complex 2/k-dimensional subspaces ($0 \leq k \leq N$) of the supertwistor space $\Sigma^N \sim C^{4/N}$. A 2/k-subspace of $C^{4/N}$ is a 2/k-dimensional plane going through the origin and thus by 2 even and (N-k) odd homogeneous equations. The set of all 2/k-subspaces in Σ^N forms the superGrassmannian manifold $G_{2/k,4/N}(\Lambda)$ (see /4/). It has the dimension $[4+k(N-k)]/2N$. Insisting that the even part should be 4 we hawe two possibilities: either to put k=0 which corresponds to the supermanifold $G_R := G_{2/0,\ 4/N}(\Lambda)$ or k = N which leads to the supermanifold $G_L := G_{2/N,\ 4/N}(\Lambda)$. Since under SU(2,2/N) any 2/k – subspace is mapped onto another, SU(2,2/N) acts on $G_{2/k,4/N}(\Lambda)$. It turns out that complex superspaces G_R and G_L of dimension 4/2N are generalizations of chiral superspaces for $N > 1$. In order to get non-chiral superspaces we have to consider more complicated manifolds than Grassmannian, namely the flag supermanifolds with the length of the flag equal to 2. Then there are more possibilities of obtaining a 4/4N-dimensional superspace (see /8/) but the usual transformation properties are obtained if we choose (see /4/):

$$\widetilde{M} := F(2/0, 2/N, C^{4/N}) .$$

Let us now briefly describe the supertwistor correspondence. Since we have more superspaces at our disposal, we can speak about the correspondence between $P(\Sigma^N)$ and $\widetilde{M}$ or between $P(\Sigma^N)$ and G_R OR BETWEEN $P(\Sigma^N)$ and G_L which are given by means of natural projections of flag supermanifold. Besides the introduced spaces G_L, G_R and $\widetilde{M}$ we define the spaces

$$P(\Sigma^N) \approx P(C^{4/N}) \text{ and } F = F(1/0, 2/0, 2/N, \Sigma^N) .$$

Then the following diagram holds

$$\begin{array}{ccc} & F & \\ \swarrow^{\mu} & & \searrow^{\nu} \\ P(\Sigma^N) & \underset{\psi}{\overset{\varphi}{\rightleftarrows}} & \widetilde{M} \\ \varphi_L, \psi_L, \varphi_R, \psi_R & G_L \swarrow^{\pi_L} & \swarrow^{\pi_R} \\ & G_R & \end{array} \qquad (3.2)$$

Here the mappings μ, ν, π_L, π_R are projections defined by

$$\begin{aligned} \mu &: (L_{1/0}, L_{2/0}, L_{2/N}) \mapsto L_{1/0}, \\ \nu &: (L_{1/0}, L_{2/0}, L_{2/N}) \to (L_{2/0}, L_{2/N}), \\ \pi_R &: (L_{2/0}, L_{2/N}) \mapsto L_{2/0} , \\ \pi_L &: (L_{2/0}, L_{2/N}) \mapsto L_{2/N} , \end{aligned} \qquad (3.3)$$

where $L_{k/m}$ denotes a k/m-dimensional subspace, $(L_{1/0}, L_{2/0}, L_{2/N}) \in F$. The supertwistor correspondence are given by the pairs of the mappings.

$$\begin{aligned} \varphi &= \nu \circ \mu^{-1} , & \psi &= \mu \circ \nu^{-1} , \\ \varphi_L &= \pi_L \circ \varphi , & \psi_L &= \psi \circ \pi_L^{-1} , \\ \varphi_R &= \pi_R \circ \varphi , & \psi_R &= \psi \circ \pi_R^{-1} . \end{aligned} \qquad (3.4)$$

The first part of the correspondence is determined by mapping ψ. For instance, to any point from G_R (i.e. to any 2/0-subspace $L_{2/0}$) there corresponds the set of all 1/0-subspaces $L_{1/0}$ (projective twistors) for which $L_{1/0} \subset L_{2/0}$, i.e. the 1/0-dimensional projective straight line in $P(\Sigma^N)$. The second part of the supertwistor correspondence is described by mapping φ and specifies which object in the considered superspace corresponds to a projective supertwistor. It turns out that these objects are generalizations of α-planes in CM; in superspaces $\widetilde{M}$, G_L and G_R we call them α-, α_L- and α_R-superplanes respectively and they are of dimensions 2/3N, 2/2N and 2/N respectively. A similar correspondence exists for dual superspaces (see /4/).

In the twistor theory, form (2.1) can be used to define first conjugation and then real noncompactified Minkowski space $M \subset CM$. Thus real Minkowski space M is associated via the twistor correspondence with the projection space $P(N) \subset P(T)$ of null twistors. Namely, a real point $x \in M$ corresponds to a line in $P(T)$ which whole lies in $P(N)$ (and does not meet a special line corresponding to the vertex of a complex light cone at infinity of CM), and a null-twistor from $P(N)$ corresponds to a null straight line in M, i.e. to a geodesic. It is defined as an intersection of the corresponding α-plane in CM with the real space M. (The α-plane corresponding to twistor Z intersect M iff $Z \in N$, i.e. Z is null.)

The same situation happens for supertwistors. If we denote by $\mathcal{N} \subset \Sigma^N$ the set of all null supertwistors (i.e. those for which $\bar{X}_A X^A = 0$) and by R a 2/0-subspace then the necessary and sufficient condition that $R \subset \mathcal{N}$ is that R belongs to 2/N subspace R^+ defined by

$$R^+ = \left\{ Y^A \in \Sigma^N \,\middle|\, \bar{Y}_A X^A = 0, \ \forall X^A \in R \right\} .$$

Thus to a real point in M (i.e. for which $Z^+ = (L^+, R^+) = Z = (R,L)$, $L \in G_L$, $R \in G_R$) there corresponds a projective line lying in $P(\mathcal{N})$. The α-superplane corresponding to a supertwistor contains real points iff the corresponding supertwistor is null. The set of real points lying in the α-superplane represents a generalization of a null line.

In the twistor theory we work with the complex compactified space $CM \sim G_{2,4}(T)$ and then by using the correspondence with null twistors we separate from it the real Minkowski space M invariant with respect to the conformal group. However, space CM also contains the real Euclidean space E or more precisely its compactification S. For this space the twistor correspondence can be also established; however, realness of S does not follow from the requirement that some form on twistors should vanish (as was the case for Minkowski space M) but from invariance with respect to an antilinear mapping $\sigma : C^4 \to C^4$, $\sigma^2 = -1$. (By introducing a quaternionic structure on $C^4 \sim H^2$, this mapping represents the multiplication by quaternionic unit j /9/.) It is clear that 4-dimensional sphere S is not invariant under SU(2,2) but under another real form of its complexified covering group SL(4,C), namely the group SL(2,H).

In the supertwistor space Σ^N there exists a mapping analogous to the mapping σ only in case N is even. In /4/ first the quaternionic structure of Σ^{2k} was introduced, then the Euclidean superspace E_R^N, E_L^N and E^N determined and transformations under which these superspaces are invariant were specified.

4. TWISTOR AND SUPERTWISTOR CONTOUR INTEGRAL METHOD

It is well known that various differential equations appear as relations when we pass, by means of an integral transform, to manifolds of higher dimensions. This is an important source for solving differential equations. For instance, in the simplest example studied by Fritz John in 1938 solutions of a ultrahyperbolic second-order differential equation were found. They are functions on the 4-dimensional space of lines obtained by integrating the function in the 3-dimensional real space along lines. Penrose, Atiyah, Drinfeld, Hitchin, Manin and their successors encounter similar effects of differential equation "disappearance" but in a more complicated complex situation /10/. It includes not only linear field equations (Dirac, Weyl, Maxwell, linearized Einstein-Hilbert) but also some non-linear ones (Yang-Mills). They must deal with cohomology instead of functions as the following simple example - free massless field equations-indicates.

Free zero-mass field equations for an arbitrary spin $s = \frac{n}{2} > 0$ can be written as

$$\nabla^{\alpha\alpha'} \psi_{\alpha'\beta'\ldots\gamma'}(x) = 0 \,, \tag{4.1}$$

where $\psi_{\alpha'\beta'\ldots\gamma'}$ are totally symmetric spinors (the special cases s = 1/2, 1,2 give the Dirac-Weyl, Maxwell and linearized Einstein-Hilbert equations respectively). Their solutions can be written by means of a contour integral expression /11/:

$$\psi_{\alpha'\beta'\ldots\gamma'}(x) = \oint_{L_x} \pi_{\alpha'} \pi_{\beta'} \ldots \pi_{\gamma'} \, \rho_x \, f_{(-n-2)}(Z^a) \, \pi_{\delta'} \, d\pi^{\delta'} \,, \tag{4.2}$$

where $Z^a = (\omega^\alpha; \pi_{\alpha'})$ is a twistor, $f(Z^a)$ is a holomorphic function of twistor Z^a of homogeneity degree (-n-2), n is the number of indices of ψ, L_x is the projective line in P(T) corresponding via twistor correspondence to point $x \in CM$ at which the wave-function is to be evaluated and ρ_x is an operator restricting function f to L_x :

$$\rho_x f(\omega^{\alpha}, \pi_{\alpha'}) = f(ix^{\alpha\alpha'}\pi_{\alpha'}, \pi_{\alpha'}) \ . \qquad (4.3)$$

For a free wave function in x-space, we have holomorphicity in the forward tube which corresponds, in P(T) space, to holomorphicity in $P(T^+)$. But (apart from polynomials, which are trivial in this context) there are no homogeneous functions that are globally holomorphic on $P(T^+)$. Thus we have to pass to something a little more subtle. Namely we cover $P(T^+)$ with open sets $\{U_i\}$ (possibly infinite in number but if so, locally finite). Now, instead of having a single holomorphic twistor function, we envisage such a function f_{ij} for every pair of sets U_i, U_j and of the same homogeneity degree (-n-2) , where the domain of f_{ij} is $U_i \cap U_j$ and where $f_{ij} = - f_{ji}$. The set of $\{f_{ij}\}$ with these properties is a 1-cochain. Denoting by ρ_i restriction of the domain of f_{jk} by **intersection** with U_i we require the condition

$$\rho_i f_{jk} + \rho_j f_{ki} + \rho_k f_{ij} = 0$$

so that set $\{f_{ij}\}$ is a 1-cocycle (for the covering $\{f_i\}$) with coefficients in a sheaf of germs of holomorphic functions on P(T). The sheaf properties that our functions must satisfy are:

i) if $\rho_i f = \rho_i g$ for all i , then $f = g$,

ii) if $\rho_i f_j = \rho_j f_i$ for all i, j, then $f_i = \rho_i g$ for some g.

Thus contour integrals of type (4.2) need some basic sheaf cohomology theory. Without going into details it turns out that a twistor wave-function for a massless free field of helicity s is an element of the 1st cohomology group

$$H^1(P(T^+),\ \mathcal{O}(-2s-2))$$

with coefficients in sheaf $\mathcal{O}(-2s-2)$ (meaning roughly a class of "functions" holomorphic but"twisted" by $-2s-2$). Thus there is the 1-1 correspondence between holomorphic solutions of the linear massless field equations with spin s in CM^+ (the space of time-like $x \in CM$) and the elements of the first cohomology group $H^1(P(T^+),\ \mathcal{O}(-2s-2))$ with coefficients in sheaf $\mathcal{O}(-2s-2)$ (for details see also /11 /. An analogous result holds for matrix analogue of Maxwell's equations, the Yang-Mills equations. By the theorem of Ward /9 / anti-self-dual holomorphic Yang-Mills fields in CM^+ are in the 1-1 correspondence with holomorphic bundle on $P(T^+)$ holomorphically trivial on projective lines in $P(T^+)$. The self-dual Yang-Mills field have a similar interpretation in terms of the dual space $P(T^*)$. However, due to the nonlinearity of the Yang-Mills equation an arbitrary Yang-Mills field cannot be decomposed into the sum of anti- and self-dual fields. The most interesting physical applications of these results are related to instantons and monopoles in the Euclidean space (for details se /10/,/12 /. It appears that under some essential restrictions the similar correspondence is valid in the presence of gravitation, i.e. when the space is curved (R.Penrose in /10/ cf. /13/).

Let us note that integral formulae similar to (4.2) hold in a supertwistor case too (see /7/, /14/, /15/, /16/ and in particular /17/). For example for superfields $\phi_{\alpha'\beta'\ldots\gamma'}$ defined on superspace G_R we have /17/:

$$\phi_{\alpha'\beta'\ldots\gamma'}(w,\bar{\theta}^i) = \oint_{L_{w\bar{\theta}}} \pi_{\alpha'}\pi_{\beta'}\ldots\pi_{\gamma'}\,\rho_{w\bar{\theta}}\, f(X^A)\,\pi_{\delta'}\,d\pi_{\delta'}\,, \quad (4.4)$$

where supertwistor $(X^A) = (\omega^\alpha, \pi_{\alpha'}, \vartheta^i)$. By expanding f in terms of anticommuting variables ϑ^i and by substituting the obtained expansion to (4.4) we get superfield $\phi_{\alpha'\beta'\ldots\gamma'}$ expressed in terms of a series in θ's with the usual fields as coefficients, i.e. we obtain the supermultiplet of usual fields associated with the superfield ϕ. Since we know the transformation properties of supertwistors under SU(2,2/N), we can determine (via (4.4)) the transformation properties of various components of the supermultiplet and find representations of SU(2,2/N). Let us remark in this context that in § 3 we have not discussed superspaces with $k \neq 0, N$ or related with more complicated flag supermanifolds which might be important for formulation of supersymmetric theories in terms of unconstrained superfields recently proposed (cf. /18/).

The other area where supertwistors and the supertwistor integral transform play a natural role is supersymmetric Yang-Mills theories in four dimensions (in particular N=4). In N=4 supersymmetric Yang-Mills theory supertwistor transforms emerge from a geometrical interpretation of the superspace constraint equations. In fact Witten /14/ observed that the constraint equations satisfied by the corresponding superconnection are precisely equivalent to the integrability on light-like lines in superspace. The statement of integrability on lines is a condition that enables us to introduce vector bundles on the space of all light-like lines which can be regarded as a surface in the supertwistor space. Thus the physical supersymmetric Yang-Mills field can be geometrically interpreted as a vector bundle with nontrivial base.

In fact it seems that the supertwistor theory can also be

used as a suitable framework not only to interpret the constraint equation but also the equations of motion of supersymmetric Yang-Mills theories and to find their solutions (see e.g. /18/, /19/), and, after a modification, those of supergravity theories /20/.

Recently, since N=4 supersymmetric Yang-Mills theory in 4-dimensions is related to supersymmetric theories in 10-dimensions and these have made a dramatic rise to prominence, a supertwistor like transforms of 10 dimensional supersymmetric Yang-Mills theory was studied by Witten /21/. According to him, since superstring theory is in some way a generalization of ten-dimensional supersymmetric field theory, it is possible that the supertwistor transformation is also the proper starting point for the understanding of the geometrical meaning of superstring theory.

REFERENCES

/1/ R.Penrose, J.Math.Phys. 8 (1967), 345
R.Penrose, Rep.Math. Phys. 12 (1977), 65 and references therein; see also references in /3/, /11/

/2/ S.G.Gindikin, Math. Intell. 5 (1983), 27

/3/ R.Penrose in Proc. of the Int.Congress of Mathematicians in 1978 (Ed. O.Lehto), Helsinki, 1980 p.189
R.Penrose in Cosmology and Gravitation; Spin, Torsion, Rotation and Supergravity (Eds. P.G.Bergmann, V.de Sabbata), Plenum Press, N.Y. 1980, p.287;
see also R.Penrose, R.S.Ward in General Relativity and Gravitation, Vol.2 (Ed.A.Held), Plenum Press, N.Y.1980, p.283

/4/ M.Kotrla, J.Niederle, Czech.J.Phys. B35 (1985), 602

/5/ E.g. R.Penrose in Quantum Gravity (Eds.C.J.Isham, R.Penrose, D.W.Sciama) Clarendon Press, Oxford 1975, p. 268

/6/ E.g.A.Salam, J.Strathdee, Nucl.Phys. B76 (1974), 477

/7/ A.Ferber, Nucl.Phys. B132 (1978), 55;
see also M.Daniel, Phys.Lett. 65B (1976), 246

/8/ Yu.Manin in Problem in High Energy Physics and Quantum Field Theory, Vol.1, Protvino, 1982 (in Russian)

/9/ R.S.Ward, Phys.Lett. A61 (1977), 81;
E.Corrigan, Phys.Rep. 49 (1979), 95;
M.F.Atiyah, Geometry of Yang-Mills fields (preprint), Scuola Normale Superiore Pisa, 1979

/10/ R.Penrose, Gen.Rel.Grav. 7 (1976), 31;
R.S.Ward, Phys.Lett. 61A (1977), 81;
M.F.Atiyah, R.S.Ward, Comm.Math.Phys. 55 (1977), 117;
M.F.Atiyah, M.J.Hitchin, V.G.Drinfeld and Yu.I.Manin, Phys. Lett. 65A (1978), 175;
see also Complex Manifold Techniques in Theoretical Physics (Ed. D.E.Lerner and P.D.Dommers), Pitman, 1979

/11/ R.Penrose, Int.J.Th.Phys. 1 (1968), 61;
R.Penrose, J.Math.Phys. 10 (1968), 38;
R.Penrose, M.A.M.Mac Callam, Phys.Rep. 6C (1972), 241,
see also 2nd and 3rd refs. in /3/

/12/ N.J.Hitchin, Comm.Math.Phys. 83 (1982), 579

/13/ N.J.Hitchin, Math.Proc.Camb.Phil.Soc. 85 (1979), 475

/14/ E.Witten, Phys.Lett. 77B (1978), 394

/15/ R.Grimm, M.Sohnius, J.Wess, Nucl.Phys. B133 (1978), 275;
M.Sohnius, Nucl.Phys. B136 (1978), 461

/16/ Y.I.Manin in Problems of High Energy Physics and Quantum Field Theory, Vol.1, Protvino 1982

/17/ M.Kotrla, J.Niederle, Czech.J.Phys. B37 (1987), 338

/18/ J.Lukierski, preprint IFT 534, Wroclaw 1981;
V.I.Ogievetsky, E.S. Sokatchev, Yadernaya Fis. 31 (1980) 205,821;
E.A.Ivanov, J.Phys. A16 (1983), 2571;
L.B.Litov, V.N.Pervushin in Proc.XII Int.Conf. on High Energy

Physics, Leipzig 1984

/19/ I.V.Volovich, Theor. i Mat.Fiz. 54 (1983), 89 (in Russian);
I.V.Volovich, Lett.Math.Phys. 7 (1983), 517;
I.V.Volovich, Phys.Lett. 129B (1983), 429

/20/Yu.I.Manin in Proc. of the Int.Seminar "Group Theiretical Methods in Physivs", Zvenigorod 1982, Vol.1, Moscow 1983 (in Russian)

/21/ E.Witten, Twistor-like transform in ten dimensions; preprint Princeton University, Princeton 1985

Spinors, Reflections and Clifford Algebras : A Review *

R. COQUEREAUX **

I.H.E.S.

35, route de Chartres
91440 Bures-sur-Yvette
France

* Based in part on preprint UTTG-21-85 (University of Texas at Austin - same title).

** On leave from Centre de Physique Théorique (section 2), CNRS, Luminy, Marseille, France.

1. INTRODUCTION

The present article is a review on the algebraic aspects of the theory of spinors in n dimensions. Many things have been written on this old subject and all physicists have learned, in their training, how to handle the usual algebra of Dirac matrices, how to deal with Dirac spinors, how to define Weyl and Majorana spinors, how to implement discrete operations like parity and time reversal, how to define the charge conjugation operator and how to define a Dirac scalar product allowing us to write down interesting Lagrangians for the spinors under study. However, it is usually impossible, in classical textbooks, to desentangle those properties which would be true for spinors in n dimensions from those which rely heavily upon the 3+1 signature of our space time metric. In the present days, the physics of "higher dimensions" has emerged and it is commonly believed that events in our usual space time are only shadows of what took place in a "bigger space". In this perspective, it is necessary, for the theoretical physicist, to go back to the beginning and to study again the theory of spinors without assuming any prior knowledge of the usual 3+1 dimensional case. Although many papers exist on this subject, both in the physical and the mathematical literature, the user may have difficulties to put "everything together" because the above mentioned papers are usually concerned with very particular questions and there is usually a considerable gap between an article and another one. The present paper tries to fill this gap. It assumes no prior knowledge on the subject and therefore could be used by students without any acquaintances of the usual Dirac algebra. The paper is self contained and all results are usually proved (sometimes, however, we only give the sketch of the proofs: writing them in full would either triple the size of the article or make it very dry). The whole subject being very old, it is very often difficult to acknowledge original references: many result often attributed to some authors were already floating around or even well known by some other people when they have been published. For this reason no attempt is made to give references in the main body of the text: they are all gathered at the end and a special section (Sect. 7) has been added to provide a guide through the literature. Also there is no claim for originality here, although some proofs are certainly new and some results perhaps also. Among other things, this paper contains a proof of the modulo 8 periodicity of Clifford Algebras for an arbitrary signature (p,q) of the space time metric, a study of their representations in relation with the existence of a charge conjugation operator c , a study of Dirac, Weyl, Majorana and Weyl-Majorana spinors, a study of Pin and Spin groups, a study of discrete reflections and a study of spin invariant scalar products of spinors to be used in constructing real Lagrangians. For more details, see the following table of contents. Many interesting topics where the algebraic role of spinors is important have been however left over and those missing entries are gathered in Sect. 7 along with appropriate references.

Table of Contents

2. GENERAL PROPERTIES OF CLIFFORD ALGEBRAS.

2.1 Real Clifford Algebras: Definitions.

2.1.1 First Definition.

The real Clifford algebra $G = G(p,q)$ is the universal real algebra generated by one unit 1 and $n = p + q$ symbols γ^μ submitted to the relations $(\gamma^\mu)^2 = 1$ for $\mu \varepsilon \{1,2,\dots,p\}$, $(\gamma^\nu)^2 = -1$ for $\nu \varepsilon \{p+1,\dots,p+q\}$ and $\gamma^\mu\gamma^\nu + \gamma^\nu\gamma^\mu = o$ for $\mu \neq \nu$. The product in G is called the Clifford product.

The notation (p,q) is not universal and is therefore also written (q,p) in half of the quoted references.

Let E be a vector space of dimension n with a non degenerate pseudo euclidean metric of g of signature (p,q); we can associate with every vector v of E an element $\not{V} = V_\mu\gamma^\mu$ of G, where (V_μ) are the components of V in a chosen orthonormal frame. In such a way we get the relation $\not{u}\not{v} + \not{v}\not{u} = 2g(u,v)$ for $u,v\varepsilon E$.

Very often it is convenient to drop the (/) notation and to think of E as a subspace of G. That is what we will do in the following.

It is clear that G is a real vector space and that a (vector) basis for G can be taken as follows: $\{1, \gamma^\mu, \gamma^\mu\gamma^\nu, \gamma^\mu\gamma^\nu\gamma^\rho, \dots\}$ with $\mu < \nu < \rho \dots$. The real dimension of G is therefore $1 + C_1^n + C_2^n + \dots + C_{n-1}^n + 1 = (1+1)^n = 2^n$.

The adjective "real" in the definition of G is quite important; for example, in 3+1 dimensions the element $i\gamma^0\gamma^1\gamma^2\gamma^3$, usually called γ^5, is not an element of G but an element of the complexified algebra $G^C = G \otimes C$.

The algebra G(p,q) is usually not isomorphic with G(q,p), as we shall see later. This is unfortunate, and for this reason we shall sometimes write (p_+,q_-) to remind the reader of our convention.

2.1.2 Second Definition.

In physics textbooks, $G(p,q)$ is often defined via an explicit faithful realization using square matrices over a field which may be R,C or H (quaternions): one has just to exhibit a set of generators that satisfy the above constraints (cf. 2.1.1). The Clifford product is then realized as a product of matrices.

Actual calculations using an explicit realization of the generators take a lot of space (specially if $n = p + q$ is big) and can always be achieved without using explicit matrices, for this reason we will deliberately avoid their use.

2.1.3 Third Definition.

Call $\Lambda = \sum_{p=0}^{n} \Lambda^p$ the algebra of differential forms over the vector space E; of course $\Lambda^0 = R$, $\Lambda^1 = E^*$ (the dual of E). Let $\{\omega^1, \omega^2, \ldots, \omega^n\}$ be an arbitrary basis of E^*, it is well known that Λ is a vector space of dimension 2^n generated by the elements $\{1, \omega^\mu, \omega^\mu_\wedge \omega^\nu, \omega^\mu_\wedge \omega^\nu_\wedge \omega^\rho, \ldots\}$ with $\mu < \nu < \rho \ldots$ E being a (pseudo) euclidean space, its metric can be used to compute scalar products of forms; by using Λ and the metric g, we will define a new product between the elements of Λ: the Clifford product.

If $u \in \Lambda^1$, define $\bar{g}(u,.)$ as follows:

For $\sigma \in \Lambda^p, \tau \in \Lambda, \bar{g}(u, \sigma_\wedge \tau) = \bar{g}(u,\sigma)_\wedge \tau + (-1)^p \sigma_\wedge \bar{g}(u,\tau)$

For $V \in \Lambda^1$ $\quad \bar{g}(u,v) = g(u,v)$

Notice $\bar{g}(u,.)$ is an antiderivation maps Λ^p into Λ^{p-1}.

If $u \in \Lambda^1$ and $\tau \in \Lambda$, define the Clifford product $u\tau$ as follows:

$$u\tau = u_\wedge \tau + \bar{g}(u,\tau)$$

Extend the above definition of the Clifford product to the whole of Λ by imposing linearity and associativity.

Λ, endowed with the Clifford product is isomorphic with the real Clifford algebra of E.

Let us work out a few examples in the case of a 4 dimensional space:

$$\omega^2\omega^3 = \omega^2 \wedge \omega^3 + g(\omega^2,\omega^3)$$

$$\omega^1\omega^2\omega^3 = \omega^1 \wedge \omega^2 \wedge \omega^3 + g(\omega^2,\omega^3)\omega^1 + g(\omega^1,\omega^2)\,\omega^3 - g(\omega^1,\omega^3)\omega^2$$

$$\begin{aligned}\omega^0\omega^1\omega^2\omega^3 = {} & \omega^0 \wedge \omega^1 \wedge \omega^2 \wedge \omega^3 + g(\omega^2,\omega^3)\omega^0 \wedge \omega^1 + g(\omega^1,\omega^2)\omega^0 \wedge \omega^3 - g(\omega^1,\omega^3)\omega^0 \wedge \omega^2 \\ & + g(\omega^0,\omega^1)\omega^2 \wedge \omega^3 - g(\omega^0\omega^2)\omega^1 \wedge \omega^3 + g(\omega^0,\omega^3)\omega^1 \wedge \omega^2 \\ & + g(\omega^0,\omega^1)\, g(\omega^2,\omega^3) + g(\omega^0,\omega^3)\, g(\omega^1,\omega^2) - g(\omega^0,\omega^2)\, g(\omega^1,\omega^3)\end{aligned}$$

Notice that the first example implies $\omega^2\omega^3 + \omega^3\omega^2 = 2g(\omega^2,\omega^3)$, as it should.

The above construction shows in particular how to express a given Clifford product as a sum of exterior products; the projection along a given Λ^p can be used to define (generalized) traces. In particular, if dim E=n, it is clear that n x (projection on Λ^0) is the usual trace that one would obtain by using γ-matrices. For example:

$$\mathrm{Tr}(\omega^0\omega^1\omega^2\omega^3) = 4\,(g^{01}g^{23} + g^{03}g^{12} - g^{02}g^{13})$$

We have seen how to express (or to define!) a Clifford product in terms of exterior products, we can also do the converse.

For $a,b \in G$, we define first

$$[a,b] = ab - ba \qquad \text{and} \qquad a \odot b = \frac{1}{2}\,(ab + ba)$$

Then for $v^1, v^2 \dots v^p \in E^*$ we define (or recover) the exterior product via the following iterating relations:

$$2.\ v^1{}_\wedge v^2{}_\wedge \dots {}_\wedge v^{2p} = [v^1{}_\wedge v^2{}_\wedge \dots {}_\wedge v^{2p-1}, v^{2p}]$$

$$v^1{}_\wedge v^2{}_\wedge \dots {}_\wedge v^{2p}{}_\wedge v^{2p+1} = (v^1{}_\wedge v^2{}_\wedge \dots {}_\wedge v^{2p}) \odot v^{2p+1}$$

Notice that if the basis (ω^p) is orthonormal, then the element $\varepsilon = \omega^1\omega^2 \dots \omega^n = \omega^1{}_\wedge\omega^2{}_\wedge \dots {}_\wedge\omega^n$ is the volume element and defines an orientation of E.

2.1.4 Fourth Definition.

The following definition is the one most often found in mathematical textbooks, it is short and nice but maybe not so easy to use. The proof that it is equivalent to the previous ones is left to the reader.

Let T be the tensor algebra over the (pseudo) euclidean vector space E (with metric g); then it is clear that the set I of all linear combinations of elements of the kind $s \otimes (u \otimes u - g(u,u)1) \otimes t$ where $u \in E$, $s,t, \in T$ is a right and left ideal of T, in other words

$$\forall\, t \in T,\ t \otimes I \subset I \text{ and } I \otimes t \subset I$$

The Clifford algebra is then defined as the coset space $G = T/I$.

2.1.5 Complex Clifford Algebra, Definition.

Although we are mainly interested in the structure of real Clifford algebra, we will need, at times, to consider its complex extensions which is of course $G^C = G(p,q) \otimes C$.

Notice that if $\{\gamma^1,\ldots,\gamma^n\}$ is a basis for $G(p,q)$, then $\{\gamma^1, i\gamma^1,\ldots,\gamma^n, i\gamma^n\}$ is a basis for G^c when considered as an algebra over R. Of course, $G(p,q)$ and $G(p',q')$ with $p + q = p' + q' = n$ have the same complex intension G^C.

2.2 Structure of Clifford Algebras

Physicists are not only interested in representations of the spin group (to be defined later) but also in representations of a bigger group (the Pin group) containing also discrete symmetries like parity or time reversal. We shall see later how to embed the Pin group into G and the spin group into G_o (the subalgebra of G generated by products of an even number of γ^μ); the whole discussion can be traced back to the properties of the Cliford algebra itself and the analysis, at this level, is much simpler. The questions to be answered are the following: is G simple, is G_o simple?; as we shall see the results are very different in even and odd dimensions, also, for a given dimension n, the results depend upon the value of the square of the element $\varepsilon = \gamma^1\gamma^2\ldots\gamma^n$.

2.2.1 Notations.

We call G_o the even subalgebra of G i.e. the subalgebra generated by linear combinations of even products of γ^μ.

For example $2 + a_{\mu\nu}\gamma^\mu\gamma^\nu \in G_o$ but $V_\mu\gamma^\mu \not\in G_o$

It is convenient to denote by a special symbol the element of G which is the product of all γ^μ symbols, we call it $\varepsilon = \gamma^1\gamma^2\ldots\gamma^p\gamma^{p+1}\ldots\gamma^{p+q}$. Notice that, due to the anticommunitativity of the γ^μ, there are two possible definitions, depending upon the chosen ordering; we have therefore to choose such an ordering (this amount to say that we have choosen a particular orientation of E). We will call ε the orientation operator. Notice that there is no $i = \sqrt{-1}$ in the definition of ε.

A subalgebra of a given algebra A is called bi-invariant if it is stable under right and left multiplication by elements of A. If A does not contain any bi-invariant subalgebra (besides 0 and itself) it is called simple.

We call Z the center of G and Z_o the center of G_o.

2.2.2 Real Clifford Algebra, n even.

From the defining properties of a Clifford algebra, it is obvious that, in this case, ε commutes with even elements of G but anticommutes with odd elements (this property is well known in the case of the usual 3+1 Dirac algebra). It is then easy to show that the center Z_o of G_o is generated by 1 and ε (obviously, $\varepsilon \in G_o$). In order to know if G (or G_o) is simple, we can try to exhibit a nontrivial projection operator P (with $P^2 = P$) commuting with the algebra G (or G_o). There are no such candidates for G, but, in the case of G_o and if $\varepsilon^2 = 1$, we can construct such an operator, namely $P = \frac{1+\varepsilon}{2}$. The whole discussion then relies upon the value of ε^2 which is found to be $(-1)^{\frac{n}{2}+q} = (-1)^{\frac{p-q}{2}}$ by a direct calculation. The results can be summarized as follows (most of the $\Rightarrow$ and $\Leftarrow$ are left to the reader).

G is simple. $\varepsilon \in G_o$, $Z = R$, $Z_o = \{a1 + b\varepsilon \; ; \; a,b \in R\}$.

$n = p_+ + q_-$

n even $\quad G_o$ simple $\Longleftrightarrow \varepsilon^2 = -1 \Longleftrightarrow p-q = 2 \bmod 4 \Longleftrightarrow Z_o \sim C$

G_o not simple $\Longleftrightarrow \varepsilon^2 = 1 \Longleftrightarrow p-q = 0 \bmod 4 \Longleftrightarrow Z_o \sim R+R$

and in this last case $G_o = \frac{1-\varepsilon}{2} G_o + \frac{1+\varepsilon}{2} G_o$

In the case $\varepsilon^2 = 1$, the two simple components of G_o are sometimes denoted by G_o^L and G_o^R (L and R standing for left and right).

Notice that in the case of the Dirac algebra ($p = 3$ and $q = 1$), $\varepsilon^2 = -1$ and G_o is simple. We would need a $i = \sqrt{-1}$ to break it into two pieces but this can only be done at the complexified level (sec. 2.2.4).

2.2.3 Real Clifford Algebra, n odd.

From the defining properties of a Clifford algebra, it is easy to see that, now, ε commutes with even and odd elements: $\varepsilon \in Z$. If $\varepsilon^2 = 1$, then we can use ε to build a projector action on G; but there is no such candidate for G_o. A direct calculation shows that $\varepsilon^2 = (-1)^{\frac{p-q+1}{2}}$. The results can be summarized as follows.

G_o is simple. $\varepsilon \in G$, $Z_o = R$, $Z = \{a1 + b\varepsilon;\ a,b\ \varepsilon R\}$.

$n = p_+ + q_-$

n odd $\quad$ G is simple $\Longleftrightarrow \varepsilon^2 = -1 \Longleftrightarrow p - q = 3 \bmod 4 \Longleftrightarrow Z \sim C$

G is not simple $\Longleftrightarrow \varepsilon^2 = 1 \Longleftrightarrow p - q = 1 \bmod 4 \Longleftrightarrow Z \sim R+R$

and in this last case $G = (\frac{1 - \varepsilon}{2})\, G + \frac{(1+\varepsilon)}{2}\, G$

Thus G is simple if and only if $p-q \neq 1 \pmod 4$ and G_o is simple if and only if $p-q \neq 0 \pmod 4$.

Also one get $G = G_o \otimes Z = G_o + G_o\varepsilon$.

Notice that the roles of G and G_o are somewhat interchanged by going from the case n even to the case n odd.

From the above results one could think that the classification of the G(p,q) algebras depends only on p−q mod 4. Actually, we will make later a more subtle analysis showing that the classification depends in fact on p−q mod 8.

2.2.4 <u>Complex Clifford Algebras.</u>

At the complexified level, things are easier since it is always possible to find and element θ (which will be ε or iε) of square 1 and therefore to construct projectors $\frac{1 \pm \theta}{2}$ which will project $G_o^C = G_o \otimes C$ on simple components when n is even and will project $G^C = G \otimes C$ on simple components when n is odd. The following summarizes the discussion.

Call $\theta = \varepsilon$ if $\varepsilon^2 = 1$, call $\theta = i\varepsilon$ if $\varepsilon^2 = -1$

n even. G^C is simple

$G_o^C = \frac{1-\theta}{2} G_o^C + \frac{1+\theta}{2} G_o^C$ is not simple

n odd $\quad G^C = \frac{1-\theta}{2} G^C + \frac{1+\theta}{2} G^C$ is not simple

G_o^C is simple

The complexified form of the usual G(3,1) is therefore simple; θ is usually called γ^5 in this case. The non simplicity of G_o^C is of course related to the splitting of a Dirac spinor into two Weyl spinors as we shall see later.

We will call θ the chirality operator. Notice that for a space (time) of given signature, θ may differ from the orientation operator ε by the presence of an i factor. It is very often claimed that the chirality operator does not exist in odd dimensions, we shall see later (sect. 4.2.1) how to make this statement meaningfull and true.

2.3 Clifford Group, Pin, Spin and Spin↑ Group.

2.3.1 First Definition (topological)

Let us remember that the full Lorentz group L associated with a pseudo euclidean vector space E of signature (p,q) has four connected components:

$$L = L_+^{\uparrow} \cup L_+^{\downarrow} \cup L_-^{\uparrow} \cup L_-^{\downarrow}$$

where the $\pm$ sign refers to those elements of determinant ± 1 and where the ↑↓ sign refers to the transformations of positive or negative temporal signature. If E is euclidean, L has only two connected components. With other notations, calling O(g) the orthogonal group of the metric g we have the isomorphisms:

$$O(g) = L \quad , \quad SO(g) = L_+^{\uparrow} \cup L_+^{\downarrow} \quad , \quad SO{\uparrow}(g) = L_+^{\uparrow}$$

It is well known that none of the connected components of L is simply connected (their first homotopy group is z_2); therefore it is natural to consider their covering groups _

Pin, Spin and Spin ↑ are the (2-fold) covering groups of $O(g)$, $SO(g)$, $SO\uparrow(g)$.

2.3.2 Second Definition (algebraic)

The Clifford group Γ is by definition the set of all elements s of the Clifford algebra G which, at the same time, are invertible and satisfy the property: $\forall x \in E \quad , \quad sxs^{-1} \in E$.

It is clear that any invertible element v of E belongs to the Clifford group Γ (as usual we think of E as a subspace of its Clifford algebra G).

$$\text{Indeed } vxv^{-1} = \frac{1}{v^2}\, vxv = \frac{1}{v^2}\left(-vvx + 2\, v\, g(v,x)\right)$$
$$= -\, x + 2g(v,x)\ v/v^2$$

i.e., the transformation $x \to vxv^{-1}$ is minus the symmetry with respect to the hyperplane conjugate to v. More generally, for $s \in \Gamma$, the transformation $\chi(s)$ defined by $\chi(s)x = sxs^{-1}$, $x \in E$ belongs to the orthogonal group $O(g)$, indeed it is well known that this group is generated by products of such symmetries.

It is clear that χ defines a representation of the Clifford group (the vectorial representation) but this representation is not faithfull, for example $\chi(s)$ and $\chi(5s)$ are the same orthogonal transformation. More precisely, we get χ^{-1} (identity) $= Z - \{0\}$; where Z, as before, denotes the center of G. A standard way to decrease the "size" of Γ is to introduce a normalization constant; for that, we need to define first the Dirac anti-involution.

It is convenient to define the maps β_+ and β_- as the anti-involutions of G whose restriction to E are respectively the identity $\left(\beta_+(\gamma^\mu) = \gamma^\mu\right)$ and minus the identity $\left(\beta_-(\gamma^\mu) = -\gamma^\mu\right)$ i.e.

$$\beta_+\,(\gamma^{i1}\gamma^{i2}\ldots\gamma^{ip}) = \gamma^{ip}\ldots\gamma^{i2}\gamma^{i1} \text{ and } \beta_-\,(\gamma^{i1}\gamma^{i2}\ldots\gamma^{ip}) = (-1)^p\,\gamma^{ip}\ldots\gamma^{i2}\gamma^{i1}$$

Depending upon the chosen convention (p,q) or (q,p) for the metric, the map β (β_- or β_+) will coincide with the usual Dirac "bar"-operation: $\beta(s) = \bar{s}$ (see Sect. 6.3.1). In the present section, we assume that $n = p+q$ is even and that the signature is hyperbolic, i.e., $(n-1,1)$ or $(1, n-1)$. Moreover we set $\bar{s} = \beta_+(s)$ if $\gamma^{02} = +1$ and $\bar{s} = \beta_-(s)$ if $\gamma^{02} = -1$. If we want to extend the definition of β at the complex level, one has to set $\beta(i) = -i$.

The map $s \in \Gamma \to N(s) = \bar{s}s \in R$ is sometimes called the spinorial norm. We can now define Pin, Spin and Spin↑ as the following subgroups of Γ:

$$\mathrm{Pin} = \{s \in \Gamma / \ |N(s)| = 1\}$$

$$\mathrm{Spin} = \mathrm{Pin} \cap G_o$$

$$\mathrm{Spin}^{\uparrow} = \{s \in \mathrm{Spin} \ / \ N(s) = 1\}$$

χ is still a representation of the above groups. Notice that $\chi(\Gamma) = \chi(\mathrm{Pin})$ and we can think of Pin as a "normalized" Γ. Moreover $\chi(\mathrm{Pin}) = L$ if n is even but $\chi(\mathrm{Pin}) = L_+$ if n is odd (it is clear, for example, that the spacetime parity $\gamma^\mu \to -\gamma^\mu$ cannot be written as $s\gamma^\mu s^{-1}$ if n is odd; it would imply $s\varepsilon s^{-1} = -\varepsilon$ but ε commutes with s for n odd therefore $\varepsilon = -\varepsilon$!). It is easy to show that $\chi(\mathrm{spin}) = L_+$ and that $\chi(\mathrm{spin}^\uparrow) = L_+^\uparrow$.

2.3.3 A Few Well Known Properties - Examples (here assume n even and the structure hyperbolic)

Every element of Pin can be written $s = u_1 u_2 \ldots u_k$ with $u_k \in E$.

If k is even and $\bar{s}s > 0$ then $\chi(s)$ belongs to $L_+^\uparrow$, for example $s = \gamma^1\gamma^2$.

If k is even and $\bar{s}s < 0$ then $\chi(s)$ belongs to $L_+^\downarrow$, for example $s = \varepsilon$ covers the PT operation: $\varepsilon\gamma^\mu\varepsilon^{-1} = -\gamma^\mu$.

If k is odd and $\bar{s}s > 0$ then $\chi(s)$ belongs to $L_-^\uparrow$

for example $s = \gamma^0$ covers the P operation: $\gamma^0\gamma^0\gamma^{0-1} = \gamma^0, \gamma^0\gamma^i\gamma^{0-1} = -\gamma^i$

If k is odd and $\bar{s}s < 0$ then $\chi(s)$ belongs to $L_-^\downarrow$

for example $s = \varepsilon\gamma^0$ $(= \pm\gamma^1\gamma^2\cdots\gamma^n)$ covers the T operation:

$(\varepsilon\gamma^0)\gamma^0(\varepsilon\gamma^0)^{-1} = -\gamma^0$, $(\varepsilon\gamma^0)\gamma^i(\varepsilon\gamma^0)^{-1} = \gamma^i$.

Warning: some authors introduce an additional twist in the definition of χ, writing $s\gamma\bar{s}^{-1}$ rather than $s\gamma s^{-1}$; although usefull in specific cases, the other convention leads to a terminology which is opposite to the one which is traditional in physics.

2.3.4 The Lie Algebra of Spin↑

We already saw how to express the groups Γ, Pin, Spin and Spin↑ as subsets of the Clifford algebra G; we will show now algebra(s) of these groups can also be realized as subsets(s) of G.

First we remark that if $x,y,z \in E$ then $[xy,z] \in E$ indeed $[xy,z] = xyz - zxy = x\,g(y,z) - g(x,z)y$. Therefore, by developing the exponential, we get $\exp(txy)\, z \exp(-txy) \in E$, which means that $\exp(txy)$ belongs to Γ and therefore $xy \in \mathrm{Lie}\,(\Gamma)$. The next remark is that $\overline{\exp(s)} = \exp(\bar{s})$ as we can see by developing the exponential. Finally $N(\exp(s)) = \exp(\bar{s})\exp(s)$.

The previous remarks show that Lie (Γ) is generated by 1 and the products xy where $x,y \in E$. This shows for example that when $x = 4$, Lie (Γ) has dimension $1 + \frac{3\times4}{2} = 7$. However Lie (Γ) is too big: Lie (Spin↑) is more interesting. Notice that, with $x,y \in E$

$$\exp(xy) \in \text{spin}\uparrow \Longleftrightarrow N(\exp xy) = 1 \Longleftrightarrow e^{\overline{xy}}\, e^{xy} = 1$$

$$\Longleftrightarrow e^{yx+xy} = 1 \Longleftrightarrow xy + yx = 0$$

$$\Longleftrightarrow g(x,y) = 0$$

Therefore x belongs to Lie (Spin↑) if it can be written as a linear combination of products xy with $x,y \in E$ and x orthogonal to y. For instance, when x = 4, Lie (Spin↑) being generated by products $\gamma^\mu\gamma^\nu$ has dimension 4x3/2 = 6 as it should be (this Lie algebra is of course isomorphic with the one of the Lorentz group).

Example. Consider the case (p,q) = (3,1) and the element $\frac{\beta}{2}\gamma^0\gamma^1 \in$ Lie (spin↑) then $s = \exp(\frac{\beta}{2}\gamma^0\gamma^1) \in$ spin↑ and a short calculation leads to

$$s = \text{ch}\beta/2 + \gamma^0\gamma^1\,\text{sh}\beta/2 \qquad ; \quad \text{also} \quad s\gamma^0 s^{-1} = \gamma^0\text{ch}\beta + \gamma^1\text{sh}\beta$$

$$s = \gamma^0(-\gamma^0\,\text{ch}\beta/2 + \gamma^1\,\text{sh}\beta/2) \qquad ; \qquad s\gamma^1 s^{-1} = \gamma^0\text{sh}\beta + \gamma^1\text{ch}\beta$$

$$s\gamma^i s^{-1} = \gamma^i \quad \text{for} \quad i = 2,3$$

This shows that the Lorentz transformation $\chi(s)$ is a boost along γ^1 and that, according to the discussion of section 2.3.2 it can be written as the product of two symmetries: the first relative to the hyperplane orthogonal to $-\gamma^0\text{ch}\beta/2 + \gamma^1\text{sh}\beta/2$, the second relative to the hyperplane orthogonal to γ^0.

Further examples. In the same way it is easy to show that $\exp(\frac{\beta}{2}\gamma^0(\vec{v}\cdot\vec{\gamma}))$ describes a boost of parameter β in the direction $\vec{v}$ and that $\exp(\frac{\theta}{2}\vec{n}\cdot\vec{\Sigma})$ where $\vec{\Sigma} = (\Sigma^1\Sigma^2\Sigma^3) = (\gamma^2\gamma^3,\ \gamma^3\gamma^1,\ \gamma^1\gamma^2)$ describes a rotation of angle θ around the direction $\vec{n}$, ($|n|^2 = 1$).

It is maybe not useless to remember that the elements of spin which are not in the connected component of the identity cannot be written as an exponential. For example, with a Lorentz signature and n even, the element $\varepsilon = \gamma^0\gamma^1\ldots\gamma^{n-1}$ belongs to Spin (it covers the PT opration, element of $L^{\downarrow}_{+}$ sec 2.3.2), not to Spin↑ be written the exp [ ] of something.

2.3.5 Isomorphisms in Low Dimension

We give, without proof a few well known isomorphisms between Spin groups and classical groups. Notice that although G(p,q) and G(g,p) are usually not isomorphic, Spin (p,q) and Spin (q,p) are the same (this comes from the fact that $G_0(p,q) = G_0(q,p)$). Below we list Spin↑ (p,q).

(p,q)	Spin↑	(p,q)	spin↑	(p,q)	spin↑
(3,0)	SU(2)	(2,1)	SL(2,R)		
(4,0)	SU(2)xSU(2)	(3,1)	SL(2,C)		
(5,0)	U(2,H)	(4,1)	U(1,1,H)	(3.2)	Sp(4,R)
(6,0)	SU(4)	(5,1)	SL(2,H)	(4.2)	SU(2,2)

In the above table H stands for the quaternions; U(n,H) is the real compact group sometimes called USp(2n) (or even Sp(n)) in the literature.

Notice that the conformal group in a space time of dimension (p,q) is locally isomorphic to spin (p+1, q+1); besides if we define SL(2,K) as the group of projective transformations of the projective line KP^1 (K being R,C,H or O (octonions)), we have SL (2,K) = SO(n+1,1), n = 1,2,4,8. This explains some of the above isomorphisms; in particular we get SL(2,O) = spin (9,1), a group which appears naturally in 10 dimensional string theories with fermions (notice that the usual definition of SL(2,k) groups does not work when K = O).

3. Classification and Periodicity

3.1 A Theorem of Dimensional Reduction

The proof of the modulo 8 periodicity of real Clifford algebras relies mostly on the following theorem:

> F being a vector space endowed with a scalar product g, and $E \subset F$ being a vector subspace of even dimension such that the restriction g_E of g to E is not singular, then, calling $E^\perp$ the orthogonal complement of E in F, one has
>
> $$G(F,g) = G(E,g_E) \otimes G(E^\perp, \varepsilon_E^2\, g_{E^\perp})$$
>
> ε_E being the ε element of E introduced in 2.2.1.

The proof is easy and goes as follows: let $\{\gamma^\mu\}, p = s,\ldots,m$ be a set of generators for $G(p_1,q_1)$, $p_1 + q_1 = m$ and $\{\gamma^\alpha\}$ $(\alpha = 1,..,s)$, a set of generators for $G(p_2,q_2)$, $p_2 + q_2 = s$; we build the following objects:

$$\Gamma^\mu = \gamma^\mu \otimes 1 \qquad \mu = 1,\ldots,m$$

$$\gamma^\alpha = \varepsilon \otimes \Gamma^\alpha \qquad \alpha = 1,\ldots,s \qquad \text{where } \varepsilon = \prod_{\mu=1}^{m} \gamma^\mu$$

Then it is obvious that

$$\Gamma^\mu\Gamma^\nu + \Gamma^\nu\Gamma^\mu = 2g^{\mu\nu} \quad , \quad \Gamma^\alpha\Gamma^\beta + \Gamma^\beta\Gamma^\alpha = \varepsilon^2 2g^{\alpha\beta}$$

and, using the fact that ε anticommutes with γ^μ for m even, $\Gamma^\mu\Gamma^\alpha + \Gamma^\alpha\Gamma^\mu = 0$. Moreover $(\Gamma^\mu)^2 = (\gamma^\mu)^2$, $(\Gamma^\alpha)^2 = \varepsilon^2.(\gamma^\alpha)^2$. This proves the above theorem.

For example using $(p_1,q_1) = (3,1)$ and $(p_2,q_2) = (o,s)$, the generators $\{\Gamma^\mu, \Gamma^\alpha\}$ generate therefore $G(3+s,1)$. (We should remember that (p,q) means (p_+, q_-)).

The above theorem may therefore be used as a tool for building "big" Clifford algebras out of "small" ones, in particular we get the following relations.

3.2 Recurrence Relations

Let us choose a two dimensional vector space E with a metric g_E, then

$\varepsilon = \gamma^1\gamma^2$,

if its signature is (2,0), then $\varepsilon^2 = -1$ and we get $G(2,0) \otimes G(p,q)$

$= G(q+2,p)$

if its signature is (0,2), then $\varepsilon^2 = -1$ and we get $G(0,2) \otimes G(p,q)$

$= G(q,p+2)$

if its signature is (1,1), then $\varepsilon^2 = +1$ and we get $G(1,1) \otimes G(p,q)$

$= G(p+1,q+1)$

These recurrence relations will allow us to find explicit faithfull realizations of the algebras $G(p,q)$ in terms of metrix algebras, but first, we need to work out a few cases of low dimension.

3.3 A Few Cases of Low Dimension

We will only need to consider the 1-dimensional and 2-dimensional cases.

3.3.1 Algebras G(1,0) and G(0,1)

$G(0,1)$ is generated by 1 and γ with $\gamma^2 = -1$, therefore this algebra is isomorphic with the complex numbers.

$G(1,0)$ is generated by 1 and γ with $\gamma^2 = +1$, therefore this algebra is isomorphic with the algebra $R + R$ of matrices of the type $\left[\begin{array}{c|c} \alpha & 0 \\ \hline 0 & \beta \end{array}\right]$, $\alpha, \beta \in R$.

3.3.2 Algebras G(2,0), G(1,1) and G(0,2)

Let us first recall the definitions of the Pauli matrices:

$$\sigma^0 = \begin{bmatrix} 1 & 0 \\ 0 & 1 \end{bmatrix}, \quad \sigma^1 = \begin{bmatrix} 0 & 1 \\ 1 & 0 \end{bmatrix}, \quad i\sigma^2 = \begin{bmatrix} 0 & 1 \\ -1 & 0 \end{bmatrix}, \quad \sigma^3 = \begin{bmatrix} 1 & 0 \\ 0 & -1 \end{bmatrix}$$

the squares of σ^0, σ^1, $i\sigma^2$ and σ^3 are respectively 1,1,-1,1.

For G(2,0) we can choose $\gamma^1 = \sigma^3$, $\gamma^2 = \sigma^1$ then $\varepsilon = \gamma^1\gamma^2 = i\sigma_2$ and clearly G(2,0) is isomorphic with the algebra R(2) of 2x2 real matrices.

For G(1,1) we can choose $\gamma^1 = \sigma^1$, $\gamma^2 = i\sigma^2$ then $\varepsilon = \gamma^1\gamma^2 = -\sigma^3$ and again G(1,1) is isomorphic with the algebra R(2).

For G(0,2) we can choose $\gamma^1 = i\sigma^1$, $\gamma^2 = i\sigma^2$, then $\varepsilon = \gamma^1\gamma^2 = -i\sigma^3$. Notice that $\varepsilon^2 = -1$. An arbitrary element of G(0,2) can therefore be written as $a1 + b\gamma^1 + c\gamma^2 + d\varepsilon = \begin{bmatrix} a-id & ib+c \\ ib-c & a+id \end{bmatrix}$ with a,b,c,d real. Since ${\gamma^1}^2 = {\gamma^2}^2 = \varepsilon^2 = -1$ and all anticommute, it is clear that G(0,2) is isomorphic with the algebra H of quaternions.

3.3.3 Comments and Summary

As already stated G(p,q) and G(q,p) are usually not isomorphic (for example G(0,1) and G(1,0)).

G(p,q) and G(p′,q′) are sometimes isomorphic (for example G(2,0) = G(1,1)). Let us summarize the previous results

$$G(1,0) = R + R \quad , \quad G(0,1) = C$$

$$G(2,0) = R(2) \quad , \quad G(1,1) = R(2) \quad , \quad G(0,2) = H$$

3.4 Tensor Product of Matrices With Coefficients in R, C, H

The last ingredients used to establish the modulo 8 periodicity theorem are the following known identities (K(n) denote the nxn matrices over the field K). The tensor products are taken over the real.

$$R(n) \otimes R(m) = R(nm) \qquad C \otimes C = C + C$$

$$R(n) \otimes C = C(n) \qquad H \otimes C = C(2)$$

$$R(n) \otimes H = H(n) \qquad H \otimes H = R(4)$$

Most of these identities are obvious, the last two (involving H) can be easily proven by representing a quaternion as a 2 x 2 complex matrix like in section 3.3.2.

The above table allow us to express any tensor product of Matrix algebras as a sum of matrix algebra over R, C or H. For example

$$H(3) \otimes C(2) \otimes C = R(3) \otimes H \otimes R(2) \otimes C \otimes C = R(6) \otimes H \otimes (C + C)$$

$$= R(6) \otimes C(2) + R(6) \otimes C(2) = R(6) \otimes R(2) \otimes C + R(6) \otimes R(2) \otimes C$$

$$= R(12) \otimes C + R(12) \otimes C = C(12) + C(12)$$

3.5 The Modulo 8 Periodicity of Real Clifford Algebras.

3.5.1 Periodicity Relations

From 3.2, 3.3 and 3.4 we get

$$G(n+8,\ 0) = G(2,0) \otimes G(0,n+6)$$

$$= G(2,0) \otimes G(0,2) \otimes G(n+4,0)$$

$$= G(2,0) \otimes G(0,2) \otimes G(2,0) \otimes G(0,n+2)$$

$$= G(2,0) \otimes G(0,2) \otimes G(2,0) \otimes G(0,2) \otimes G(n,0)$$

$$= R(2) \otimes H \otimes R(2) \otimes H \otimes G(n,0)$$

$$= R(4) \otimes R(4) \otimes G(n,0) = R(16) \otimes G(n,0)$$

Therefore G(n+8,0) is an algebra of the same type as G(n,0), only the dimensions differ. It is worthwhile to write

$$G(p+4,\ q-4) = G(p,q)$$

$$G(p+8,q) = G(p,q) \otimes R^{16} = G(p,q+8)$$

In the same way, we get

$$\begin{aligned} G(p,q) &= G(1,1) \otimes G(p-1,\ q-1) \qquad \text{(suppose } q < p) \\ &= G(1,1) \otimes G(1,1) \otimes G(p-2,\ q-2) = \ldots \\ &= \underbrace{G(1,1) \otimes \ldots \otimes G(1,1)}_{q} \otimes G(p-q,0) = R(2^q) \otimes G(p-q,\ 0) \end{aligned}$$

This last relation shows that it is enough to consider the purely euclidean case and that the whole classification depends upon the value of p-q mod 8.

3.5.2 <u>The Classification of G(p,q).</u>

We have only to consider the G(n,0) for n=1 to 8 and apply the above periodicity relations. Let us work out for example G(5,0);

$$\begin{aligned} G(5,0) &= G(2,0) \otimes G(0,3) = G(2,0) \otimes G(0,2) \otimes G(1,0) \\ &= R(2) \otimes H \otimes (R + R) = H(2) + H(2) \end{aligned}$$

Then, the periodicity relations tell us that

$$G(p,q) = H(2^{\frac{n-1}{2}-1}) + H(2^{\frac{n-1}{2}-1}) \quad \text{when } p-q = 5 \text{ mod } 8.$$

In this way we construct the following table.

p-q mod 8	0	1	2	3	4	5	6	7
G(p,q)	$R(2^{n/2})$	$R(2^{\frac{n-1}{2}}) + R(2^{\frac{n-1}{2}})$	$R(2^{n/2})$	$C(2^{\frac{n-1}{2}})$	$H(2^{\frac{n}{2}-1})$	$H(2^{\frac{n-1}{2}}) + H(2^{\frac{n-1}{2}})$	$H(2^{n/2-1})$	$C(2^{\frac{n-1}{2}})$
n = p + q								

Although everything is contained in the above table, we also give the following tables which are more handy to use since we are mainly interested in euclidean signatures or in hyperbolic signatures. In the hyperbolic case, the table contains many entries which will be explained and used later. In both cases, we also give the structure of the even subalgebras $G_o(p,q)$, it can easily be shown that

$$G_o(p,q) = G(p,q-1) = G(q,p-1) = G_o(q,p)$$

In both cases, we also list the complex extension G^C of $G(p,q)$.

The euclidean case. We list only the first 8 cases. Notice that $G_o(o,n) = G_o(n,o)$

$n = p + q$	$G_o(o,n)$	$G(o,n)$	$G(n,o)$	$G^C(n)$
1	R	C	R+R	C + C
2	C	H	R(2)	C(2)
3	H	H + H	C(2)	C(2) + C(2)
4	H + H	H(2)	H(2)	C(4)
5	H(2)	C(4)	H(2) + H(2)	C(4) + C(4)
6	C(4)	R(8)	H(4)	C(8)
7	R(8)	R(8) + R(8)	C(8)	C(8) + C(8)
8	R(8) + R(8)	R(16)	R(16)	C(16)

The hyperbolic case. Notice $G_o(n-1,1) = G_o(1,n-1)$

n	4	5	6	7	8	9	10	11
$G^C(n)$	C(4)	C(4)+C(4)	C(8)	C(8)+C(8)	C(16)	C(16)+C(16)	C(32)	C(32)+(32)
G(n-1,1)	R(4)	C(4)	H(4)	H(4)+H(4)	H(8)	C(16)	R(32)	R(32)+R(32)
G(1,n-1)	H(2)	H(2)+H(2)	H(4)	C(8)	R(16)	R(16)+R(16)	R(32)	C(32)
$G_o(n-1,1)$	C(2)	H(2)	H(2)+H(2)	H(4)	C(8)	R(16)	R(16)+R(16)	R(32)
θ	iε	/	ε	/	iε	/	ε	/
$c^2=1$, cγ = cγ	yes	no	no	no	yes	yes	yes	yes
	cθ=−θc		cθ=θc		cθ=−θc		cθ=θc	
Dirac	C^4	C^4	C^8	C^8	C^{16}	C^{16}	C^{32}	C^{32}
Weyl	C^2	/	C^4	/	C^8	/	C^{16}	/
Majorana	R^4	/	/	/	R^{16}	R^{16}	R^{32}	R^{32}
Weyl-Majorana	/	/	/	/	/	/	R^{16}	/
β(θ)	−θ	/	−θ	/	−θ	/	−θ	/

In the above table, ε as usual is the orientation operator (sect. 2.1.3)
θ is the chirality operator (sect. 2.2.4 and sect. 4.2) and $\theta^2 = 1$
c is an antilinear operator (sect. 4.1.2), 4.13 and 4.14 called charge conjugation
β denotes β_+ or β_- (see section 2.3.2) i.e., the Dirac adjoint;

* Notice finally that the entry "yes" in the table means that there exists an antilinear operator c with $c^2 = 1$ and cγ = γc ; the entry "No" means that there is no such c .

4. Spinors

4.1 Dirac Spinors.

4.1.1 Definition

We know that the complex extension $G^C(n)$ of the Clifford algebra is simple when n is even and not simple when n is odd. The periodicity, at the complex level is only mod. 2. We indeed saw that

$$\text{for n even,} \quad G^C(n) = C(2^{n/2})$$

$$\text{for n odd,} \quad G^C(n) = C(2^{\frac{n-1}{2}}) + C(2^{\frac{n-1}{2}}) \quad , \quad G^C_o(n) = C(2^{\frac{n-1}{2}})$$

We define Dirac spinors as an irreducible representation of G on a complex vector space D. Then, automatically it gives a representation of G^C. If n is even we have only one such representation, we have two if n is odd. For a Clifford algebra $G^C(n)$, we get in both cases $D = C^f$, $f = 2^{[\frac{n}{2}]}$, $[\frac{n}{2}]$ being the integer part of $\frac{n}{2}$; for example, if n = 4 or 5, the space of Dirac spinors is the same, namely C^4. In both cases D carries an irreducible representation of $G^C(n)$ but it is not faithfull when n is odd.

4.1.2 The Two Types of Dirac Representation in the even case - Majorana Spinors.

As we shall see below, when n is even, there exist two kinds of Dirac spinors: those of real type and those of quaternionic type (there are no Dirac spinors of complex type when n is even. Let us call A = G(p,q) if n = p+q is even, then we get $A^C = G^C(n)$. The space D of Dirac spinors carries an irreducible representation of the complex algebra A^C (there is only one such irreducible representation, up to equivalence, since A^C is simple). We will show that two situations (and two only) may occur:

Situation 1. There exist an A^C - isomorphism c of D on $\bar{D}$ (i.e. an antilinear operator of D commuting with the γ^μ symbols) with $c^2 = 1$. In this case the vector space $M = \{\psi \in D / c\psi = \psi\}$ is a real vector space and carries a R-irreducible representation of the real algebra A. Moreover, D, considered as a real vector space decomposes into a direct sum of two equivalent irreducible representations of A. M is called the space of Majorana spinors. In this case, we say that the representation of A^C on the space of Dirac spinors is of real type.

Situation 2. There exist an A^C isomorphism c of D on $\bar{D}$ (i.e. an antilinear operator of D commuting with the γ^μ symbols) with $c^2 = -1$. Therefore, there are no Majorana spinors and the space of Dirac spinors D, considered as a real vector space, also carries an irreducible representation of the real algebra A. In this case, we say that the representation of A^C on the space of Dirac spinors is of quaternionic type.

Notice that there always exist a change conjugation operator c but, in the second case it does not possess non zero eigeinstates with real eigenvalues (imposing $\psi = c\psi$ would imply $\psi = -\psi$ if $c^2 = -1$).

Let us now prove the above. Let c be a one to one antilinear operator of D commuting with the action of the γ^μ symbols, we could also say that c is an intertwining operator of D to $\bar{D}$ (we have obviously a representation of A^C on the vector space $\bar{D}$ (i.e. D with the conjugated external law). But D and $\bar{D}$ are equivalent since A^C is simple, therefore the space of such operators is of dimension 1 (Schur lemma). This implies that c^2 is a linear operator of D proportional to the unit matrix: $c^2 = \alpha 1$. Let us show that α is real: for any $\psi \in D$ we have $\alpha c\psi = c^2 c\psi = cc^2\psi = c\alpha\psi$ therefore $\alpha c = c\bar{\alpha}$ and α is real. We can now normalize c by introducing $c' = \lambda c$, $\lambda \in C$ but then $c'^2 = \lambda\bar{\lambda}c^2 = |\lambda|^2\alpha^2$ which shows that by rescaling, we cannot change the sign of c^2. We can therefore after assume that $c^2 = 1$ or that $c^2 = -1$. Notice that D and $\bar{D}$ being equivalent, we can always find such a c (Dirac spinors are never of complex type).

4.1.3 The three types of Dirac representations in the odd case - Majorana spinors

When n is odd, the orientation operator ε belongs to the center of the algebra, it will be represented by a number ± 1 or $\pm i$ on the space of Dirac spinors; in both cases ($\varepsilon^2 = 1$ or $\varepsilon^2 = -1$) we have two representations and we may choose one. Now, it is clear from the tables of section 3.5.2 that all three types of representation (real, complex, quaternionic) occur. If $\varepsilon^2 = 1$, by an analysis analog to the one of the previous section, we see that again, two situations may occur: either there exists an antilinear operator c with $c\gamma = \gamma c$ and $c^2 = 1$, the representation is of real type and we may represent the algebra on the real vector space of Majorana spinors $M = \{\psi \in D / c\psi = \psi\}$. Or there exist c with $c\gamma = \gamma c$ and $c^2 = -1$, the representation is then of quaternionic type.

If $\varepsilon^2 = -1$, the representation is always of the complex type but again two situations may occur; it is a priori clear that there is no antilinear c commuting with the γ^μ since it would then commute with ε (which is impossible because $\varepsilon = \pm i$). If $\varepsilon^2 = -1$, we have therefore the following: either we can find c, antilinear with $c\gamma = -\gamma c$ and $c^2 = 1$, in this case, by going from G(p,q) to G(q,p) we would find an algebra whose representation is of quaternionic type;

or we can find c with $c\gamma = -\gamma c$ and $c^2 = -1$ and in this case, by going from G(p,q) to G(q,p) we would find an algebra whose representation is of real type.

From the physical point of view, since it is commonly believed that the choice of (p,q) or (q,p) is a matter of conventions (although the algebra **are** not isomorphic), we can always choose the signature which leads to an **algebra** which has $\varepsilon^2 = 1$ and we are back to the previous discussion with only two **kinds** of Dirac spinors.

4.1.4 <u>On the definition of the charge conjugation operator.</u>

An antilinear operator commuting with the γ's and of square 1 is "usually" called a "charge conjugation operator". However, the previous discussion shows that one can sometimes find antilinear operators of square -1, commuting or not with the γ's even if we prefer not to call them "charge conjugations

4.1.5 <u>Examples.</u>

Tables of sect. 3.5.2 show without any more calculations that Majorana spinors exist in

dimensions 1,2,6,7,8 mod 8 if the signature is euclidean

dimensions 1,2,3,4,8 mod 8 if the signature is hyperbolic

For the hyperbolic case, we therefore get in particular $c^2 = 1$ in dimensions $n = 4,8,9,10,11$.

Notice that Majorana spinors do not provide a faithfull representation of the Clifford algebra when n is odd; this explains why the dimensions 1 and 3 mod 8 have sometimes been excluded in some older articles like [14']. The terminology (real, quaternionic) comes from the fact that, in the first situation, the commutant of the space carrying the irreducible representation of the algebra A is just $R \cdot 1$ whereas in the second situtation, the commutant is $C \cdot 1 + cC \simeq H$. Let us give a few examples which can be extracted directly from the table in sec. 3.5.2 (hyperbolic case)

$G^C(5) = C(4)+C(4)$, $G_o^C = C(4)$, the space of Dirac spinors is C^4

$G^C(10) = C(32)$, the space of Dirac spinors is C^{32}

$G(3,1) = R(4)$, $G(1,3) = H(2)$, $G^C(4) = C(4)$; the space of Dirac spinors is C^4, it is of real type and $M = R^4$ is the space of Majorana spinors.

$G(5,1) = G(1,5) = H(4)$, $G^C(6) = C(8)$, the space of Dirac spinors is C^8, it is of quaternionic type: there are no Majorana spinors, C^8 itself provides an irreducible representation space for the real Clifford algebra H(4) (indeed $H^4 \sim C^8$).

4.2 Weyl Spinors and Weyl-Majorana Spinors.

4.2.1 The Odd Dimensional Case

When n is odd, we can write (see sect. 2.2.3) $G = G_o + G_o\varepsilon$ and therefore $G^C = G_o^C + G_o^C\theta$. Recall that the orientation operator is $\varepsilon = \prod_{\mu=1}^{n} \gamma^\mu$ and the chirality operator θ (which has square + 1) is ε if $p-q = 1 \bmod 4$ and is $i\varepsilon$ if $p-q = 3 \bmod 4$. However, when n is odd, ε is represented by ± 1 or $\pm i$ in the space of Dirac spinors, this explains why it is commonly claimed that the helicity operator θ does not exist in odd dimensions. Would we consider a faithfull representation of G^C, like it is sometimes done in the mathematical literature (the elements of such a representation still being called spinors!), we should then study the action of θ.

4.2.2 The Even Dimensional Case.

When n is even, we know (see Sect. 2.2.2) that G^C is simple, however the complex extension G_o^C of G_o is not simple since we can always write

$$G_o^C = \frac{(1-\theta)}{2} G_o^C + \frac{(1+\theta)}{2} G_o^C \quad , \quad \theta^2 = 1$$

with $\theta = \varepsilon$ if $(p-q) = 0 \bmod 4$ and $\theta = i\varepsilon$ if $p-q = 2 \bmod 4$. Therefore the representation of G^C on the space D of Dirac spinors induces a representation of the subalgebra G_o^C which is always reducible.

We define $D = W^L + W^R$; $\dim W^L = \dim W^R = \dim D/2$.

where $W^L = \{\psi \in D \ / \ \theta\psi = -\psi\}$ = space of left Weyl spinors

and $W^R = \{\psi \in D \ / \ \theta\psi = +\psi\}$ = space of right Weyl spinors

The space W^L of left Weyl spinors (for example) carries and irreducible representation of the complex algebra G_o^C; in the same way as we studied the passage complex → real and obtained two possible types of Dirac representations (those of quaternionic type and those of real type leading to the definition of Majorana spinors), we will now study this for Weyl spinors and obtain three possible types of Weyl representation (quaternionic, complex and real), the last type leading to the definition of Weyl-Majorana spinors.

4.2.3 The Three Types of Weyl Representations.

Let us now consider the real Clifford algebra $G_o(p,q)$ and its complex extension $G_o^C(n)$, n = p + q even, let us also choose a space W of Weyl spinors, for example $W = W_L$. Three situations (and only three) may occur:

Situation 1. There exists a G_o^C isomorphism c from W to $\bar{W}$ (i.e., an antilinear operator of W) with $c^2 = 1$. In this case the vector space $V = \{\psi \in W/c\psi = \psi\}$ is a real vector space and carries a R-irreducible representation of the real algebra G_o. Moreover, W, considered as a real vector space decomposes into a direct sum of two equivalent representations of G_o. If $W = W_L$, then $V = V_L$ is called the space of left Weyl Majorana spinors. In this case we say that the representation of G_o^C on the space W of Weyl spinors is of real type.

Situation 2. There exists a G_o^C isomorphism c from W to $\bar{W}$ which $c^2 = -1$. There are no Weyl Majorana spinors and the space W, considered as a real vector space, also carries and irreducible representation of the real algebra G_o. In this case the representation of G_o^C on the space W of Weyl spinors is of quaternionic type.

Situation 3. This is the case where W and $\bar{W}$, as representaion of G_o^C, are unequivalent, c does not leave W invariant, this happens when $c\theta = -\theta c$. W, considered as a real vector space carries also an irreducible representation of the real algebra G_o. The representation of G_o^C on the space W is of complex type.

The proof of the above goes along the same lines as the one given in sect. 4.1.2, only the names are different ; in that case, W and $\bar{W}$ are inequivalent and there is no charge conjugation operator acting on W .

We could of course start directly with Dirac spinors; in order to get a Weyl Majorana spinor (say right), we have to impose the conditions $c\psi = \psi$ and $\theta\psi = \psi$. This is only possible if $c^2 = \theta^2 = 1$ and if $c\theta = \theta c$ (if not, we would get $\psi = -\psi$). This last constraint is only possible if $\theta = \varepsilon$ (and not $i\varepsilon$) since c is antilinear.

4.2.4 Examples.

Here again we use the table in sect. 3.5.2 (hyperbolic case)

$G_o(9,1) = R(16) + R(16)$, $G_o^C = C(16) + C(16)$; the spaces of Weyl spinors are $W_L = W_R = C^{16}$. These representation are of real type and we get a space $V_L = R^{16}$ of left Weyl Majorana spinors and a space V_R of right Weyl Majorana spinors. There is a change conjugation operator c with $c^2 = 1$.

$G_o(5,1) = H(2) + H(2)$, $G_o^C = C(4) + C(4)$; the space of Weyl spinors are $W_L = W_R = C^4$. These representations are of quaternionic type, they are no Weyl Majorana spinors but notice that C^4 provides an irreducible representation of the real Clifford alebra (indeed $H^2 = C^4$). There is an antilinear operator c leaving W (W_L or W_R) invariant but its square $c^2 = -1$.

$G_o(3,1) = C(2)$, $G_o^C = C(2) + C(2)$; the spaces of Weyl spinors are $W_L = W_R = C^2$. These representation are of complex type: there is no antilinear operator c leaving a space W (W_L or W_R) invariant. Notice that C^2 also provides a representation for the real Clifford algebra. Of course, there are no Weyl Majorana spinors.

4.3 Summary of This Section.

For explicit results, one can use the tables of section 3.5.2.

The following chart illustrates the logic followed in the present section (for n, even and choosing, for example $W = W_R$).

Dirac, $c^2 = -1$ —— $\theta\psi = \psi$ ——> Weyl, quaternionic type

Dirac, $c^2 = 1$ —— $c\theta = -\theta c$ ——> Weyl, complex type

$c\theta = \theta c$ ——> Weyl, real type —— $c\psi = \psi$ ——> Weyl Majorana

$c\psi = \psi$ ——> Majorana, $c\theta = \theta c$ —— $\theta\psi = \psi$ ——> Weyl Majorana

——> Majorana, $c\theta = -\theta c$

For a space-time (hyperbolic signature), the tables of sect. 3.5.2 shows that

Majorana spinors exist in dimensions 4,8,9,10,11 mod 8

Weyl Majorana spinors exist in dimensions 2 mod 8, in particular when n = 2,10,18,26.

For a purely euclidean space, we find that

Majorana spinors exist in dimensions 1,2,6,7,8 mod 8.

Weyl Majorana spinors exist in dimension 0 mod 8, in particular n = 8.

More generally, for a metric of signature (p,q), Weyl Majorana spinors exist if and only if when p−q = 0 mod 8.

5. REFLECTIONS

In this section, we use a hyperbolic signature with the $(\gamma^o)^2 = -1$ convention; in order to obtain the corresponding results for the other convention, we would just have to replace γ^μ by $i\gamma^\mu$. The expression of the reflection operators for other space-time signatures can be obtained easily from what follows. We consider separately the case of the "one-particle theory" and of "field theory" where spinor fields become anticommuting objects, indeed the definitions of Parity, Time reversal and charge conjugation are different in these two cases (compare for instance Bjorken and Drell tome 1 vs tome 2).

5.1 Reflections in one particle theory

5.1.1 Parity

For "classical" spinor fields, the parity operator P is a linear operator such that $\psi(u^o,\vec{u}) \rightsquigarrow \psi^P(u^o, -\vec{u}) = P\,\psi\,(u^o, \vec{u})$; this relation makes sense only if space time is flat, but, more generally we can define P as follows:

P is an element of the group Pin which covers the usual Lorentz space reflection, i.e., $P\,\gamma^\mu P^{-1} = \gamma^o$ if $\mu = 0$

$= \gamma^i$ if $\mu = i$

obviously $\pm\,\gamma^o$ is solution; we choose $P = \gamma^o$. Notice that at the complexified level we could multiply by an arbitrary phase.

5.1.2 Total reflection

For "classical" spinor fields, the total reflection operator ε is a linear operator such that $\psi(u^o, \vec{u}) \rightsquigarrow \psi^\varepsilon\,(-u^o, -\vec{u}) = \varepsilon\,\psi(u^o, \vec{u})$ in flat space. More generally, we will define ε as follows:

ε is an element of the group Pin which covers the Lorentz Total Reflection, i.e., $\varepsilon\gamma^\mu\varepsilon^{-1} = -\,\gamma^\mu$.

Obviously $\pm\,\gamma^o\gamma^1\ldots\gamma^n$ is solution and we choose $\varepsilon = \gamma^o\gamma^1\ldots\gamma^n$.

The total reflection operator therefore coincides with the orientation operator introduced in 2.2.2. When the space time is even dimensional, ε is related to the chirality operator θ (generalised "γ^5", with $\theta^2 = 1$) by $\theta = \varepsilon$ or $\theta = i\varepsilon$, depending upon the square of ε (see last table of section 3.5.2).

5.1.3 Charge conjugation

We already introduced and discussed the properties of this operator in sect. 4.

It is always possible to construct an antilinear operator c ($ic = -ci$), $c\gamma = \gamma c$ acting on the space of spinors : $\psi(u) \rightsquigarrow \psi^c(u) = c\psi(u)$. when $c^2 = 1$, this is a charge conjugation in the usual sense (it is possible to define Majorana spinors why satisfy $\psi = \psi^c$).

When $c^2 = -1$, the situation is different but c is nevertheless a "symmetry" of the Dirac equation if ($\partial\!\!\!/\psi = 0$ then $\partial\!\!\!/\psi^c = 0$ as well); (the symmetries of the massive Dirac equation are studied in sect. 6.5.4).

It is usual in standard textbooks to introduce a matrix C, therefore a linear operator, such that $\psi^c(u) = [c\psi](x) = C\gamma^o\psi^*(u)$; the * opeation is antilinear as it should: $^*i\psi = -i\psi^*$, however the map $\psi \to \psi^*$ has no intrinsic meaning in a complex vector space, therefore the matrix C is representation dependent as well: the composition of the operators * and C is representation independent, as the charge conjugation operator c itself.

5.1.4 Time reversal

For "classical" spinor fields, we define the Time reversal operator τ as follows:

τ is defined by $\varepsilon = Pc\tau$

τ is antilinear since P, ε are linear and c antilinear.

In flat space, τ is such that $\psi(u^o, \vec{u}) \rightsquigarrow \psi^\tau(-u^o, \vec{u}) = \tau\psi(u^o, \vec{u})$. In standard textbooks, it is standard to introduce a matrix T, therefore a linear operator, such that $\psi^\tau(-u^o, \vec{u}) = T\psi^*(u^o, \vec{u})$; here again, the matrix T as well as the * operator are representation dependent, however τ is representation independent.

5.1.5 Combined $c\tau$ operation

It is clear from the definition of τ and $P = \gamma^o$ that $c\tau = \pm \gamma^1\gamma^2..\gamma^n = \gamma^o\varepsilon$.

$$\text{Therefore } c\tau\, \gamma^\mu (c\tau)^{-1} = \begin{cases} -\gamma^o & \text{if } \mu = 0 \\ +\gamma^i & \text{if } \mu = i \end{cases}$$

$c\tau$ is therefore a linear operator coinciding with the element of the Pin group which covers the usual Lorentz time reflection, in particular it is worth to notice that this role is not played by the Time reversal operator τ (despite its name).

Notice that, in flat space time, $c\tau\psi(u^o, \vec{u}) = c\psi^\tau(-u^o, \vec{u}) = \psi^{c\tau}(-u^o, \vec{u})$; moreover, although the matrices C and T are separately representation dependent, their product CT is not and is equal to ε (up to a phase).

5.1.6. Remarks

The action of the previous operators is clearly well defined on a faithfull representation of the Clifford algebra; however, when n is odd, ε does not act on Dirac spinors (sect. 4.1.1, 4.2.1) and some of the previous properties may become unclear: in the last case, we extend the representation of G_o^C to the full of G^C, this amounts either to double the number of components of ψ and fill them with zeros or, more simply, to keep the same spinors and set $\varepsilon = 1$ (identity) for Dirac spinors in odd dimensions.

The representation dependent matrices T, C and the operation * have been introduced for the convenience of the reader who wants to establish a link between our definitons and some traditional notations; however it should be stressed that the use of T, C and * is usually unnecessary in actual calculations and is dangerous from the conceptual point of view.

5.2 Reflections in field theory

In quantum field theory, spinors fields become anticommuting objects and the definitions of the reflections operators are different. A careful analysis of these operators would require the introduction of the mathematical apparatus of field theory; however, as it very often happens, we can mimic the results of QFT by using anticommuting classical spinors, i.e., by making the tensor product of our spinor space by the odd part of a Grassman algebra. Then we have to define the reflections on the Grassman part as well. A consistent choice is to take $\underline{P} = P \otimes 1$, $\underline{C} = c \otimes \tilde{c}$, $\underline{T} = \tau \otimes 1$ where $\tilde{c}$ is a conjugation on the Grassman part. Then

P is linear, c is antilinear, τ is antilinear, $\varepsilon = Pc\tau$ is linear

$\underline{P}$ is linear, $\underline{C}$ is linear, $\underline{T}$ is antilinear, $\underline{E} = \underline{PCT}$ is antilinear

The above is enough to compute the various transformation properties of quadratic forms under discrete symmetries.

6. Scalar products of spinors and Lagrangian densities

Ultimately, we want to construct a real Lagrangians (and a real action) involving spinor fields coupled via. a differential operator and possible mass terms. This Lagrangian will have to be invariant under the Spin group and we have therefore to see what are the possible scalar products that can be used to construct such a Lagrangian. By "scalar product" we mean "bilinear form" (it can be symmetric or not).

6.1. The adjoints β_+ and β_-: the scalar products types $\langle\ ,\ \rangle_+$ and $\langle\ ,\ \rangle_-$

Let S be a space of spinors and $\langle\ ,\ \rangle$ be a scalar product (possibly complex valued) in S, invariant under the group Spin. Let s be an element of the Clifford algebra G and $\beta(s)$ its adjoint for the scalar product $\langle\ ,\ \rangle$, i.e., such that

$$\forall v,w \in S \quad \langle v, sw \rangle = \langle \beta(s)v, w \rangle$$

This implies in particular that $\beta(s_1 s_2) = \beta(s_2)\beta(s_1)$ and $\beta(\beta(s)) = s$, i.e., $\beta^2 = 1$. Therefore β has to be an anti-involution of the Clifford algebra G. Also, we want $\langle\ ,\ \rangle$ to be invariant under the spin group, i.e., we want $s \in \mathrm{Spin} \subset G$ to be an isometry of $\langle\ ,\ \rangle$.

$$[s \text{ is an isometry}] \Longleftrightarrow \forall v \in S \quad \langle v,v \rangle = \langle sv, sv \rangle$$

$$\Longleftrightarrow \forall v \in S \quad \langle v,v \rangle = \langle \beta(s)sv, v \rangle$$

$$\Longleftrightarrow \beta(s)s = 1 \qquad \beta(s) = s^{-1}$$

Therefore, β is such that

$$\forall s \in \mathrm{Spin} \subset G \quad , \quad \beta(s)s = 1 \quad .$$

It is clear that β has to coincide with one of the two antiinvolutions β_+ and β_- already introduced in section 2.3.2. We will therefore say that a scalar product is of β_+ type (resp. β_- type) and denote it by $\langle\ ,\ \rangle_+$ (resp. $\langle\ ,\ \rangle_-$) when the adjoint is obtained via the antiinvolution β_+ (resp. β_-).

The reader may wonder why we do not use (for the moment) the Dirac "bar" notation; the reason is that, for specific signatures of spacetime, it is sometimes possible to define up to 8 possible kinds of real valued Spin invariant scalar products of spinors (4 of them of β_+ type and 4 of them of β_- type); only one of them is called "$\bar{\psi}\psi$". The reason why the others do not enter in traditional Lagrangians is either because they do not contribute to field equations or because they violate sacrosanct conservation laws (in some cases also, the different definitions may collapse for the spinors under study).

Finally, notice that here, we have only shown that Spin invariant scalar products fall into two families, we have not explained how to construct them: this will be done in sect. 6.3.

6.2 Complex valued versus real valued scalar products

Let us remind the reader that, in this paragraph, components of spinors are treated as commuting numbers: conclusions will be trivially modified when we consider anticommuting spinors (the adjectives "symmetric", "antisymmetric" becoming "graded symmetric" and "graded antisymmetric").

Let $\langle\ ,\ \rangle$ be a complex valued scalar product, then we will say that

it is of type
- C_a if it is antisymmetric $\langle v,w\rangle = -\langle w,v\rangle$
- C_s if it is symmetric $\langle v,w\rangle = \langle w,v\rangle$
- C^*_s if it is sesquilinear symmetric (hermitian) $\left\{ \begin{array}{l} \langle v,w\rangle = \langle w,v\rangle^* \\ \langle \alpha v,w\rangle = \alpha^*\langle v,w\rangle \\ \langle v,\alpha w\rangle = \alpha\langle v,w\rangle \end{array}\right.$

If $\langle\ ,\ \rangle$ is antisymmetric sesquilinear, i.e. if $\langle v,w\rangle^* = -\langle v,w\rangle$ then the new scalar product

$$\langle v,w\rangle_{new} = \langle iv,w\rangle \text{ is hermitian.}$$

Notice that if s is a <u>linear</u> operator and $\langle\ ,\ \rangle$ a scalar product, we have $\langle v,sw\rangle = \langle\beta(s)v,w\rangle$ where $\beta(s)$ is the adjoint of s but if c is an <u>antilinear</u> operator then we define its adjoint $\beta(c)$ as the operator satisfying $\langle v,cw\rangle = \langle\beta(c)v,\ w\rangle^*$.

We will ultimately need to take the real part or the imaginary part of the C-valued scalar product (for instance to construct a Lagrangian), it should be noticed that, calling real antisymmetric by the symbol R_A and real symmetric by the symbol R_s, we have, with obvious notations:

$$\mathrm{Re}\,(C_a) = R_a \qquad \mathrm{Im}\,(C_a) = R_a$$

$$\mathrm{Re}\,(C_s) = R_s \qquad \mathrm{Im}\,(C_s) = R_s$$

$$\mathrm{Re}\,(C^*_s) = R_s \qquad \mathrm{Im}\,(C^*_s) = R_a$$

When $\langle\ ,\ \rangle$ is real symmetric, it can be put into diagonal form and it becomes interesting to know its signature. Take $\langle\ \rangle = \langle\ \rangle_-$; the γ^μ are then antisymmetric since $\langle v,\gamma^\mu w\rangle = \langle \beta_-(\gamma^\mu)v,w\rangle = -\langle\gamma^\mu v,w\rangle$, let $\{v_j\}$ be an orthogonal basis: $|\langle v_j,v_j\rangle| = 1$, choose $\langle v_i,v_i\rangle = 1$, then

$$\langle\gamma^\mu v_i,\gamma^\mu v_i\rangle = -\langle\gamma^{\mu 2}v_i v_i\rangle = \begin{cases} +1 & \text{if } \gamma^{\mu 2} = -1 \\ -1 & \text{if } \gamma^{\mu 2} = 1 \end{cases}$$

In case there exist one γ^μ of square +1, it becomes possible to each basis spinor v_i of square (1) to associate a spinor of square (-1): this shows that the scalar product is neutral (as many + signs as - signs). Along the same lines, one can prove that:

If E is a vector space of signature (p,q), (with p + sign any q - signs) and $\langle\ ,\ \rangle_\pm$ are spin invariant scalar products in a space of spinors for E, then

$$\begin{cases} \langle\ ,\ \rangle_- \text{ is positive definite} & \text{if } (p,q) = (o,q) \\ \langle\ ,\ \rangle_- \text{ is neutral} & \text{if } (p,q) \neq (o,q) \end{cases}$$

$$\begin{cases} \langle\ ,\ \rangle_+ \text{ is positive definite} & \text{if } (p,q) = (p,o) \\ \langle\ ,\ \rangle_+ \text{ is neutral} & \text{if } (p,q) \neq (p,o) \end{cases}$$

It may be useful to recall that the invariance groups of the scalar products of type C_s, C_a, C^*_s, R_s, R_a are respectively $O(2m,C)$, $Sp(2m,C)$, $U(2m,C)$ ($U(m,m,C)$ if neutral), $O(2m,R)$($O(m,m,R)$ if neutral)), $Sp(2m,R)$. Notice that $\langle\ ,\ \rangle_+$ being invariant under a group I_+ (being one of the above), and $\langle\ ,\ \rangle_-$ being invariant under a group of I_- (being one of the above), we have Spin $\subset I_+$ Spin $\subset I_-$.

6.3 Construction of the scalar products

6.3.1 The Dirac scalar product

We now show how to construct all scalar products of β_+ and β_- type. Take $F = C^f$ as the space of Dirac spinors of a given Clifford algebra, this space F carries also a representation of the finite group generated by the $\pm\gamma^\mu$ and their products. It is well known that there exist one (and only one) symmetric positive definite sesquilinear form on F for which the γ^μ are isometries; call this form $(\ ,\ )$. (ψ,φ) is usually written as $\psi^*\varphi$ in the literature (but remember that ψ^* has no invariant meaning). Let us call, as usual, +, the adjunction with respect to $(\ ,\ )$. The γ^μ being isometries, we have $(\gamma^\mu\psi,\gamma^\mu\varphi) = (\gamma^{\mu+}\gamma^\mu\psi,\varphi) = (\psi,\varphi)$ and $\gamma^{\mu+}\gamma^\mu = 1$, therefore $\gamma^{\mu+} = \gamma^\mu$ if $\gamma^{\mu 2} = 1$ and $\gamma^{\mu+} = -\gamma^\mu$ if $\gamma^{\mu 2} = -1$. The hermitian scalar product $(\ ,\ )$ has however no reason to be invariant under the Spin group. Let us order our basis

$(\gamma^\mu) = \{\gamma^{\mu 1},\gamma^{\mu 2}, — \gamma^{\mu p},\gamma^{\mu p+1}\ldots,\ \gamma^{\mu p+q}\}$ with $\begin{cases}\gamma^{\mu i^2} = +1 & \text{if } i \le p \text{ and}\\ \gamma^{\mu i^2} = -1 & \text{if } i > p\end{cases}$

define $\gamma^o = \gamma^{\mu p+1}\gamma^{\mu p+2}\ldots\gamma^{\mu p+q}$

then

$$\psi,\varphi \in F \longrightarrow \frac{i^{2_1[q+2]}}{i^q} \qquad (\psi,\gamma^o\varphi)$$

where $[q/2]$ denotes the integer part, defines a spin invariant and hermitian symmetric scalar product (the rather awkward coefficient is there to make it symmetric). It is sesquilinear i.e., antilinear in the first variable since $(\ ,\ )$ is already sesquilinear. Using the above properties of the adjoint + of the γ^μ for (,) one can compute $\beta(s)$, the adjoint of s for the new product and check that $\beta(s)s = 1$. A technical exercise shows that $\beta(\gamma_\mu) = (-1)^q \gamma_\mu$ therefore this scalar product is of β_+ type if q is even and of β_- type if q is odd — see also [12]. Being of $\beta_\pm$ type, this scalar product is spin invariant.

In the following, we assume that our n-dimensional space-time has only one kind of "time" i.e., p = 1 or n − 1 depending upon ones favorite conventions. Call γ^o the time like Clifford symbol; then

If $\gamma^{02} = -1$, $\gamma^{i2} = 1$, the above definition leads to the scalar product

$\langle\psi, \varphi\rangle_- \doteq i(\psi, \gamma^o \varphi)$ which is of the β_- type: $\beta(\gamma^\mu) \doteq \bar{\gamma}^\mu = -\gamma^\mu$

as already stated. In this case, a Lagrangian written as

$\mathcal{L} = \langle\psi, \partial\!\!\!/\psi\rangle_- + m \langle\psi,\psi\rangle_-$ leads to the correct Dirac and Klein Gordon equations.

If $\gamma^{02} = +1$, $\gamma^{i2} = -1$, the above prescription leads to a scalar product

defined by $\dfrac{i^{2}i^{[(n-1)/2]}}{i^{n-1}}\,(\psi, \gamma^1\gamma^2 \ldots \gamma^{n-1}\varphi)$ - let us take n = 4, for example: the factor in front drops out and we get the scalar product $\langle\psi, \varphi\rangle_- = (\psi, \gamma^1\gamma^2\gamma^3\varphi) = (\psi, \gamma_o \varepsilon \varphi)$ which is of β_- type according

to the above results. It is then useful to introduce a new scalar product $\langle\psi,\varphi\rangle_+ \doteq \langle\psi, \varepsilon\varphi\rangle_- = (\psi, \gamma^o\varphi)$ which is readily seen to be of β_+ type, i.e., $\beta(\gamma^\mu) \doteq \bar{\gamma}^+ = \gamma^\mu$ and symmetric hermitian as well. But then, the Dirac Lagrangian should be written

$\mathcal{L} = \langle\psi, i\partial\!\!\!/\psi\rangle_+ - m \langle\psi,\psi\rangle_+$ in order to lead to the usual Dirac and Klein Gordon equation.

We assume in the following that such a conventional choice has been made and we call $\langle\psi,\varphi\rangle$ the Dirac product (also $\bar{\psi}\varphi$) keeping in mind that is of β_- type if $\gamma_o^2 = -1$ but of β_+ type if $\gamma_o^2 = +1$, that it is symmetric hermitian and that its full invariance group is the non compact U(m,m,C), see also sect 2.3.2.

6.3.2 The scalar product $\psi\varphi + i\langle\psi,\theta\varphi\rangle = i\bar{\psi}\theta\varphi$

θ being as usual the chirality operator (with $\theta^2 = 1$ and $\beta(\theta) = \bar{\theta} = -\theta$), we introduce for n even, the scalar product

$$\psi,\varphi \longrightarrow \langle\langle \psi,\varphi\rangle\rangle \equiv i\,\langle\psi,\theta\varphi\rangle$$

It is easy to show that it is also sesquilinear symmetric but of opposite β type as the Dirac product, i.e.

$$\langle\langle \psi,\varphi\rangle\rangle^* = \langle\langle \varphi,\psi\rangle\rangle$$

but $\langle\langle \psi,\gamma^\mu\varphi\rangle\rangle = -\langle\langle\overline{\gamma^\mu}\psi,\varphi\rangle\rangle$ where, for example $\bar{\gamma}^\mu = \beta_+(\gamma^\mu) = \gamma^\mu$ if $\gamma_o^2 = +1$.

It is, of course, invariant under the group Spin (its full invariance group being still U(m,m,C)) and can be used in a Lagrangian (but it can give a zero contribution, see sect. 6.5.1).

6.3.3 The adjoint of the charge conjugation operator

Let $\langle\ ,\ \rangle$ a hermitian scalar product, then, if s is a linear operator, its adjoint $\beta(s)$ satisfies, as we know the relation $\langle\psi,s\varphi\rangle = \langle\psi(s)\psi,\varphi\rangle$; however, the definition of the adjoint of an antilinear operator (c in our case) is different:

$$\langle c\psi,\varphi\rangle = \langle\psi,\beta(c)\varphi\rangle^* \quad \text{by definition}$$

$$= \langle\beta(c)\varphi,\psi\rangle \quad \text{by hermiticity of } \langle\ ,\ \rangle$$

The adjoint $\beta(c)$ can be computed as follows (we do it on an example and will use a positivity argument). Take $\langle\phi,\phi\rangle = (\phi,\ \gamma^o\phi)$ with (,) positive, and suppose $\gamma^{02} = +1$. Then

$$(\phi,\gamma^o\phi) = \langle\phi,\phi\rangle \qquad >0$$

Also $\qquad (c\phi,\gamma^o c\phi) = \langle c\phi,c\phi\rangle \qquad >0$

Assume $\gamma^o c = c\gamma^o$, then $\qquad (c\phi,c\gamma^o\phi) = \langle c\phi,c\phi\rangle \qquad >0$

but $\beta(c)c = \pm 1$ $\qquad (\beta(c)c\gamma^o\phi,\phi) = \beta(c)c \quad (\gamma^o\phi,\phi)$

$$= \beta(c)c\ (\phi,\phi) \qquad >0$$

Therefore $\beta(c)c = +1$ and we know $\beta(c)$ if we know c^2. We leave to the reader the task of computing $\beta(c)$ in each case. For example, in the case of signature (3,1) we get $c^2 = 1 \Rightarrow \beta(c) = c$.

The value of the adjoint $\beta(c)$ of the charge conjugation operator c relies therefore on the value of $c^2 = \pm 1$ and on the relations $\gamma^\mu c = \pm c\gamma^\mu$.

6.3.4 The scalar product $\psi,\varphi \to \overline{\psi}^c\varphi$

The Dirac product $\langle\ ,\ \rangle$ was hermitian (and, in particular, antilinear in the first variable) but the new scalar product $\psi,\varphi \to \langle c\psi,\varphi\rangle$ is no longer hermitian (since c anticommutes with i): it is bilinear.

$$\langle c\lambda\psi,\varphi\rangle = \langle\lambda^* c\psi,\varphi\rangle = \lambda\ \langle c\psi,\varphi\rangle$$

$$\langle c\psi,\lambda\varphi\rangle = \lambda\ \langle c\psi,\varphi\rangle$$

We should now study if it is symmetric or antisymmetric; notice first that

$\langle c\psi,\varphi\rangle = \langle\psi,\beta(c)\varphi\rangle^*$, by definition of the adjoint $\beta(c) = \bar{c}$ of the antilinear operator c

$= \langle\beta(c)\ \varphi,\psi\rangle$, by hermiticity of $\langle\ ,\ \rangle$

Therefore, this scalar product is

symmetric (C_S) if $\beta(c) = c$

antisymmetric (C_A) if $\beta(c) = -c$

If we take for example n = 4 mod 8, this scalar product becomes graded antisymmetric for anticommuting spinors and $\bar{\psi}^c \psi$ does not vanish. However, a Lagrangian containing such a term violates the global invariance $\psi \to e^{i\alpha}\psi$ associated conventionally with lepton number conservation, indeed (take $\alpha = \pi/2$),

$$\langle ci\psi, i\psi\rangle = \langle -ic\psi, i\psi\rangle = +i\ \langle c\psi, i\psi\rangle = -\ \langle c\psi, \psi\rangle$$

For this reason, it is not usually taken into consideration (but for giving a mass to the neutrino).

6.3.5 The scalar product $\psi, \varphi \to i\bar{\psi}^c \theta \varphi$

When n is even, we can use again the chirality operator θ, to define the scalar product $\psi, \varphi \to i\ \langle c\psi, \theta\varphi\rangle$. Being obviously Spin invariant, it can again be used in the construction of Lagrangians, however, as the previous one, it violates the global transformation $\psi \to e^{i\alpha}\psi$.

It is first clear that its β type is opposite to the previous one and that its symmetry properties are determined from those of $\langle c\psi, \varphi\rangle$:

$$\langle c\psi, \theta\varphi\rangle = \langle \beta(\theta)c\psi, \varphi\rangle = -\ \langle \theta c\psi, \varphi\rangle, \text{ therefore}$$

$$\text{if } \theta c = c\theta,\ \langle c\psi, \theta\varphi\rangle = -\ \langle c\theta\psi, \varphi\rangle = -\ \langle \theta\psi, \beta(c)\varphi\rangle^* = -\ \langle \beta(c)\varphi, \theta\psi\rangle$$

$$\text{if } \theta c = -\ c\theta,\ \langle c\psi, \theta\varphi\rangle = +\ \langle c\theta\psi, \varphi\rangle = +\ \langle \theta\psi, \beta(c)\varphi\rangle^* = +\ \langle \beta(c)\varphi, \theta\psi\rangle$$

This last scalar product is therefore not hermitian, it is of type C_S or C_A.

6.3.6 Summary of section 6.3

The construction has been the following

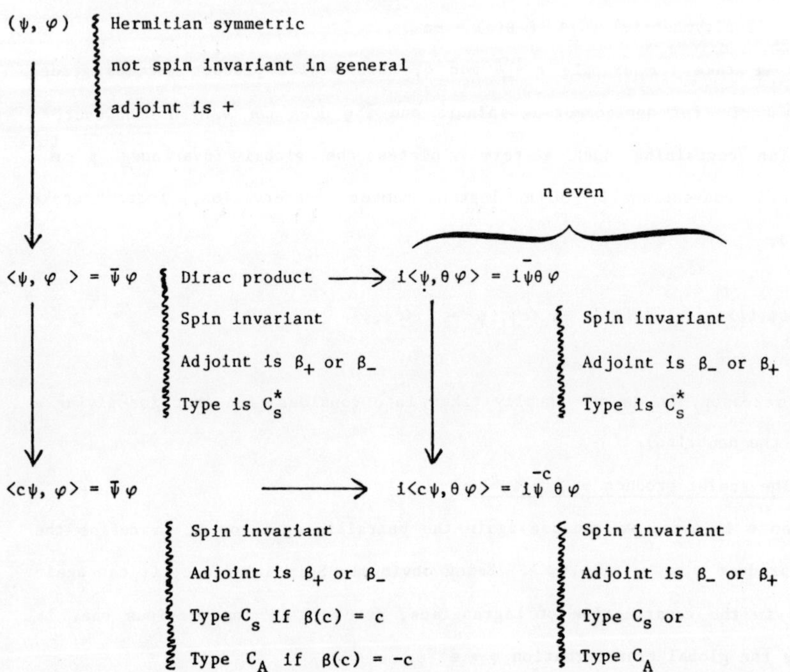

Moreover, with n=4 and hyperbolic signature* (see sect 6.3.1 for other cases):

---

* In most textbooks of particle physics, the (+---) convention is chosen. One then has only to remember that the adjoint of the Dirac scalar product $\langle\psi,\varphi\rangle \equiv \bar{\psi}\varphi = \psi^{+}\gamma^{o}\varphi$ is given by β_{+} ; also, if Γ is an element of the Clifford algebra, one usually writes $\bar{\Gamma} = \beta_{+}(\Gamma)$. Besides, in this case $\varepsilon^{2} = -1$, $\gamma_5 \equiv \theta = i\varepsilon$, $\gamma_5^{2} = 1$, $\beta_{+}(\gamma_5) = \bar{\gamma}_5 = -\gamma_5$, $\beta_{+}(\varepsilon) = \bar{\varepsilon} = +\varepsilon$, also $c\gamma^{\mu} = \gamma^{\mu}c$, $c^{2} = 1$ and $\beta_{+}(c) = c$ (cf. 6.3.3).

$$\langle\psi,\varphi\rangle \doteq i(\psi,\gamma^{o}\varphi) \quad \text{and} \quad \beta=\beta_- \quad : \quad \beta(\gamma^\mu) = \bar{\gamma}^\mu = -\gamma^\mu \quad \text{IF } \gamma^{02} = -1$$

$$\langle\psi,\varphi\rangle = (\psi,\gamma^{o}\varphi) \quad \text{and} \quad \beta = \beta_+ \quad : \quad \beta(\gamma^\mu) = \bar{\gamma}^\mu = +\gamma^\mu \quad \text{IF } \gamma^{02} = +1$$

The four spin invariant complex valued scalar products defined above give rise to eight spin invariant real valued scalar products by taking their real and imaginary parts.

Recall that the symbols C_A, C_S meaning antisymmetric and symmetric become graded antisymmetric and graded symmetric when applied to anticommuting spinors.

6.4 Bilinear forms valued in R,C,H, R ⊕ R, H ⊕ H

It is clear from the classification of Clifford algebra given in sect. 3.5.2 that the commutant K of G(p,q) is one of the following algebra: R,C,H, R ⊕ R, H ⊕ H. One sometimes defines spin invariant "scalar products" of spinors valued in this commutant K (even if K is not a field). (See ref. [14]). For a given value of (p,q), we have one such scalar product of β_+ type and another of β_- type. By decomposing the commutant on a complex (or real) basis, we of course recover the more familiar scalar products (C or R valued) discussed previously.

6.5 Lagrangians for spinor fields

6.5.1 Kinetic terms and mass terms

We now go back to the general situation where E is a pseudo Riemannian manifold of signature (p,q). Locally Spinor fields are maps from E to a vector space F. Let $\langle\langle\ ,\ \rangle\rangle$ denote some real scalar product (symmetric or antisymmetric, positive or not), invariant under the group Spin and for which the adjoinction operator is β. We want to build a Lagrangian density and in particular a kinetic energy term; such a term will contain a piece of the kind $\langle\langle\phi,\gamma^\mu\partial_\mu\phi\rangle\rangle$; by definition we get

$$<<\phi,\gamma^\mu\partial_\mu\phi>> = <<\beta(\gamma^\mu)\phi,\partial_\mu\phi>>$$
$$= \partial_\mu <<\beta(\gamma^\mu)\phi,\phi>> - <<\partial_\mu\beta(\gamma^\mu)\phi,\phi>>$$

Let us assume for a while that $<< , >>$ is symmetric and that $\beta=\beta_+$ i.e., $\beta(\gamma^\mu) = \gamma^\mu$; we find by integrating over E that

$$2\int<<\phi,\gamma^\mu\partial_\mu\phi>> \text{ dvol} = \int\partial_\mu<<\phi,\gamma^\mu\phi>> \text{ dvol}$$

Our expected kinetic term becomes a pure divergency: it does not contribute to the equations of motion!. Of course if $\beta=\beta_-$ i.e., $\beta(\gamma^\mu) = -\gamma^\mu$ the conclusions are not the same and we get a non trivial kinetic term. Therefore, it is clear that not any real, Spin invariant scalar product in the space F can be used to build a non trivial kinetic term. There is a further subtlety: in the above, we assumed that ϕ and $\partial_\mu\phi$ were commuting (we wrote $<<\phi,\gamma^\mu\partial_\mu\phi>> = <<\gamma^\mu\partial_\mu\phi,\phi>>$ assuming $<< , >>$ symmetric; the conclusions would have been just reversed if we had used anticommuting spinors (something we have to do anyhow): in order $<<\phi,\gamma^\mu\partial_\mu\phi>>$ to be a non trivial kinetic term for anticommuting spinors and if the adjoint β is β_+ we need $<< , >>$ to be real valued (graded) symmetric.

For anticommuting spinors, it is also clear that, while discussing possible mass terms (of the kind m $<<\phi,\phi>>$), we get a zero contribution from (graded) symmetric scalar products (whether $\beta = \beta_+$ or β_- is here irrelevant).

The following summarizes the possibilities:

Mass terms : use real valued graded symmetric scalar products

Kinetic terms: use real valued graded symmetric scalar products if $\beta=\beta_+$

or use real valued graded antisymmetric scalar products if $\beta=\beta_-$

For example, in the usual case of a Minkowski spacetime, if we choose the $\gamma_o^2 = -1$ convention, we know that the Dirac product $\bar{\psi}\psi$ is hermitian symmetric and of β_- type ($\bar{\gamma}_\mu = -\gamma_\mu$), therefore $\mathrm{Im}(\bar{\psi}\partial\!\!\!/\psi)$ being real graded antisymmetric leads to a non trivial kinetic term whereas $\mathrm{Re}(\bar{\psi}\partial\!\!\!/\psi)$ leads to a total divergency.

6.5.2 Special cases of Weyl, Majorana and Weyl-Majorana spinors.

When we want to build Lagrangians for spinor fields satisfying constraints like $\theta\psi = \pm\,\psi$ or $c\psi = \psi$, two kinds of simplification occur:

Scalar products which are independent for Dirac spinors may become dependent for spinors with constraints; for example $\langle c\psi,\theta\psi\rangle$ is the same as $\langle c\psi,\psi\rangle$ if $\theta\psi = \psi$.

Scalar products which are usually non vanishing may vanish automatically because of the constraints; for example, if $\theta\psi = \pm\,\psi$ then

$$\langle\psi,\psi\rangle = \langle\theta\psi,\theta\psi\rangle = \langle\beta(\theta)\theta\psi,\psi\rangle = -\,\langle\theta^2\psi,\psi\rangle = -\langle\psi,\psi\rangle \Longrightarrow \langle\psi,\psi\rangle = 0.$$

6.5.3 Reflections and equations of motion. (hyperbolic signature).

Dirac spinors - It is easy to prove that P and τ are symmetries of the massive Dirac equation in all cases (for example, if $\gamma^\mu\partial_\mu\psi = -m\psi$ then $\gamma^\mu\partial_\mu{}^P\psi^P = \psi^o\partial_o\gamma^o\psi - \partial_i\gamma^i\gamma^o\psi = \gamma^o(\gamma^\mu\partial_\mu\psi) = -\,m\gamma^o\psi = -m\psi^P$). As we saw in previous chapters, it is always possible to construct an antilinear operator c acting on Dirac spinors; when $c^2 = 1$, it is a charge conjugation in the usual sense but when $c^2 = -1$ it is not possible to define real eigenstates of c (no Majorana spinors); c is nevertheless, in all cases, a symmetry of the massless Dirac equation: if $\partial\!\!\!/\psi = 0$ then $\psi^c = c\psi$ obeys $\partial\!\!\!/\psi^c = 0$ as well.

Weyl spinors - If $n = 4$ or $8 \bmod 8$ i.e., $4 \bmod 4$ then $c\theta = -\theta c$. If ψ is a Weyl spinor with right helicity ($\theta\psi = \psi$) then $\psi^c = c\psi$ will be of opposite helicity ($\theta\psi^c = \theta c\psi = -c\theta\psi = - c\psi = -\psi^c$). Also $P = \gamma^o$ connects the two kinds of helicity: if $\theta\psi = \psi$ then $\theta(P\psi) = \theta\gamma^o\psi = \gamma^o\theta\psi = - \gamma^o\psi = - (P\psi)$ since θ anticommutes with the γ^μ in all even dimensions. Therefore in those dimensions the Weyl equation is left unchanged by the product cP (or by τ) but not separately by c and P.

If $n = 2,6 \bmod 8$ i.e., $2 \bmod 4$ then $c\theta = \theta c$ and the conclusions are very different: P is violated (as usual) but c is conserved therefore τ is violated and $P\tau$ is conserved.

Notice that in all cases if ψ is a Dirac spinor then $\psi + \Theta\psi$ is a Weyl spinor.

Majorana spinors - Here c acts trivally therefore $P\tau$ is conserved. Notice that if ψ is a Dirac spinor then $\psi + c\psi$ is a Majorana spinor. If $n = 2 \bmod 8$, we know there exist Weyl Majorana spinors (satisfying both $c\psi = \psi$ and $\theta\psi = \psi$, or $\theta\psi = - \psi$); being in particular W is violated therefore τ is violated as well but $P\tau$ is conserved.

7. Guide for references and other topics

It is usually impossible to put everything in a review...many aspects of spinors and Clifford algebras have not been discussed here, in particular those aspects which belong to the realm of differential geometry or differential topology (possibility of defining spinors on a Riemannian manifold, definition and properties of the Dirac equation on curved spaces, etc...), this would require a book by itself and no references will be given. Even from the algebraic point of view, we have left over many things, the following tries to guide the reader through the existing literature, it also can be used as a guide for the present paper.

General study of Clifford algebras → [2][3][7][11][13][14][14'][19][0][20]

Periodicity relations → [2][5][14]

Spinors (representations) → [5][7][8][17][19]

Scalar products of spinors → [10][12][14][16]

Spinor and exceptional groups → [15]

Spinors in differential geometry → [4][6][18]

Acknowledgements

The present study originated while A. Jadczyk and myself were working on an article entitled "Harmonic expansion and dimensional reduction of G/H Kaluza Klein theories"[1] and was originally conceived as an appendix; the size of the appendix becoming bigger than the size of the paper itself, it has been purely discarded. Many thanks are therefore due to A. Jadczyk with whom I discussed many of the topics presented here. This set of lectures was then written and typed at the Theory group, Department of Physics at the University of Austin, Texas and I would like to thank Professor S. Weinberg for giving me the opportunity of writing this review (UTTG-21-85 supported in part by Robert Welch Foundation and NSF PHY 8304629). Last modifications to the Austin preprint were finally made during my stay at IHES whose hospitality I want to acknowledge. Finally, I am glad to thank Professor G. Furlan and A. Trautman for editing this review in a volume dedicated to Professor Paolo Budinich.

[0] E. Cartan. Leçon sur la théorie des spineurs I, II. (Exposés de géométrie; XI, Paris-Hermann)-1938.

[1] R. Coquereaux, A. Jadczyk. Class. Quantum Grav. 3 (1986) 29-42.

[2] M.F. Atiyah, R. Bott and A. Shapiro. Topology, Vol. 3, Sup. 1 (1969).

[3] N. Bourbaki. Algebra Ch. 9 (Hermann, Paris, 1958).

[4] N. Hitchin. Adv. Math 14 (1979) 1.

[5] A.A. Kirillov. Elements of the theory of representations. (Springer, Berlin, 1976).

[6] R. Bott. Lectures on K[x]. Benjamin, New York, 1969).

[7] R. Brauer, H. Weyl. Amer. J. Math 57 (1935) 425.

[8] J. Scherk. Cargèse Lecture. (Plenum, New York, 1979).

[9] C. Chevalley. The algebraic theory of spinors. (Columbia, New York 1954).

[10] P. Lounesto. Ann. Inst. H. Poincaré A XXXIII no. 1 (1980) 53.

[11] N. Salingaros. J. Math Phys. 23 (1) 1982.

[12] D. Kastler. Electrodynamique Quantique. (Dunod, Paris, 1961).

[13] J.M. Souriau. Geometric et Relativité. (Hermann, Paris, 1964).

[14] Porteous. Topological geometry. (VNR, London, 1969).

[14'] R. Coquereaux. Physics Lett., Vol. 115B, n°5, p.389-395.

[15] A Sudbery. J. Math. Gen. 17 (1984) 939-955.

[16] P. Lounesto. Foundations of Physics. Vol. 11, no. 9/10, 1981.

[17] N. Bourbaki. Algebra de Lie. Chap. 9, 10 (Masson, Paris, 1980).

[18] D. Husemoller. Fiber bundles. (Springer Verlag, 1966).

[19] M. Karoubi. "Algèbre de Clifford et K-théorie", Thèse de Doctorat, Université de Paris (1967).

[20] C. Wetterich. Discrete symmetries in Kaluza-Klein theories, Nuclear Physics B234 (1984) 413.

$\overline{SL}(n,R)$ SPINORS FOR PARTICLES, GRAVITY AND SUPERSTRINGS

Dj. Šijački

Institute of Physics, P. O. Box 57, Belgrade, Yugoslavia

ABSTRACT

A brief account of the spinors of the double covering $\overline{SL}(n,R)$ of the SL(n,R) groups, n=2,3,4,10, in the fields of Particle Physics and Gravity is given. The basic notions and results due to Harish-Chandra on the noncompact group representations in the maximal compact subgroup basis are presented and applied to the $\overline{SL}(n,R)$, n=2,3,4 cases. A construction of all unitary irreducible representations (unirreps) of these groups is achieved by making use of the most general scalar product kernel operator. The $\overline{SL}(n,R)$, n=2,3,4 unirrep labels and the noncompact operators matrix elements for an arbitrary unirrep are given. The decontraction formula method for obtaining spinorial $\overline{SL}(10,R)$ unirreps is outlined, and the $\overline{SO}(10)$ content of the simplest such representations is given.

INTRODUCTION

Current knowledge about the unitary irreducible representations (unirreps) of the SL(n,R) groups is incomplete.[1)] In particular very little is known about the unirreps of the double-covering groups $\overline{SL}(n,R)$. The cases n=2,3,4,10 are important in physics.

SL(2,R)≃SU(1,1) and its multiple coverings are of significance for classical and quantized strings,[2)] projective transformations, etc. The SL(3,R) group representations have been applied to classify the excitations of deformed nuclei[3)] and of hadronic states lying along Regge trajectories.[4)] The SL(4,R) group is a relativistic extension of the hadronic SL(3,R), and is generated by the following charge operators[5)]

$$Q^a{}_b = \int d\sigma_\mu \left(x^a \Theta_b{}^\mu(x) - \frac{1}{4}\delta^a_b (x^c \Theta_c{}^\mu(x)) + \text{intrinsic part} \right),$$

where $\Theta_a{}^\mu$ is the local symmetric stress-energy-momentum tensor for hadrons, and its components obey some equal-time commutation relations. We have shown recently that the complete spectrum of baryon and meson resonances for each flavour is given by spinor and tensor infinite-component systems based on $\overline{SL}(4,R)$ unirreps.[6)]

In the standard approach to General Relativity one starts with the group of general coordinate transformations GCT, i.e. the group of diffeomorphisms DiffR^4. The theory is set up upon the principle of general covariance. A unified description of both tensors and spinors would require the existence of respectively tensorial and (double valued) spinorial representations of the GCT group. In other words one is interested in the corresponding single valued representations of the double covering

$\overline{\mathrm{GCT}}$ of the GCT group, since the topology of GCT is given by the topology of its linear compact subgroup. It is well known that the finite-dimensional representations of $\overline{\mathrm{GCT}}$ are characterized by the corresponding ones of the $\overline{\mathrm{GL}}(4,R) \supset \overline{\mathrm{SL}}(4,R)$ group, and $\overline{\mathrm{SL}}(4,R)$ does not have finite spinorial representations. However there are infinite-dimensional spinors of $\overline{\mathrm{SL}}(4,R)$ which are the true "world" (holonomic) spinors.[7)] There are two ways to introduce finite spinors: i) One can make use of the nonlinear representations of the $\overline{\mathrm{GCT}}$ group, which are linear when restricted to the Poincaré subgroup.[8)] ii) One can introduce a bundle of cotangent frames, i.e. a set of 1-forms e^a (tetrads; a=0,.. .,3 the anholonomic indices) and define in this space an action of a physically distinct local Lorentz group. Owing to this Lorentz group one can introduce finite spinors, which behave as scalars w.r.t. $\overline{\mathrm{GCT}}$. The bundle of cotangent frames represents an additional geometrical construction corresponding to the physical constraints of a local gauge group of the Yang-Mills type, in which the gauge group is the isotropy group of the space-time base manifold. One is now naturally led to enlarge the local Lorentz group to the whole linear group $\overline{\mathrm{GL}}(4,R)$, and together with translations one obtaines the affine group $\overline{\mathrm{GA}}(4,R)$. The affine group translates and deforms the tetrads of the locally Minkowskian space-time,[9)] and provides one with either infinite-dimensional linear or finite-dimensional nonlinear spinorial representations.[10)]

It has been realized recently that superstrings may provide both the first mathematically consistent (renormalizable or even finite) quantum theory of gravity and the first truly unified theory of all fundamental interactions and matter.[11)] As for the quantum gravity, the crucial observation was that $\alpha' \to 0$ limit of string scattering amplitudes corresponds to appropriate scattering amplitudes of a point particle (string excitation) field theory.[12] In the conventional lagrangian formulation for superstrings, the 2-dimensional curved (locally reparametrizable) string world-sheet R^2 is embedded in a flat 10-dimensional Minkowski space-time. On the other hand, macroscopic gravity is described classically by Einstein's theory, corresponding to a curved Riemannian R^4 manifold. It is now clearly necessary first to find a generic curved space formulation for superstring, and then to study the theory on manifolds with some of the dimensions compactified.

The attempts to embed the superstring in a curved manifold[13] have encountered three fundamental difficulties: i) The fermionic $\Theta(\zeta)$ frame-fields required by supersymmetry and constructed at any point ζ^μ, $\mu=0,1$ of the world-sheet as spinors in $M^{1,9}$, cannot be embedded in a curved generic Riemannian, i.e. Riemann-Cartan, manifold R^{10}, since there exist no finite-dimensional spinorial representations of $\overline{SL}(10,R)$ on the one hand, and on the other hand one cannot apply the usual tetrad formulation, as indicated below. ii) The critical dimensionality of the embedding space is modified[14] in the presence of generic curvature, thus destroying the essential condition for a ghost-free finite-theory. iii) Without spinors we also lose supersymmetry, essential as a constraint for the removal of a tachyon whose unwanted presence now makes the theory unphysical.

We have suggested recently solutions for all three difficulties, based upon the application of the doubly-covered groups of Diffeomorphisms and Superdiffeomorphisms in the "tangent' at ζ^μ (Ref. 15,16). We have constructed explicitly the (infinite) spinorial and tensorial representations of the double covering $\overline{SL}(10,R)$ group[15] thus answering the quest in (i), while realizing the Principle of General Covariance. Further, we have solved problems (ii) and (iii) by embedding $\overline{GL}(10,R)$ in the double-covered real-form $\overline{GQ}(10,R)$, a supergroup generated by the classical hyperexceptional superalgebra q(10).[16] This way the curving of $M^{1,9} \to R^{10}$ preserves supersymmetry (i.e. no ghosts) both on and off mass shell.

In the Polyakov formulation[17] of the Green-Schwarz quantized superstring, every point ζ of the evolving string carries a global 10-space-time-dimensional Poincaré supersymmetric frame $(X^m(\zeta),\ \theta^{a\alpha}(\zeta))$, m=0,1,...,9 denoting the components of a bosonic 10-vector frame, a=1,2,...,16 standing for the fermionic components of a real (Majorana) chiral (Weyl) spinor frame, and α=1,2.

In going from $M^{1,9}$ to R^{10} one replaces $\eta_{mn} \to g_{\hat{m}\hat{n}}$, a curved metric, and so on for all bosonic quantities. However, the method fails for the spinors. In the usual technique in General Relativity, a spinor would be defined in the local tangent (flat embedding, coordinate x^m) space at $x^{\hat{m}}$ (curved embedding), where a local Lorentz group can act on it. This is achieved by introducing the "decad" frames

$$e^m_{\hat{m}}(\hat{x}) = \partial x^m/\partial x^{\hat{m}}, \qquad \psi^a_{\hat{m}}(\hat{x}) = \partial\theta^a/\partial x^{\hat{m}}.$$

In the string formalism the original spinor field is defined as a local (fiber) frame-field on the tangent to the curved string (the base manifold) at the string coordinate ζ^μ. Indeed, the coordinates x^m (flat) or $x^{\hat{m}}$ (curved) do not appear in the formalism. As a result, the usual transition of the gamma matrices $\gamma^n \to \gamma^{\hat{n}}$ cannot be performed by an ordinary tetrad-like (decad) matrix. Our $X^m(\zeta)$ are in the flat tangent at ζ^μ, and there are no "tangents to the tangents", frames over frames. Were it not for the spinors, generic curving could have been achieved by replacing $X^m(\zeta)$ by $X^{\hat{m}}(\zeta)$, a world-vector carrying finite linear representations of GL(10,R) and nonlinear representations of the GCT group. In other words changing the structure group of the bundle from SO(1,9) to GL(10,R) and GCT. For spinors, where the double-covering is required, one has only the infinite representations of $\overline{GL}(10,R)$ and $\overline{GCT}$. We thus replace the $\theta^a(\xi)$ by $\Psi^A(\xi)$, infinite frames transforming w.r.t. the unirreps of the linear subgroup $\overline{SL}(10,R)$. In order to preserve supersymmetry we additionally replace $X^m(\xi)$ by an appropriate infinite-component $\overline{SL}(10,R)$ bosonic frames $X^M(\xi)$.

$\overline{SL}(n,R)$, n=2,3,4,10 UNIRREPS

SL(n,R) is the group of linear unimodular transformations in an n-dimensional real vector space. The group is a simple and noncompact Lie group. The space of the group parameters is not simply connected. The maximal compact subgroup of SL(n,R) is SO(n). The double covering (the universal covering for n>2) group of SL(n,R) we denote by $\overline{SL}(n,R)$. Its maximal compact subgroup is $\overline{SO}(n)\simeq Spin(n)$, the covering group of SO(n).

$$\overline{SL}(n,R)/Z_2 \simeq SL(n,R), \qquad \overline{SO}(n)/Z_2 \simeq SO(n).$$

A convenient way to parametrize any noncompact semisimple Lie group is given by means of the Iwasawa decomposition according to which the group G can be written as a product G=NAK, where N is a nilpotent subgroup of G , and its elements are upper triangular matrices with ones on the diagonal, A is an Abelian subgroup of G, and for SL(n,R) we take its elements to be of the form $a=\mathrm{diag}(e^{\lambda}, e^{\mu}, e^{\nu},\ldots,e^{-(\lambda+\mu+\nu+\ldots)})$, and finally K is the maximal compact subgroup of G. For $\overline{SL}(3,R)$, K=SU(2), and for $\overline{SL}(4,R)$, K=SU(2)⊗SU(2), and we will parametrize them in terms of the Euler angles α,β,γ and $\alpha_1, \beta_1, \gamma_1, \alpha_2, \beta_2, \gamma_2$, respectively. An element $g\in G$ can thus be written as a product g=nak, where $n\in N$, $a\in A$, $k\in K$. The Iwasawa decomposition is unique and the product of some element $k\in K$ and an arbitrary element $g\in G$ is in general an arbitrary element of G which can be uniquely written as $kg=na(k,g)k\cdot g$, where $n\in N$, $a(k,g)\in A$ and $k\cdot g\in K$. Owing to the Iwasawa decomposition every element $g \in SL(n,R)$ can be uniquely written as $g=ne^{h}k$. The Abelian subgroup of SL(n,R) has n-1 generators $A_1, A_2, \ldots$, and if $\lambda, \mu, \nu, \ldots$ are the corresponding group parameters, respectively, one has $h=\lambda A_1+\mu A_2+\nu A_3+\ldots$ Let α be a linear, in general complex, function such that $\alpha(h)=\lambda\alpha(A_1)+\mu\alpha(A_2)+\nu\alpha(A_3)+\ldots$, and let us denote $\alpha(A_1)$, $\alpha(A_2)$, $\alpha(A_3)$ ... by a,b,c,..., respectively. Exstence of the mapping α is guaranteed by the 1-dimensionality of the irreducible representations of the Abelian subgroup A. The mapping α can be extended in a natural way to a mapping from the group NA into the complex numbers since N is an invariant subgroup in NA.

The set of cosets $\overline{SL}(n,R)/NA$ is in one-to-one correspondence with the group K=SO(n) and can be parametrized by the elemnts of K. In the coset space $\overline{SL}(n,R)/NA$ one has as well a measure, which we choose to be the invariant measure dk on K. Let $H=L^2(K)$ be the separable Hilbert space of functions on K which are square integrable with respect to the invariant measure on K, i.e. $H=\{f(k)\,|\,k\in K\}$, such that $\int dk f^*(k)f(k)<\infty$, and let $\int dk=1$.

Every nontrivial unitary representation of a noncompact group is necessarily infinite-dimensional and this partly accounts for the complexity which occures when one deals with unitary representations. The class of real semisimple Lie groups is especially complex. Harish-Chandra[18)] defines a representation U(g) of $G=\overline{SL}(n,R)$ on H in the following way: U(g) is a homomorphic continuous mapping from G into the set of linear transfotmations on H given by

$$(U(g)f)(k)=e^{(h(k,g))}f(k\cdot g),$$

where $g\in G$, $f\in H$, $k\in K$, $e^{h}\in A$ and where $(U(g)f)(k)$ denotes the value of U(g)f at the point k. Harish-Chandra now defines the concept of infinitesimal equivalence of two representations in the following way: Two representations are infinitesimally equivalent if there exists a similarity transformation of one representation into the other, with a nonsingular, not necessarily unitary operator. In the case of equivalence there exist a unitary operator by means of which the transformation between the two representations is carried out. If both of two infinitesimally equivalent representations are unitary, then they are equivalent. Suppose now that U(g) is a representation of a group G on a Hilbert space H. Suppose further that H_1 and H_2 are the two closed invariant subspaces of H, such that $H_2 \subset H_1 \subset H$, and $H_1 \neq H_2$. Then U(g) induces a representation U'(g) on the quotient H_1/H_2 in a natural way. The representation U'(g) is said to be deducible from the representation U(g). Harish-Chandra has proved that every unirrep is infinitesimally equivalent to some irreducible representation deducible from some representation U(g) of the above form. Thus it is always possible to construct a bilinear form $(\tilde{f},\tilde{g})$ in some quotient space H_1/H_2, where $\tilde{f},\tilde{g} \in H_1/H_2$. One can extend the domain of this bilinear form to all H_1 uniquely by defining (,) to vanish on H_2. Unitarity now means that $(U(g)f,U(g)f)=(f,f)$, $f \in H_1$, $g \in G$, and the additional conditions that the bilinear form is a scalar product are hermiticity and positive definiteness $(f,g)=(g,f)^*$ and $(f,f) \geq 0$ $\forall f,g \in H_1$. It is convenient to extend the domain of the scalar product to the whole space H. Being interested in obtaining all unirreps of $\overline{SL}(n,R)$, n=2,3,4, we will start with the most general scalar product: $(f,g)=\iint dk_1 dk_2 f^*(k_1)\varkappa(k_1,k_2)g(k_2)$, $f,g \in H$, where $\varkappa(k_1,k_2)$ is a kernel, the integration is over K and dk is an invariant measure. The problem of finding all unirreps of $\overline{SL}(n,R)$ becomes now the problem of finding all scalar products, i.e. kernels for which the representation U(g) is irreducible and unitary. It is straight-forward to evaluate the infinitesimal operator of a 1-parameter subgroup of $\overline{SL}(n,R)$ which corresponds to an arbitrary group parameter ε. We find

It is interesting to point out that X_ε is not a differential operator in the group parameters corresponding to the nilpotent subgroup. It is also clear that the Casimir operators of the $\overline{SL}(n,R)$ group are of the form $C=C(a,b,c,\ldots)$.

In order to present the explicit forms of the $\overline{SL}(n,R)$ generators, n=2,3,4, we first separate them according to compactness and it is most convenient to take them in the spherical basis. We will list the minimal set of the commutation relations. The remaining ones can be obtained by means of the Jacobi identity.

The $\overline{SL}(2,R)$ generators are J_0, $T_\pm$. The commutation relations are:

$$[J_0, T_\pm] = \pm T_\pm , \quad [T_+, T_-] = -2 J_0$$

The $\overline{SL}(3,R)$ generators are J_0, $J_\pm$, T_μ, $\mu=0,\pm1,\pm2$. J_0 and $J_\pm$ generate the SU(2) subgroup, while T_μ form an SU(2) second rank irreducible tensor operator. The commutation relations are:

$$[J_0, J_\pm] = \pm J_\pm, \quad [J_+, J_-] = 2 J_0, \quad [J_0, T_\mu] = \mu T_\mu ,$$
$$[J_\pm, T_\mu] = \sqrt{6-\mu(\mu\pm1)}\, T_{\mu\pm1} \quad [T_{+2}, T_{-2}] = -4 J_0.$$

The $\overline{SL}(4,R)$ generators are $J_0^{(i)}$, $J_\pm^{(i)}$, $Z_{\alpha\beta}$, i=1,2; $\alpha,\beta=0,\pm1$. $J_0^{(i)}$ and $J_\pm^{(i)}$ generate the SU(2)⊗SU(2) subgroup, while $Z_{\alpha\beta}$ form, w.r.t. SU(2)⊗SU(2), a (1,1)-irreducible tensor operator. The commutation relations are:

$$[J_0^{(i)}, J_\pm^{(j)}] = \pm \delta_{ij} J_\pm^{(i)}, \quad [J_+^{(i)}, J_-^{(j)}] = 2\delta_{ij} J_0^{(i)}, \quad [J_0^{(i)}, J_0^{(j)}] = 0$$
$$[J_0^{(1)}, Z_{\alpha\beta}] = \alpha Z_{\alpha\beta}, \quad [J_0^{(2)}, Z_{\alpha\beta}] = \beta Z_{\alpha\beta}$$
$$[J_\pm^{(1)}, Z_{\alpha\beta}] = \sqrt{2-\alpha(\alpha\pm1)}\, Z_{\alpha\pm1,\beta}, \quad [J_\pm^{(2)}, Z_{\alpha\beta}] = \sqrt{2-\beta(\beta\pm1)}\, Z_{\alpha,\beta\pm1},$$
$$[Z_{+1,+1}, Z_{-1,-1}] = -(J_0^{(1)} + J_0^{(2)}).$$

An explicite evaluation yields the following expressions for the $\overline{SL}(n,R)$ generators.

$\underline{n=2}$: $\quad J_0 = i\frac{\partial}{\partial\gamma}, \quad T_\pm = e^{\mp i\gamma}\left(\pm a + i\frac{\partial}{\partial\gamma}\right).$

$\underline{n=3}$: $\quad J_0 = i\frac{\partial}{\partial\gamma}, \quad J_\pm = e^{\mp i\gamma}\left(\frac{-i}{\sin\beta}\frac{\partial}{\partial\alpha} \mp \frac{\partial}{\partial\beta} + i\,\mathrm{ctg}\,\beta\frac{\partial}{\partial\gamma}\right),$

$$T_0 = i\left[\sqrt{\tfrac{2}{3}}\,(3\cos^2\alpha\sin^2\beta-1)a + \sqrt{\tfrac{2}{3}}\,(3\sin^2\alpha\sin^2\beta-1)b + \sqrt{6}\sin\alpha\cos\alpha\sin^2\beta\frac{\partial}{\partial\alpha} - \sqrt{6}\sin\beta\cos\beta\frac{\partial}{\partial\beta}\right].$$

$\underline{n=4}$: $\quad J_0^{(i)} = J_0(\gamma_i), \quad J_\pm^{(i)} = J_\pm(\alpha_i,\beta_i,\gamma_i),$

$$Z_{0,0} = i\Big[\tfrac{1}{2}\cos\beta_1\cos\beta_2(c-a-b) + \tfrac{1}{2}\cos(\alpha_1-\alpha_2)\sin\beta_1\sin\beta_2\, c + \tfrac{1}{2}\cos(\alpha_1+\alpha_2)\sin\beta_1\sin\beta_2(a-b)$$
$$+ \sin\alpha_1\cos\alpha_2\sin\beta_1\sin\beta_2\frac{\partial}{\partial\alpha_1} + \cos\alpha_1\sin\alpha_2\sin\beta_1\sin\beta_2\frac{\partial}{\partial\alpha_2}$$
$$- \sin\beta_1\cos\beta_2\frac{\partial}{\partial\beta_1} - \cos\beta_1\sin\beta_2\frac{\partial}{\partial\beta_2}\Big].$$

The remaining generators are most easily obtained by making use of the commutation relations. In order to analyse the representations it is convenient to have the matrix elements of the group generators. Also, in this case the task of determining the scalar products of the unitary representations is considerably simplified. The most general results are obtained in the $\left|{j \atop k\,m}\right\rangle$, $\left|{j_1 \atop k_1 m_1}\,{j_2 \atop k_2 m_2}\right\rangle$ basis of the SU(2), SU(2)⊗SU(2) representations respectively, $j, j_1, j_2 = 0, 1/2, 1, \ldots$ The matrix elements of the compact generators are well known, and we list only the matrix elements of the noncompact generators.

$\underline{n=2}$: $\quad J_0|m\rangle = m|m\rangle, \quad T_\pm|m\rangle = (m \mp a)|m\pm 1\rangle.$

$\underline{n=3}$:

$$\left\langle{j' \atop k'm'}\right|T_\mu\left|{j \atop k\,m}\right\rangle = (-)^{j'-m'}\begin{pmatrix} j' & 2 & j \\ -m' & \mu & m\end{pmatrix}\left\langle{j' \atop k'}\right\|T\left\|{j \atop k}\right\rangle, \qquad \mu = 0, \pm1, \pm2$$

$$\left\langle{j' \atop k'}\right\|T\left\|{j \atop k}\right\rangle = (-)^{j'-k'}\sqrt{(2j'+1)(2j+1)}\,\Big\{\frac{-i}{\sqrt{6}}\left[2\sigma - j'(j'+1) + j(j+1)\right]\begin{pmatrix} j' & 2 & j \\ -k' & 0 & k\end{pmatrix} +$$
$$+ i(\delta+k+1)\begin{pmatrix} j' & 2 & j \\ -k' & 2 & k\end{pmatrix} + i(\delta-k+1)\begin{pmatrix} j' & 2 & j \\ -k' & -2 & k\end{pmatrix}\Big\}$$

$\sigma = a+b\,,\ \delta = a-b.$

$\underline{n=4}$:

$$\left\langle{j_1' \atop k_1'm_1'}\,{j_2' \atop k_2'm_2'}\right|Z_{\alpha\beta}\left|{j_1 \atop k_1 m_1}\,{j_2 \atop k_2 m_2}\right\rangle = (-)^{j_1'-m_1'}(-)^{j_2'-m_2'}\begin{pmatrix} j_1' & 1 & j_1 \\ -m_1' & \alpha & m_1\end{pmatrix}\begin{pmatrix} j_2' & 1 & j_2 \\ -m_2' & \beta & m_2\end{pmatrix}\left\langle{j_1' \atop k_1'}\,{j_2' \atop k_2'}\right\|Z\left\|{j_1 \atop k_1}\,{j_2 \atop k_2}\right\rangle$$

$$\left\langle \begin{matrix} j_1' & j_2' \\ k_1' & k_2' \end{matrix} \right\| Z \left\| \begin{matrix} j_1 & j_2 \\ k_1 & k_2 \end{matrix} \right\rangle = (-)^{j_1'-k_1'} (-)^{j_2'-k_2'} \frac{i}{2} \sqrt{(2j_1'+1)(2j_2'+1)(2j_1+1)(2j_2+1)} \times$$

$$\times \left\{ \left[e+4-j_1'(j_1'+1)+j_1(j_1+1)-j_2'(j_2'+1)+j_2(j_2+1) \right] \begin{pmatrix} j_1' & 1 & j_1 \\ -k_1' & 0 & k_1 \end{pmatrix} \begin{pmatrix} j_2' & 1 & j_2 \\ -k_2' & 0 & k_2 \end{pmatrix} \right.$$

$$- (c+k_1-k_2) \begin{pmatrix} j_1' & 1 & j_1 \\ -k_1' & 1 & k_1 \end{pmatrix} \begin{pmatrix} j_2' & 1 & j_2 \\ -k_2' & -1 & k_2 \end{pmatrix} - (c-k_1+k_2) \begin{pmatrix} j_1' & 1 & j_1 \\ -k_1' & -1 & k_1 \end{pmatrix} \begin{pmatrix} j_2' & 1 & j_2 \\ -k_2' & 1 & k_2 \end{pmatrix}$$

$$\left. + (d+k_1+k_2) \begin{pmatrix} j_1' & 1 & j_1 \\ -k_1' & 1 & k_1 \end{pmatrix} \begin{pmatrix} j_2' & 1 & j_2 \\ -k_2' & 1 & k_2 \end{pmatrix} + (d-k_1-k_2) \begin{pmatrix} j_1' & 1 & j_1 \\ -k_1' & -1 & k_1 \end{pmatrix} \begin{pmatrix} j_2' & 1 & j_2 \\ -k_2' & -1 & k_2 \end{pmatrix} \right\},$$

$$e = c - a - b, \quad d = a - b.$$

The representation labels a; σ, δ; and c,d,e are at present arbitrary complex numbers and will be determined from the unitarity requirement.

We start with the most general scalar product of the Hilbert space. We find, by making use of the fact that dk is an invariant measure and of the additivity properties of Wigner's functions, the following expressions for the scalar product in terms of the matrix elements of the kernel and the expansion coefficients

$$(f,g) = \sum_{(j)(k)(k')(m)} f^{(j)*}_{(k')(m)} \, g^{(j)}_{(k)(m)} \, \varkappa^{(j)}_{(k')(k)} .$$

Here e.g. (j) stands for j and j_1, j_2 in the $\overline{SL}(3,R)$ and $\overline{SL}(4,R)$ case respectively. The hermiticity of the scalar product yields

$$\varkappa^{(j)*}_{(k')(k)} = \varkappa^{(j)}_{(k)(k')} .$$

Therefore $\varkappa$ is a hermitian matrix and can be diagonalized. Thus without any loss of generality we write $\varkappa$ in the form $\varkappa(j;k)$ and $\varkappa(j_1,j_2;k_1,k_2)$ for $\overline{SL}(3,R)$ and $\overline{SL}(4,R)$ respectively. The positive definiteness of the scalar product yields

$$\varkappa(j;k) \geq 0 , \quad \varkappa(j_1,j_2;k_1,k_2) \geq 0 , \quad \forall j, j_1, j_2, k, k_1, k_2 .$$

Finally we find that the hermiticity condition of an arbitrary group generator X, i.e. the unitarity of the representation, $(f,Xg)=(g,Xf)^*$ reads

$$\mathcal{H}(j_1', j_2'; k_1', k_2')\left\langle \begin{matrix} j_1' & j_2' \\ k_1' m_1' & k_2' m_2' \end{matrix} \middle| X \middle| \begin{matrix} j_1 & j_2 \\ k_1 m_1 & k_2 m_2 \end{matrix} \right\rangle = \mathcal{H}(j_1, j_2; k_1, k_2)\left\langle \begin{matrix} j_1 & j_2 \\ k_1 m_1 & k_2 m_2 \end{matrix} \middle| X \middle| \begin{matrix} j_1' & j_2' \\ k_1' m_1' & k_2' m_2' \end{matrix} \right\rangle^*$$

We now substitute in this equations the explicite expressions for the noncompact generators, and allow the representation labels values to be arbitrary complex numbers, e.g. $a=a_1+ia_2$, $a_1,a_2\in R$. We proceed to determine all solutions for the representation labels a; σ,δ; c,d,e and the corresponding kernels of the scalar products, thus determining all unirreps of $\overline{SL}(n,R)$, n=2,3,4. The irreducibility of the representations is achieved by usind the little group technique.

All unirreps of $\overline{SL}(2,R)$ are well known.[19] In terms of our label a they are as follows:

Principal series: $a_1=\frac{1}{2}$, $a_2\in R$; $\{m\}=\{0,\pm1,\pm2,\dots\}$, $\{m\}=\{\pm\frac{1}{2},\pm\frac{3}{2},\dots\}$; $\mathcal{H}(m)=1$.

Supplementary series: $|a_1-\frac{1}{2}|<\frac{1}{2}$, $a_2=0$; $\{m\}=\{0,\pm1,\pm2,\dots\}$,

$$\mathcal{H}(m)=\Gamma(1+a_1+m)\Gamma(2-a_1)/\Gamma(2-a_1+m)\Gamma(1+a_1).$$

Discrete series: $a_1=\underline{m}-1$, $a_2=0$; $\{m\}=\{\underline{m},\underline{m}+1,\underline{m}+2,\dots\}$, $\underline{m}=\frac{1}{2},1,\frac{3}{2},\dots$;

$$\mathcal{H}(\underline{m}+n)=\binom{2a_1+n}{n}.$$

$$a_1=-\overline{m}-1,\ a_2=0;\ \{m\}=\{\overline{m},\overline{m}-1,\overline{m}-2,\dots\},\ \overline{m}=-\tfrac{1}{2},-1,\dots$$

We list now all unirreps of the $\overline{SL}(3,R)$ group.[20]

Principal series: $\sigma_1=\delta_1=0$, $\sigma_2,\delta_2\in R$

$$(\varepsilon,\varepsilon')=(+1,+1):\ \{j\}=\{0^1,2^2,3^1,4^3,5^2,\dots\}$$

$$(\varepsilon,\varepsilon')=(+1,-1),(-1,\pm1):\ \{j\}=\{1^1,2^1,3^2,4^2,5^3,\dots\},\ \{\tfrac{1}{2}^1,\tfrac{3}{2}^2,\tfrac{5}{2}^3,\dots\}.$$

Supplementary series: $\sigma_1=\delta_2=0$, $\sigma_2\in R$

$$0<\delta_1<1,\ (\varepsilon,\varepsilon')=(+1,+1):\ \{j\}=\{0^1,2^2,3^1,4^3,5^2,\dots\}$$

$$(\varepsilon,\varepsilon')=(+1,-1):\ \{j\}=\{1^1,2^1,3^2,4^2,5^3,\dots\}$$

$$0<\delta_1\leq\tfrac{1}{2},\ \{j\}=\{\tfrac{1}{2}^1,\tfrac{3}{2}^2,\tfrac{5}{2}^3,\dots\}$$

Discrete series: $\sigma_1=\delta_2=0$, $\sigma_2\in R$, $\delta_1=1-\underline{j}$; $\underline{j}=\frac{3}{2},2,\frac{5}{2},3,\dots\}$

$$\{j\}=\{\underline{j}^1,(\underline{j}+1)^1,(\underline{j}+2)^2,(\underline{j}+3)^2,(\underline{j}+4)^3,\dots\}$$

Multiplicity free (ladder) series: $\sigma_1=\delta_2=0$, $\delta_1=1$,

$$\sigma_2\in R,\ \{j\}=\{0,2,4,\dots\},\ \{j\}=\{1,3,5,\dots\}$$

$$\sigma_2=0,\ \{j\}=\{\tfrac{1}{2},\tfrac{5}{2},\tfrac{9}{2},\dots\}.$$

For the general case of the $\overline{SL}(4,R)$ unirreps we present here only the labels,[21] which are in agreement with the ones of the $\overline{SO}(3,3)$ unirreps,[22] and the SL(4,R) unirreps.[1] For the general (multiplicity non free) case, we have

A) $e_1=0,\ e_2\in R,$

B_1) $d_1=0,\ d_2\in R,$

B_2) $d_1=\underline{k}_1+\underline{k}_2,\ d_2=0\ ;\quad \underline{k}_1+\underline{k}_2=\frac{1}{2},1,\frac{3}{2},\ldots,$

B_3) $0<d_1<1,\ d_2=0\ ;\quad k_1+k_2=0,\pm 2,\pm 4,\ldots,$

B_4) $0<d_1<\frac{1}{2},\ d_2=0\ ;\quad k_1+k_2\equiv\frac{1}{2}(\mathrm{mod}\,2)$ or $\frac{3}{2}(\mathrm{mod}\,2),$

C_1) $c_1=0,\ c_2\in R,$

C_2) $c_1=\underline{k}_1-\underline{k}_2,\ c_2=0\ ;\quad \underline{k}_1-\underline{k}_2=\frac{1}{2},1,\frac{3}{2},\ldots,$

C_3) $0<c_1<1,\ c_2=0\ ;\quad k_1-k_2=0,\pm 2,\pm 4,\ldots,$

C_4) $0<c_1<\frac{1}{2},\ c_2=0\ ;\quad k_1-k_2=\frac{1}{2}(\mathrm{mod}\,2)$ or $\frac{3}{2}(\mathrm{mod}\,2).$

Any combination of (A) with one (B) and one (C) determines a series of $\overline{SL}(4,R)$ unirreps. For these series $j_1\geq|k_1|,\ j_2\geq|k_2|$. There are four series of multiplicity free $\overline{SL}(4,R)$ unirreps.[23]

Principal series: $e_1=0,\ e_2\in R\ ;\quad j_1+j_2\equiv 0(\mathrm{mod}\,2)$ or $1(\mathrm{mod}\,2),$

Supplementary series: $0<e_1<1,\ e_2=0\ ;\quad j_1+j_2\equiv 1\,(\mathrm{mod}\,2),$

Discrete series: $e_1=1-\underline{j},\ e_2=0\ ;\quad \underline{j}=\frac{1}{2},1,\frac{3}{2},\ldots,|j_1-j_2|\geq\underline{j},$

Ladder series: $e_1=0,\ e_2\in R\ ;\quad j_1=j_2=j,\ \{j\}=\{0,1,3,\ldots\},\ \{j\}=\{\frac{1}{2},\frac{3}{2},\frac{5}{2}\ldots\}.$

We shall consider now briefly the $\overline{SL}(10,R)$ spinorial unirreps case. Let Q_{mn}, $m,n=0,1,\ldots,9$ be the $\overline{SL}(10,R)$ generators. The corresponding commutation relations read

$$[Q_{mn},Q_{kl}]=i(\eta_{nk}Q_{ml}-\eta_{ml}Q_{kn}),$$

where $\eta_{mn}=\mathrm{diag}(+1,-1,\ldots,-1)$. The maximal compact subgroup $\overline{SO}(10)=\mathrm{Spin}(10)$ of the $\overline{SL}(10,R)$ group is generated by $L_{mn}=\frac{1}{2}(Q_{mn}-Q_{nm})$. The remaining generators $T_{mn}=\frac{1}{2}(Q_{mn}+Q_{nm})$ are noncompact, and transform as a 54-dimensional $\overline{SO}(10)$ irreducible tensor operator, i.e. as (20000) in the Dynkin notation.

The Wigner-Inönü contraction of the $\overline{SL}(10,R)$ group w.r.t. its $\overline{SO}(10)$ subgroup is the $T_{54}\circledS\overline{SO}(10)$ group, with mutually commuting noncompact generators U_{mn},

$$U_{mn} = \lim_{\varepsilon \to 0} \varepsilon T_{mn}, \quad [U_{mn}, U_{kl}] = 0.$$

We construct the $\overline{SL}(10,R)$ unirreps in the $\overline{SO}(10)$ basis. A very convenient way to carry out an explicit $\overline{SL}(10,R)$ unirrep construction consists in i) the construction of the $T_{54} \circledS \overline{SO}(10)$ unirreps and ii) the lifting of these unirreps to the $\overline{SL}(10,R)$ case by making use of the decontraction formula.[24)]

Let L_{mn} and U_{mn} be the generators of a given $T_{54} \circledS \overline{SO}(10)$ unirrep. The generators of the corresponding (decontracted) $\overline{SL}(10,R)$ unirrep are given by L_{mn} and T_{mn}, where

$$T_{mn} = p U_{mn} + \frac{i}{2}(U \cdot U)^{-\frac{1}{2}} [L^2, U_{mn}], \qquad p \in R.$$

It is rather straight-forward to determine explicit forms of the U_{mn} operators, and thus by making use of the decontraction formula to determine the corresponding representations of the T_{mn} operators. The simplest spinorial and tensorial $\overline{SL}(10,R)$ unirreps belong to the set of multiplicity free representations. They contain in the $\overline{SO}(10)$ subgroup decomposition each $\overline{SO}(10)$ representation at most once. We list some $\overline{SL}(10,R)$ unirreps and indicate their $\overline{SO}(10)$ unirrep content. The spinorial unirrep "discrete series" contain:

$D(16) = \{16, 144, 720, 2640, 7920, \ldots\}$,

$D(560) = \{560, 3696, 8800, 15120, \ldots\}$,

$D(672) = \{672, 1440, 11088, \ldots\}$,

....

and their conjugated unirreps

$D(\overline{16}) = \{\overline{16}, \overline{144}, \overline{720}, \overline{2640}, \overline{7920}, \ldots\}$,

....

The tensorial "ladder series" are:

$D(1) = \{1, 54, 660, 4290, 19305, \ldots\}$,

$D(10) = \{10, 210, 1782, 9438, 37180, \ldots\}$.

REFERENCES

1. I. M. Gel'fand and M. I. Graev, Am. Math. Soc. Transl. 2(1956)147;
B. Speh, Math. Ann. 258(1981)113.
2. J. H. Schwarz, ed., Superstrings (World Scientific, 1985); M. Jacob, ed., Dual Theory (North-Holland, 1974).
3. L. Weaver and L. C. Biedenharn, Nucl. Phys. A 185(1972)1.
4. Y. Dothan, M. Gell-Mann and Y. Ne'eman, Phys. Lett. 17(1965)148.
5. Dj. Šijački, Ph. D. Thesis, Duke University (1974).
6. Y. Ne'eman and Dj. Šijački, Phys. Lett. 157B(1985)267.
7. Dj. Šijački, in Group Theoretical Methods in Physics, L. P. Horowitz and Y. Ne'eman eds., Ann. Isr. Phys. Soc. 3(1980)35;
Y. Ne'eman and Dj. Šijački, Phys. Lett. 157B(1985)275.
8. V. I. Ogievetsky and I. V. Polubarinov, Soviet Phys. JETP 21(1965)1093.
9. F. W. Hehl, G. D. Kerlick and P. von der Heyde, Phys. Lett. 63B(1976)446.
10. Y. Ne'eman and Dj. Šijački, Ann. Phys. (N.Y.) 120(1979)292.
11. M. B. Green and J. H. Schwarz, Phys. Lett. 136B(1984)367, 149B(1984)117, 151B(1985)21.
12. J. Scherk and J. H. Schwarz, Nucl. Phys. B81(1974)118, Phys. Lett. 52B(1974)374;
T. Yoneya, Progr. Theor. Phys. 51(1974)1907.
13. C. Lovelace, Phys. Lett. 135B(1984)75;
E. S. Fradkin and A. A. Tseytlin, Phys. Lett. 158B(1985)316.
E. Witten, Nucl. Phys. B266(1986)245.
14. D. Nemeschansky and S. Yankielowicz, Phys. Rev. Lett. 54(1985)620.
15. Y. Ne'eman and Dj. Šijački, Phys. Lett. 174B(1986)165.
16. Y. Ne'eman and Dj. Šijački, Phys. Lett. 174B(1986)171.
17. A. M. Poliakov, Phys. Lett. 103B(1981)207,211.
18. Harish-Chandra, Proc. Natl. Acad. Sci. USA 37(1951)170,362, 366,691.
19. V. Bargmann, Ann. Math. 48(1947)568.
20. Dj. Šijački, J. Math. Phys. 16(1975)298.
21. Dj. Šijački, "The Continuous Unitary Irreducible Representations of SL(4,R)"
22. A. Kihlberg, Ark. Fys. 32(1966)241.
23. Dj. Šijački and Y. Ne'eman, J. Math. Phys. 26(1985)2457.
24. Y. Dothan and Y. Ne'eman, in Symmetry Groups in Nuclear and Particle Physics, F. J. Dyson ed. (Benjamin, 1966).

SPINORS ON COMPACT RIEMANN SURFACES

C. Reina

Scuola Internazionale Superiore di Studi Avanzati (SISSA/ISAS)

Strada Costiera 11 - 34014 TRIESTE

1. Some of the results concerning spin structures on compact Riemann surfaces have been recently applied in both superstring theory and in the study of completely integrable systems. The basic technical fact in these applications is that compact Riemann surfaces are algebraic curves, (i.e. can be realized as the zero locus of homogeneous polynomial equations in complex projective space $\mathbb{C}P_N$) and spin structures on them are algebraic objects as well. Dealing with such objects is particularly pleasant, since much can be said in an algebraic geometrical set up. In this paper we shall briefly review some of the results which may be relevant in physical applications.

2. We shall identify a compact Riemann surface with an algebraic curve C. The canonical bundle K of C is simply the holomorphic cotangent bundle, with local sections of the form $f(z,\bar{z})dz$. Chiral (i.e. Weyl) local spinor fields are simply "half-forms" $\psi(z,\bar{z})(dz)^{1/2}$, and thus are sections of holomorphic square roots of the canonical bundle; in other words they are sections of the line bundles L on C such that $L^2=K$. Such a square root is called a θ-characteristics in algebraic geometry, and coincides precisely with a spin structure on C.

3. A way of showing that there are solutions to the equations $L^2=K$ is to notice that transition functions for L are square roots of the transition function (dz_α/dz_β) of K, and this involves a sign ambiguity.

A choice $S_{\alpha\beta}$ for the square roots, will define on $U_\alpha \cap U_\beta \cap U_\gamma$ a 2-cocycle $\mathcal{W}$ with $\mathbb{Z}_2$ values

$$S_{\alpha\beta} \cdot S_{\beta\gamma} \cdot S_{\gamma\alpha} = \mathcal{W}_{\alpha\beta\gamma}$$

representing the 2-nd Stieffel Witney class of C; the (mod 2)-reduction of the Euler class. Since this is even, we can conclude that $\mathcal{W}$ is a coboundary, $\mathcal{W} = \delta\eta$ and the new choice $S'_{\alpha\beta} = \eta_{\alpha\beta} S_{\alpha\beta}$ now satisfies the cocycle condition and defines a line bundle L. Notice that degL = ½ degK = g-1.

4. Next question is how many inequivalent spin structures we have on a curve C of genus g. If L_1 is one holomorphic square root of K, then any other L s.t. $L^2 = K$ can be written as $L = L_1 \otimes N$, for some degree zero line bundle N such that $L^2 = L_1^2 \otimes N^2 = K$. This N has degree zero and its square is the trivial line bundle. It is well known that degree zero line bundles are parametrized by points of a complex g-dimensional torus $H^1(C,\mathcal{O})/H^1(C,\mathbb{Z})$ which coincides with the Jacobian of C, J(C). Since J(C)=U(1)x...U(1)(2g-factors) there are 2^{2g} points $N \in J(C)$ s.t. $N^2=1$. Thus we get that many inequivalent spin structures on C.

5. The chiral (Weyl) operator acting on sections of L reduces to the Cauchy-Riemann operator $\bar\partial$. This gives us an exact sequence

$$O \to H^o(C,L) \to \Gamma(L) \xrightarrow{\bar\partial} \Gamma(L\otimes\bar{K}) \to H^1(C,L) \to O$$

where Γ(...) denotes the space of C^∞ sections of ... .
In other words $H^o(C,L) = \ker\bar\partial$ is the space of holomorphic sections of L, and $H^1(C,L) = \operatorname{coker}\bar\partial = \Gamma(L\otimes\bar{K})/\operatorname{Im}\bar\partial$. Riemann Roch theorem (i.e. the index theorem for ∂-operators on line bundles over curves) tells us that

$$\dim_{\mathbb{C}} H^o(C,L) - \dim_{\mathbb{C}} H^1(C,L) = \deg L - g + 1.$$

Since for spin structures degL = g-1, kernel and cokernel have the same dimensions. The parity of this common dimension is another invariant, so we can talk about odd and even spin structures. It turns out that out of the 2^{2g} spin structures $2^{g-1}(2^g-1)$ are odd and $2^{g-1}(2^g+1)$ are even. For instance on the Riemann sphere g=0, there is a unique spin structure of degree -1 which is necessarily even, because a negative line bundle cannot have non vanishing holomorphic sections.

6. The construction of the action for closed superstrings requires the following data

i) A conformal class of metrics {h} on a compact 2-real dimensional oriented surface S of genus g. As is well known {h} induces a complex structure turning the couple (S,{h}) into an algebraic curve C.
ii) A spin structure $L \to C$, whose sections will represent fermionic degrees of freedom on the world-sheet.

We say that two spin structure L_1, L_2 are isomorphic whenever there is a holomorphic isomorphism $\psi: L_1 \to L_2$. Clearly ψ covers an isomorphism of the bases C_1, C_2, i.e. there exists a biholomorphic map φ s.t. the diagram

$$\begin{array}{ccc} L_1 & \xrightarrow{\psi} & L_2 \\ \downarrow & & \downarrow \\ C_1 & \xrightarrow{\varphi} & C_2 \end{array}$$

commutes. The converse may be very well not true, i.e. if $\varphi : C_1 \to C_2$ is an isomorphism L_2 and $\varphi^* L_2$ will be in general non isomorphic spin structures. The set of isomorphism classes of spin structures on smooth curves of genus g is called the moduli space $\mathcal{S}_g$ of θ-characteristics. One can show that

i) $\mathcal{S}_g$ is a quasiprojective V-manifold of complex dimension 3g-3;
ii) It is disconnected, with components $\mathcal{S}_g^{\pm}$ corresponding to even and odd spin structures;
iii) $\mathcal{S}_g^{\pm}$ is a covering of the moduli space $\mathcal{M}_g$ of curves with $2^{g-1}(2g \pm 1)$ sheets.

7. Supermoduli spaces require one more bit of data, namely we want to parametrize gravitino fields up to local supersymmetries. A gravitino field χ is a section of $L^{-1} \otimes \overline{K}$ and transforms under local supersymmetry generated by a section ε of L^{-1} as $\delta\chi = \overline{\partial}\varepsilon$. As usual we have an exact sequence

$$O \to H^o(C,L^{-1}) \to \Gamma(C,L^{-1}) \xrightarrow{\partial} \Gamma(L^{-1} \otimes \overline{K}) \to H^1(C,L^{-1}) \to O$$

which identifies for us the space $H^o(C,L^{-1})$ with the space of Killing spinors, i.e. holomorphic ε's. On the other hand elements of $H^1(C,L^{-1}) = \Gamma(L^{-1} \otimes \overline{K}) / \mathrm{Im}\overline{\partial}$ are the gravitino's degrees of freedom which cannot be gauged away. From R.R. theorem we have

$$\dim_{\mathbb{C}} H^o(C,L^{-1}) - \dim_{\mathbb{C}} H^1(C,L^{-1}) = \deg L^{-1} - g+1 - = 1-g-g+1 = 2-2g$$

For g = 0, $H^1(C,L^{-1})=0$ and we have two Killing spinors. For g=1 both $H^o(C,L^{-1}) = H^1(C,L^{-1}) = \mathbb{C}$ or O while at g ≥2, $H^o(C,L^{-1}) = 0$ and $H^1(C,L^{-1}) = \mathbb{C}^{2g-2}$. These are called supermodular parameters in the physical literature.

8. To take into account these extra degrees of freedom we work on the universal curve with θ-characteristics $\mathcal{L}\to\mathcal{S}_g$; this is a fibered complex manifold over $\mathcal{S}_g$, whose fibre at s is a copy of the spin structure $\mathcal{L}$ parametrized by s itself. $\mathcal{L}$ is clearly a line bundle over the universal $\mathcal{L}\to\mathcal{C}$, $\mathcal{C}$ being itself the family of curves parametrized by $s\in\mathcal{S}$. Now the restriction of $\mathcal{L}$ to each fibre $\mathcal{C}_s$ is simply a spin structure $\mathcal{L}_s$ on $\mathcal{C}_s$, everything depending holomorphically on $s\in\mathcal{S}$.

For g>2 the direct image $\pi_!\mathcal{L} \equiv \cup_s H^1(C_s, \mathcal{L}_s)$ is a vector bundle over $\mathcal{S}$, whose fibre at s coincides with the linear space of non trivial gravitino modes, a good candidate for supermoduli space of genus g.

As a final remark, notice that we have been working holomorphically, thus neglecting anticommutativity of fermions. It should not be difficult to turn each L into a superiemann surface, of dimension (1,1), and $\pi_!\mathcal{L}$ into a supermanifold of dimension (3g-3, 2g-2) .

References

1) E. Arbarello, M. Cornalba, P.A. Griffiths and J. Harris *"Geometry of algebraic curves"* Springer-Verlag (1985), Appendix B and references quoted therein.

Simple spinors as Urfelder

by

E.R. Caianiello

Dipartimento di Fisica Teorica e S.M.S.A.

Università di Salerno

It is convenient, before addressing the subject of this talk, to establish notations and, though most briefly, remind some basic relations connecting determinants, pfaffians and anticommuting operators.Though less known than determinants, pfaffians play in our context a more fundamental role. It is determinants that descend, in several ways, from pfaffians, and not conversely;for instance, v.e.v.'s of Clifford products of fields describing both particles and antiparticles are pfaffians that degenerate into determinants. This degeneration does not occur for Majorana fields, for which they are therefore the basic, irreducible algorithm. It is convenient to use the shortened notation due to Cayley for determinants and to Jacobi (their inventor) for pfaffians. Then:

$$(1)\quad \det A = \begin{pmatrix} 1\,2\ldots n \\ 1\,2\ldots n \end{pmatrix} = \begin{vmatrix} a_{11} & \ldots & a_{1n} \\ \vdots & & \\ a_{n1} & \ldots & a_{nn} \end{vmatrix}$$

where the first row denotes row, the second column indices. If $n = 2m$ and A is antisymmetric $(hk) = a_{hk} = -a_{kh}; (h<k)$

$$(2)\quad \begin{pmatrix} 1\,2\ldots n \\ 1\,2\ldots n \end{pmatrix} = (1\;2\;\ldots\;2m)^2$$

where the pfaffian (1...2m) is represented by a triangular array and expanded, in the simplest form, as:

$$(3)\quad (1\,2\ldots 2m) = \begin{vmatrix} (12) & (13)\ldots & (1\,2m) \\ & (23)\cdots & (2\,2m) \\ & & \vdots \\ & & (2m-1,2m) \end{vmatrix} = \sum_{h=2}^{2m} (-1)^h (1h)(2,\ldots,h-1,h+1,$$

It has only "lines", line j being formed by all elements with j as first or second index. Denote the ordinary (Clifford) product of two anticommuting fields with the symbol $\wedge\!\!\!\!\cdot\;$ and their antisymmetrized (Grassmann) product with the symbol $\wedge$ (we keep, for convenience, the notation used in the references cited). Then

$$(4)\quad \psi^{(1)}(x_1)\wedge\psi^{(2)}(x_2) = \psi^{(1)}(x_1)\wedge\psi^{(2)}(x_2) + (12),$$

where (12) is the <u>contraction</u> of the two fields; time ordering would give the original form od Wick's theorem, $\psi^{(1)}\wedge\psi^{(2)}$ being then the "normal product". Then:

$$(5)\quad \psi^{(1)}(x_1)\wedge\dots\wedge\psi^{(k)}(x_k) = \sum_{r=0}^{[k/2]}\sum_{C_r}(-1)^P(h_1\dots h_{2r})\,\psi^{(e_1)}(x_{e_1})\wedge\dots\wedge\psi^{(e_{k-2r})}(x_{e_{k-2r}})$$

where $h_1<\dots<h_{2r}$, $e_1<\dots e_{k-2r}$ any combination of 2r indices h_i out of k; e_j the remaining ones, in natural order.

The coefficients of expansion (5) are pfaffians, with elements (hk) defined by (4) (according to the situation, free propagators of a type or another, or pair correlation functions). Taking the v.e.v. of (5) destroys at its r.h.s., because of the antisymmetry of Grassmann products, all terms of the expansion with more than one field; thus (omitting from now on the symbol $\wedge$):

$$(6)\quad \langle 0|\psi^{(1)}(x_1)\dots\psi^{(2m)}(x_{2m})|0\rangle = (12\dots 2m)$$

$$(7)\quad \langle 0|\psi^{(1)}(x_1)\dots\psi^{(2m-1)}(x_{2m-1})|0\rangle = (012\dots 2m-1)$$

where, by definition:

$$(8)\quad \langle 0|\psi^{(h)}(x_h)|0\rangle = (0h)$$

* * *

Work done in the early fifties [1,2] on the problem of finding compact perturbative expansions in configuration space of propagators, or Green functions, for any theory coupling anticommmuting and/or commuting fields required the evaluation of expectation values of ordinary (Clifford) products of free Fermi or Majorana fields; any such v.e.v. is a pfaffian whose elements are 2-point free propagators: this was found by expressing Clifford products as linear combinations of the corresponding Grassmann products, whose coefficients are these pfaffians (a general property that contained, as a special case, the well known Wick theorem). A concise account of relevant properties then found for pfaffians and of various applications of them,with special regard to field theory is contained in ref. (2) (analogous tools for treating commuting fields were also invented, but do not interest us here); in it and in the references quoted therein further details may be found.

The formalism then developed, which permitted the complete solution of that problem as reported in the references, has quite recently become again an object of interest, because it exactly coincides with that which emerged from the structure of "simple spinors". Our interest in them [3] was due to the fact that ordinary spinors, such as we naturally take as wave functions (e.g. satisfying Dirac equations in any number of dimensions) and think of as non reducible to simpler objects (vectors come from spinors, but not conversely) and forming a linear space, can always be expressed as linear combinations of "simple spinors", as was shown by Cartan's classic work. Simple spinors do not form a linear space (the sum of two simple spinors is not simple), and are subject to constraints that reduce the number of independent components much below that which is standard with spinors; these constraints were explicitly given by Cartan as recursion formulae.[4]

Our approach to quantum mechanics as a geometry in 8-dimensional relativistic phase space [5] led us to pay particular attention to the rôle of spinors in the context that then arises, because, as Cartan pointed out, the dimension "eight" is unique, since it is the only one in which vectors and semispinors of first and second kind have the same number of components, behave in the same way under rotations and, under operations of G_2, transform into one another: the triality principle of Cartan, i.e. a supersymmetry often referred to in the recent literature (to these features several others should be added, e.g. the direct connection of our quantum geometry with that of information theory). These things came out, quite unexpectedly, from the sole requirement of phase-space geometrization of quantum mechanics. A closer look at simple spinors seemed called for, and yielded the result, surprising for us, on which we wish to comment here: the components of simple spinors are pfaffians.[3]

This automatically implies that they, as will be apparent, are structured as expectation values of products of "basic" 2-component Weyl spinors, kinematically independent from one another and each living in a space of its own. A simple spinor appears in this perspective as a "mean field"; standard spinors are linear combinations of them in the same way as a tensor is a linear combination of multivectors. This fact holds also if one forgets about phase-space quantum mechanics, in any number of dimensions. The difference with ordinary spinors starts when the number of dimensions is greater than five, and dramatically increases with it. It appears therefore that basic 2-component spinors, which recall to mind Heisenberg's Urfelder, do not have to be postulated, but have been there all the time, whenever one has dealt with spinors, especially of high dimensions. (The 8-dimensional case has one more appeal, because then semispinors are octonions and, as already recalled, triality links them with vectors). Basic fields of this kind are not, so to say, "beyond" ordinary space, but "behind" it.

Since components of simple spinors in dimensions greater than five are functionally linked among themselves, one might want to find, starting from a Dirac equation, equations that contain only the independent ones; the outcome can only be highly non linear equations. We shall try to test here, instead, the possibility that such "basic" fields satisfy per se linear equations, each in its own space, and the consequences that this approach has on the simple spinors which derive from them.

Simple spinors satisfy, in $N=2\nu$ or $2\nu+1$ dimensions, the equation[4]

$$X\xi = 0 \quad , \quad X = \gamma_\mu x_\mu \tag{9}$$

or equivalently

$$\xi^T C X_{(p)} \xi = 0 \quad ; \quad p<\nu \, , \; \nu - p \equiv 0 \text{ or } 3 \pmod{4} \tag{10}$$

where $X_{(p)}$ denotes an arbitrary p-vector and

$$C\,\xi_{i_1\ldots i_p} = (-1)^{\frac{p(p+1)}{2}} \xi_{i_{p+1},\ldots, i_\nu} \tag{11}$$

The supernumerary conditions (9) lead to the relations

$$\xi_0 \xi_{i_1\ldots i_p} = \begin{cases} \sum_{k=1}^{p-1} (-1)^{k-1} \xi_{i_k i_p} \xi_{i_1 \ldots i_{k-1} i_{k+1} \ldots i_{p-1}} & (p \text{ even}) \\ \sum_{k=1}^{p} (-1)^{k-1} \xi_{i_k} \xi_{i_1 \ldots i_{k-1} i_{k+1} \ldots i_p} & (p \text{ odd}) \end{cases} \tag{12}$$

(which define a maximal isotropic ν -plane; all this is masterly explained in Cartan's cited works)

By setting

$$(0h) = \xi_h / \xi_0 \quad , \quad (hk) = \xi_{hk} / \xi_0 \tag{13}$$

one finds then[3]

$$\begin{cases} \xi_{h_1 \ldots h_{2k}} = \xi_0 \cdot (h_1 \ldots h_{2k}) \\ \xi_{h_1 \ldots h_{2k-1}} = \xi_0 \cdot (0 h_1 \ldots h_{2k-1}) \end{cases} \tag{14}$$

In particular, when $\nu = 4$, Cartan's conditions for semispinors to be simple, which read in our notation, with $(\ddot{0}\ddot{0}) = 0$

$$\begin{cases} \xi_{1234} = \xi_0 \cdot (1234) & \text{(I- type)} \\ (\ddot{0}\ddot{0}1234) = 0 & \text{(II- type)} \end{cases} \tag{15}$$

are identically satisfied by (1). Semispinors of first (second) type have an even (odd) number of indices. The "coincidence" which it is our purpose to point out here is the fact that <u>expression (14) is the exact counterpart of</u> (7) (identical if $\xi_0 = 1$)

These fields bring to mind Heisenberg's Urfelder, but differ in fact from them in essential ways. Heisenberg's fields refer to ordinary spacetime; their cartesian products can be made to correspond to anything wanted. The fields we are discussing here share with them only the property of being 2-component spinors. They live each in a space of its own, of dimension 3 (which may reduce to 2); these spaces are kinematically disjoint; their number is ν if the space used as the stage for the physical theory or model has dimension $N = 2\nu$ or $2\nu+1$. The simple spinors that will derive from them by taking v.e.v.'s of their Clifford products are "mean fields" and form a non linear manifold of dimension $\binom{\nu+1}{2}$. With them one can construct all linear 2^{ν}-component spinors, customarily considered as the basic primitive objects. The most important fact is, in our perspective, that they are a necessary, though unexpected, consequence of our geometrical approach to quantum mechanics, and not the result of postulation; each semispinor is then an octonion (Hurwitz's theorem on norm may be taken as a no-go theorem against further extensions).

We shall construct spinors of higher dimensions by taking direct products, rather than direct sums as is more common; this procedure offers a simple recipe for obtaining a number of unorthodox factorizations of Dirac equations and their spinor solutions, of which a few examples will be briefly shown. This fact may be of interest *per se*, and the object of future work, as it affects the standard 4-dimensional Dirac or Majorana equations to begin with. In the next Section it will be shown how to operate on basic spinors, when considered as q-numbers, to generate from them "simple" spinors which, as mean fields on a "vacuum", still obey Dirac equations

  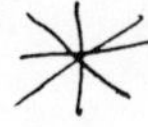

Consider a sequence of 2-component spinors, in disjoint and kinematically independent spaces; variables (coordinates and arguments of functions) are mere mathematical symbols, whose meanings are to be decided *a posteriori*, when "system identification" is performed. "Vectors" denote likewise, as long as convenient, symbols that transform like vectors. Space no. 1 plays a privileged role; the 2-component spinor is defined in it as satisfying the equation:

$$\sigma_h q^h_{(1)} \psi^{(1)} = q^4_{(1)} \mathbb{1} \psi^{(1)}, \tag{16}$$

where the full Pauli algebra is used. Next come spaces in each of which the corresponding 2-spinor satisfies an equation that contains only the generators of the Pauli algebra:

$$\sigma_3 q^3_{(2)} \psi^{(2)} = (\sigma_1 q^1_{(2)} + \sigma_2 q^2_{(2)}) \psi^{(2)}, \tag{17}$$

$$\sigma_3 q^3_{(3)} \psi^{(3)} = (\sigma_1 q^1_{(3)} + \sigma_2 q^2_{(3)}) \psi^{(3)}, \tag{18}$$

etc. The direct product of (16) and (17) gives

$$\sigma_h \otimes \sigma_3 q^h_{(1)} q^3_{(2)} \psi^{(1)} \otimes \psi^{(2)} = \left[\mathbb{1} \otimes \sigma_1 q^4_{(1)} q^1_{(2)} + \mathbb{1} \otimes \sigma_2 q^4_{(1)} q^2_{(2)} \right] \psi^{(1)} \otimes \psi^{(2)}; \tag{19}$$

clearly, $\sigma_i \otimes \sigma_3$, $\mathbb{1} \otimes \sigma_1$ and $\mathbb{1} \otimes \sigma_2$ in the 2x2 space spanned by the 4- spinor $\psi^{(1)} \otimes \psi^{(2)}$ (incorporating according to need signs and factors "i" into coefficients) yield all, and only, the generators $\gamma_1, \gamma_2, \gamma_3, \gamma_4$ of the ordinary 4-dimensional Clifford algebra, plus γ_5. To within changes of indices or reversals of direct products, all classic representations of Dirac matrices can be obtained in this way [6]. Iteration can go on indefinitely; we stop at $\nu = 4$, since our phase space quantization requires just four basic spinors (or conjugate spinors).

Space 1 differs from the others because it is the only one in which the spinor, as defined by eq. (16), is <u>not</u> simple in Cartan's sense,i.e. does not correspond to an isotropic vector $q^2=0$ (with suitable adjustments of signs and factors). Unless, of course, one of its components vanishes; the same pattern can be followed, at any level of iteration ($\sigma_3 = i\sigma_1\sigma_2$). One sees that to obtain a mass term, or equivalently a γ_5, in a Dirac equation for a 2^ν-spinor thus constructed, it is required that one,and only one, of the factors be not a simple 2-spinor. The issue is not very relevant, because (see below), the sum of two simple massless spinors gives a massive non simple spinor; keeping space 1 so privileged is only a convenient option. Simplicity of basic spinors (excepting the first) is required for the constructions of this Section, but not for the next one.

<u>Examples</u>

a) <u>Mass from massless simple spinors</u>. Consider eq.'s (16) and (17). With easy <u>ad hoc</u> adjustments they yield the free Dirac equation (covariant derivatives would change nothing in the argument)

$$\gamma_\mu \partial_\mu \psi = i m \psi$$

Take

$$\psi(x) = \exp[i(k_1+k_2)x]\cdot[\varphi_1(k_1,k_2)+\varphi_2(k_1,k_2)],$$

so that

$$(\gamma k_1 + \gamma k_2)(\varphi_1+\varphi_2) = m\varphi_1 + m\varphi_2$$

it suffices to choose

$$\gamma k_1 \varphi_1 = m \varphi_2 \quad , \quad \gamma k_1 \varphi_2 = 0$$

$$\gamma k_2 \varphi_1 = 0 \quad , \quad \gamma k_2 \varphi_2 = m \varphi_1$$

to have: φ_2 simple with respect to the isotropic vector k_1, φ_1 with respect to isotropic k_1 and $m^2 = 2k_1 \cdot k_2$. Conversely, given any simple spinor, it is immediate to associate to it, by reversing the above procedure, another one such their sum satisfies a Dirac equation with any wanted mass.

b) A Majorana equation. Again, we might write (16) and (17) as

$$\sigma_h \partial_h \psi^{(1)} = -\partial_t \psi^{(1)}$$

$$\sigma_3 \psi^{(2)} = \pm i \sigma_1 \psi^{(2)}$$

then
$$\varphi = \psi^{(1)} \otimes \psi^{(2)}$$

satisfies the Majorana conditions

$$\begin{cases} \gamma_\mu \partial_\mu \varphi = 0 \\ \dfrac{1 \pm \gamma_5}{2} \varphi = 0 \end{cases}$$

c) "Forward" and "backward" factorization. Another example obtains by taking in (16)

$$q^1_{(1)} = \partial_u \quad , \quad q^2_{(1)} = \partial_{v_+} \quad , \quad q^3_{(1)} \sim m$$

and in (17)

$$q^3_{(2)} = 1 \quad , \quad q^1_{(2)} = \partial_{\bar{u}} \quad , \quad q^{(2)}_{(2)} = \partial_{v_-}$$

with

$$u = x+iy\ ,\ \bar{u} = x-iy\ ,\ v_+ = z-t\ ,\ v_- = z+t\ .$$

A little adjusting of numerical factors and passing to the conjugate representation $\gamma_\mu \to i\gamma_\mu\gamma_5$ yields again the free massive equation

$$\gamma_\mu \partial_\mu \varphi = im\varphi$$

with $\quad \varphi = \varphi^{(1)}(x+iy, z-t) \otimes \varphi^{(2)}(x-iy, z+t)\ .$

This type of decomposition should be of interest in phase space, where it might apply to (q,p) coordinates.

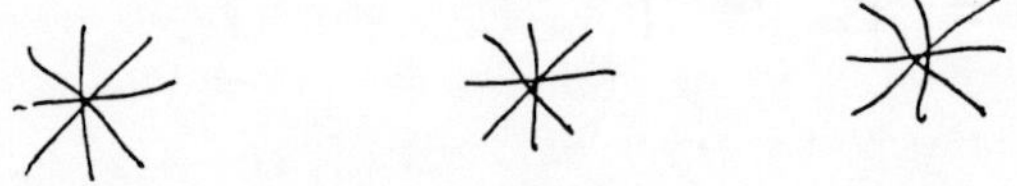

The factorizations of a Dirac equation into corresponding ones for independent, basic 2-fields, of which we have given examples, remain valid when the latter are considered as operators representing "free" fields, i.e. having gaussian distributions. We consider here only the case of Fermi statistics, with N=8, as an example to exhibit the "engineering" of matching expressions (7) and (14). The key to it is, of course, the fact that the contraction of two such fields is a c-number, as in (5). It must be emphasized that many ways are open; the two components $\varphi_0^{(k)}$ and $\varphi_1^{(k)}$ of each basic spinor can be given any wanted behaviour in passing to a q-theory: thus, $\varphi_0^{(k)}$ can be supposed to commute or anticommute with $\varphi_1^{(k)}$, to remain a c-number, to be an operator of which the "vacuum" is an eigenstate as with overcomplete fields... Our purpose here is only to show a working example, crude and simple as possible; we use the fact that spinors (whether basic or 2^ν- simple spinors) are <u>projective</u> objects, which permits to choose a basis (which may changed at will by multiplying by any wanted factor) that greatly expedites our proposed identification: we take, that is, as primary objects, instead of $(\varphi_0^{(k)}, \varphi_1^{(k)})$, the others $\left(1, \xi^{(k)} = \frac{\varphi_1^{(k)}}{\varphi_0^{(k)}}\right)$, i.e. the unity and and complex operators $\xi^{(k)}$ of which we assume that they satisfy eq.s (7) and (8). This gives us, conveniently, $\xi_0 = 1$ in (14) and (15) securing thus their complete identification; a quite suggestive feature emerges out of this choice, as it would make applicable to general cases the interesting remark by P. Budinich[7] that 2-spinors offer a formalism <u>equivalent to that of strings</u>, in a context of extreme geometrical clarity, where "mathematical points" do not any more stand for physical particles. We hope to expand on this issue in the future, and note here only that, should this conjecture prove acceptable, it would again appear as a straightforward consequence of phase space geometrization, and be necessarily congruent with other consequences of that approach, e.g. mass spectra for bosons and fermions in excellent agreement with available data[8].

With the choice made here for basic spinors, the wanted results are read out immediately from the cited formulae. The 16-component pure spinor of 8-dimensional phase space consists of two semispinors, as follows (we have taken $\xi_0 = 1$):

I -type: components $(\xi_0, \xi_{12}, \xi_{13}, \xi_{14}, \xi_{23}, \xi_{24}, \xi_{34}, \xi_{1234})$

II -type: components $(\xi_1, \xi_2, \xi_3, \xi_4, \xi_{123}, \xi_{124}, \xi_{134}, \xi_{234})$

where: $$\xi_{1234} = \xi_{12}\xi_{34} - \xi_{13}\xi_{24} + \xi_{14}\xi_{23}$$

$$\xi_{123} = \xi_1\xi_{23} - \xi_2\xi_{13} + \xi_3\xi_{12}$$

etc. Altogether, 4 + 7 independent components; the remaining ones have the direct meaning of <u>volumes</u> (being square roots of determinants). It is possible to choose a special representation that annihilates one semispinor, e.g. as in (9) or by arranging that in (8) all $\langle 0|\varphi^{(4)}|0\rangle = 0$, one then would of course obtain further reductions on the number of truly independent components of simple spinors. This does not seem necessary at this point, only further analyses can tell how things should stand in the present perspective.

In conclusion, the author wishes to express his warmest thanks to Prof. P. Budinich for interesting discussions on Cartan's simple spinors; he is also indebted to Prof.s P. Furlan, G. Marmo. A. Trautman. J.A. Wheeler and many others.

1 - E.R Caianiello, Nuovo Cimento 10,1634 (1953);11,492 (1954)

2 - E.R. Caianiello,"Combinatorics and renormalization in Quantum field Theory", Benjamin,Mass. 1973

3 - E.R. Caianiello, A. Giovannini,Lett. Nuovo Cimento, 34,301 (1982)

4 - E. Cartan, "La theorie des Spineurs", Hermann, Paris (1938) .Note that Cartan' denomination "spineurs simples" is badly translated in the customary "pure spinors": we retain Cartan's, and speak of "simple spinors".

5 - E.R. Caianiello, Nuovo Cimento B,59,350 (1980);with G. Marmo and G. Scarpetta,Nuovo Cimento A,86,337 (1985); with G. Di Genova, "Some consequences of phase space quantum geometry" in "Quantum Field Theory",F. Mancini ed.,Elsevier,1986. And references quoted.

6 - E. Corinaldesi and F. Strocchi, "Relativistic wave mechanics",North Holland,Amsterdam,1963

7 - P. Budinich, ISAS Rep. 10/86/E.P.

8 - E.R. Caianiello and W. Guz, Lett. Nuovo Cimento,43 (1985)

9 - P. Budinich, Phys. Reports,137,35 (1986)

Applications of Cartan Spinors to Differential Geometry in Higher Dimensions

by

L.P. Hughston

Lincoln College, Oxford OX1 3DR, England

May 1987

1. Introduction

Higher dimensional spinors, particularly the 'pure' spinors of E. Cartan (1937), are useful as a tool in the study of geometrical problems in higher dimensional spaces. This is on account of the intimate relationship between spinors and the conformal geometry of such spaces.

As an illustration I propose to show how the theorems of Robinson (1961) and Kerr (cf. Penrose 1967, and Penrose & Rindler 1986, §7.4) on null solutions of Maxwell's equations can be generalized to any even-dimensional flat space-time.

2. Twistor Geometry in Four Dimensions

Let us consider first the case of four dimensional flat complex space-time, which may be regarded as a quadric surface Q^4 in CP^5 (see e.g. Hughston & Hurd 1983). The natural conformal metric on Q^4 is built up from the geometry of the quadric: two points are null-separated if and only if each lies on the tangent plane of the other.

In Q^4 there are two families of totally null two-planes, called α-planes and β-planes. The aggregate of all α-planes forms a manifold CP^3 which is the *twistor space* associated with Q^4. By virtue of this association structures in Q^4 may be correlated with corresponding structures in CP^3.

For example, a point x in Q^4 corresponds to a complex projective line (one-dimensional linear space) in CP^3, whereas each point on that line corresponds to an α-plane in Q^4 that passes through x. The one-dimensional system of α-planes through a given point x in Q^4 is the *projective spin space* associated with that point. A *projective spinor field* on a region R of space-time consists of an assignment to each point $x \in R$ of an α-plane through x.

3. Robinson's Theorem in Four Dimensions

This theorem is concerned with the construction of null solutions of Maxwell's equations — whereas Robinson's original result is for an arbitrary normal hyperbolic pseudo-Riemannian manifold, we shall consider the corresponding result in a flat complex manifold. Robinson's theorem shows in effect that any analytic null solution of Maxwell's equation can be represented by a symmetric spinor field $\phi_{A'B'} = e^{\psi}\xi_{A'}\xi_{B'}$, for a suitable choice of ψ, where $\xi_{A'}$ satisfies the equation

$$\xi^{A'}\xi^{B'}\nabla_{AA'}\xi_{B'} = 0 \ . \qquad (3.1)$$

It is a straightforward matter to verify that this equation is invariant under a rescaling of the form $\xi_{A'} \to \lambda(x)\xi_{A'}$. Thus (3.1) represents a differential equation for a projective spinor field in the sense described above in §2.

4. The Kerr Theorem in Four Dimensions

This theorem provides a local construction for all analytic solutions of equation (3.1) by consideration of the zero set of a homogeneous holomorphic function $F(Z^{\alpha})$, where Z^{α} represent homogeneous coordinates for CP^3. Thus solutions of (3.1) are represented locally by *analytic varieties* in CP^3.

According to standard conventions (cf. Penrose & MacCallum 1972, Penrose 1975, Penrose & Rindler 1976) we may represent Z^{α} by the spinor pair $(\omega^A, \pi_{A'})$. Then according to the Kerr theorem solutions of (3.1) are spinor fields $\xi_{A'}(x)$ which satisfy

$$F(ix^{AA'}\xi_{A'}(x), \xi_{A'}(x)) = 0 \qquad (4.1)$$

for some $F(\omega^A, \pi_{A'})$. It is a straightforward exercise to verify that if $\xi_{A'}(x)$ satisfies (4.1), then it also satisfies (3.1) - see, for example, Hughston 1979, §10.1; conversely, any spinor field $\xi_{A'}(x)$ which satisfies (3.1) can be represented locally as a solution of (4.1) for a suitable choice of $F(Z^{\alpha})$.

In geometrical terms the space-time point $x^{AA'}$ is represented by a complex projective line L in CP^3. The vanishing of F determines a variety V in CP^3. The line L intersects V at a point on L which represents a primed projective spinor in the tangent space of $x^{AA'}$. In this way the variety V determines a primed projective spinor field over a region of Q^4.

5. Robinson's Theorem in Six Dimensions

Before going on to consider the case of general even dimension it will be instructive to consider first the case of six dimensions in some detail.

In this dimension a self-dual three-form corresponds to a symmetric spinor ϕ_{AB} (A,B = 1...4). The three-form is *closed and co-closed* if and only if $\nabla^{AB}\phi_{BC} = 0$, where $\nabla^{AB} = \partial/\partial X_{AB}$. Here $X_{AB} = -X_{BA}$ (six independent components) represent coordinates for the six-dimensional space-time. On the other hand the three-form is *null* if and only if $\phi_{AB} = \xi_A \xi_B$ for some spinor field $\xi_A(X)$. If the three-form is therefore both null and closed it follows, after a short calculation, that

$$(\xi_A \nabla^{AB} \xi_{[C}) \xi_{D]} = 0 \ . \tag{5.1}$$

Conversely given any analytic solution of (5.1), the scale of ξ_A may be chosen such that $\phi_{AB} = \xi_A \xi_B$ satisfies $\nabla^{AB}\phi_{BC} = 0$. Such a solution will be called a *null Maxwell field* in six dimensions.

6. The Kerr Theorem in Six Dimensions

Now we wish to consider the problem of finding the general solution of (5.1) in six dimensions. Thus we must consider the twistor space appropriate for six dimensions. We represent space-time by a quadric Q^6, in which there are two families of self-dual three-planes, which again we shall call α-planes and β-planes. The aggregate of all α-planes constitutes a six-dimensional quadric Q^6_α. This is the twistor space for Q^6. A twistor for Q^6 can be represented by a pair of spinors (ω^A, π_A) satisfying $\omega^A \pi_A = 0$. We may think of (ω^A, π_A) as homogeneous coordinates for CP^7 (recall $A = 1 \ldots 4$), and $\omega^A \pi_A = 0$ then determines the quadric Q^6_α in CP^7. Each analytic solution of (5.1) corresponds locally to an analytic variety V^3 of dimension 3 in Q^6_α.

THEOREM *Let* $F^r(\omega^A, \pi_A) = 0$ $(r = 1,2,3)$ *determine an analytic variety of dimension* 3 *in the projective quadric* $\omega^A \pi_A = 0$, *and suppose the spinor field* $\xi_A(X)$ *defined on a region* R *in* Q^6 *satisfies*

$$F^r(X^{AB}\xi_B, \xi_B) = 0 \tag{6.1}$$

for each $X^{AB} \in R$. *Then* ξ_B *satisfies* (5.1).

Proof. Since $F^r(\omega^A, \pi_A)$ is by hypothesis homogeneous of degree (say) n_r in (ω^A, π_A) we have

$$(\omega^A \hat{\omega}_A + \pi_A \hat{\pi}^A) F^r = n_r F^r = 0$$

on the variety $(\hat{\omega}_A \equiv \partial/\partial\omega^A,\ \hat{\pi}^A \equiv \partial/\partial\pi_A)$; whence

$$\xi_A(-X^{AB}\hat{\omega}_B + \hat{\pi}^A)F^r = 0$$

by the substitution $(\omega^A, \pi_A) = (X^{AB}\xi_B, \xi_A)$. If we write

$$R^{rA} = (-X^{AB}\hat{\omega}_B + \hat{\pi}^A)F^r$$

we have

$$\xi_A R^{rA} = 0 \qquad (r = 1,2,3) \ . \tag{6.2}$$

Furthermore, by differentiation of (6.1) we obtain

$$\nabla^{RS}F^r = 0$$

whence

$$\nabla^{RS}(X^{AB}\xi_B)\hat{\omega}_A F^r + \nabla^{RS}(\xi_A)\hat{\pi}^A F^r = 0 \ ,$$

which, after some simplification by use of $\nabla^{AB}X^{CD} = \varepsilon^{ABCD}$, yields

$$(\nabla^{RS}\xi_A)R^{rA} + \varepsilon^{RSAB}\xi_B\hat{\omega}_A F^r = 0 \ ;$$

and by transvection of this relation with ξ_R we get:

$$(\xi_R\nabla^{RS}\xi_A)R^{rA} = 0 \qquad (r = 1,2,3) \tag{6.3}$$

Now since R^{rA} $(r = 1,2,3)$ are linearly independent at generic points of the variety it follows from (6.2) and (6.3) that

$$\xi_R\nabla^{RS}\xi_A = \lambda^S\xi_A$$

for some λ^S, from which it follows at once that ξ_A satisfies (5.1). □

Geometrically the point X^{AB} in Q^6 corresponds to a CP^3 in Q^6_α (see e.g. Hughston 1987 for details of this correspondence). The CP^3 thereby determined intersects V^3 at a point which in turn corresponds to a null three-plane back in Q^6. This null three-plane goes through the point X^{AB} and hence determines a projective spinor at that point. In this way V^3 generates a spinor field on a region of Q^6.

7. The Kerr-Robinson Theorem in 2n Dimensions

Remarkably a similar result can be formulated in any even dimension. By a null Maxwell field in dimension 2n we mean a self-dual n-form which is closed and totally null. Such a field can be represented by a spinor field $\xi^{A'}(X^i)$, where $i = 1...2n$ and $A' = 1...m$ where $m = 2^{n-1}$. The spinor field $\xi^{A'}$ must satisfy Cartan's 'purity' condition

$$\Gamma^{iA}_{A'}\Gamma^{B}_{iB'}\xi^{A'}\xi^{B'} = 0 \tag{7.1}$$

and the 'foliation' condition

$$(\xi^{A'}\Gamma^{iA}_{A'}\nabla_i\xi^{[B'})\xi^{C']} = 0 \;, \tag{7.2}$$

where $\Gamma^{iA}_{A'}$ is the reduced gamma matrix (Van de Waerden symbol) appropriate to dimension 2n. (My notational conventions here are essentially those of Penrose & Rindler 1986, in their appendix on higher dimensional spinors.)

The space-time can be represented by a quadric Q^{2n}, and the associated 'pure twistor space' consists of the aggregate of α-planes (totally null self-dual n-planes) in Q^{2n}. These constitute a manifold of dimension $n(n+1)/2$ which we shall denote $PT^{n(n+1)/2}$ (projective twistor space).

Solutions of (7.2), with $\xi^{A'}$ subject to the purity condition (7.1), are determined by analytic varieties of dimension n in $PT^{n(n+1)/2}$. The construction of $\xi^{A'}$ is essentially as follows. A point X in Q^{2n} is represented in the pure twistor space $PT^{n(n+1)/2}$ by an algebraic manifold $X^{n(n-1)/2}$ which corresponds to the set S_X of all α-planes through X in Q^{2n}. (The manifold $X^{n(n-1)/2}$ has in fact the structure of the twistor space $PT^{n(n-1)/2}$ for the space Q^{2n-2}. This is because Q^{2n-2} has the structure of the projective light-cone for a point in Q^{2n}.) The manifold $X^{n(n-1)/2}$ intersects the analytic variety V^n at a point ξ which in turn determines a self-dual null n-plane through the original point X. In this way for each point X in the relevant region we obtain an associated α-plane, and hence a projective spinor field. For a proof that the resulting spinor field satisfies (7.2) see, for example, Hughston & Mason (1987), which contains also a more general discussion of spinors in higher dimensions and their properties.

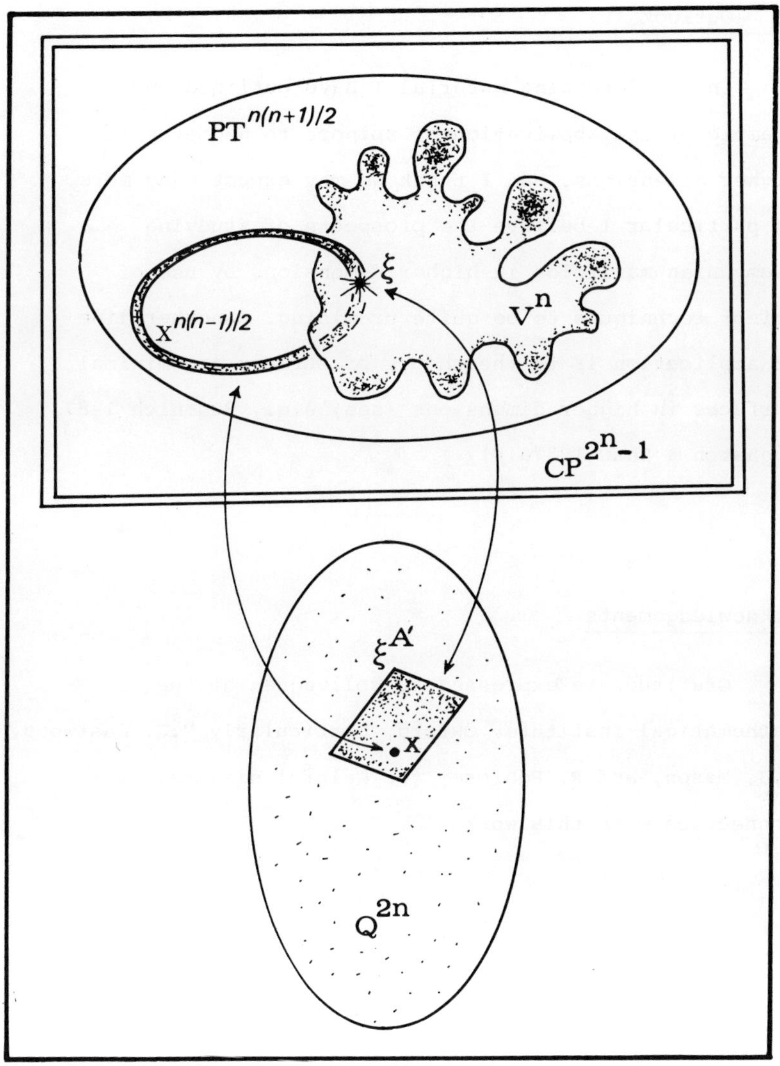

Figure: The Kerr-Robinson Theorem in Dimension 2n. *A point X in space-time Q^{2n} is represented by an algebraic manifold $X^{n(n-1)/2}$ in the pure projective twistor space $PT^{n(n+1)/2}$, which is itself an algebraic manifold in the general projective twistor space CP^m where $m = 2^n-1$. The intersection of $X^{n(n-1)/2}$ with the analytic variety V^n determines a point ξ of $X^{n(n-1)/2}$ and hence an alpha-plane $\xi^{A'}$ through the original point X of Q^{2n}.*

8. Outlook

In the foregoing material I have outlined one example of the application of spinors to geometry in higher dimensions, and I think we may expect many more. In particular I believe the prospects of studying Riemannian manifolds in higher dimensions by use of spinor techniques to be quite promising. Another line of application is to the theory of strings and minimal surfaces in higher dimensions (see, e.g., Budinich 1987, Hughston & Shaw 1987a,b).

Acknowledgements

Gratitude is expressed to colleagues at the Mathematical Institute, Oxford, particularly M.G. Eastwood, L.J. Mason, and R. Penrose, for helpful discussions in connection with this work.

References

Budinich, P. 1987, Null Vectors, Spinors, and Strings, *Commun. Math. Phys.* 107, 455-465.

Cartan, E. 1937, *Leçons sur la théorie des spineurs*, trans. 1966, *The theory of spinors*, Hermann, Paris. (Reprinted by Dover Publications, New York, 1981.)

Hughston, L.P. 1979, *Twistors and Particles*, Springer Lecture Notes on Physics, Vol. 97.

Hughston, L.P. 1987, Applications of SO(8) Spinors. In *Gravitation and Geometry, a volume in honour of I. Robinson* (ed. Rindler, W. & Trautman, A.) Bibliopolis, Naples.

Hughston, L.P. & Hurd, T.R. 1983, A CP^5 Calculus for Space-Time Fields, *Phys. Repts.* 100, 273-326.

Hughston, L.P. & Mason, L.J. 1987, A Generalized Kerr-Robinson Theorem.

Hughston, L.P. & Shaw, W.T. 1987a, Minimal Curves in Six Dimensions, *Class. Quantum Grav.* 4, no. 4.

Hughston, L.P. & Shaw, W.T. 1987b, Classical Strings in Ten Dimensions.

Penrose, R. 1967, Twistor Algebra, *J. Math. Phys.* 8, 345-366.

Penrose, R. & MacCallum, M.A.H. 1972, Twistor Theory: An Approach to the Quantization of Fields and Space-Time, *Physics Reports* Vol. 6C, No. 4.

Penrose, R. 1975, Twistor Theory: Its Aims and Achievements. In *Quantum Gravity: An Oxford Symposium* (ed. C.J. Isham, R. Penrose and D.W. Sciama). Oxford University Press.

Penrose, R. & Rindler, W. 1986, *Spinors and space-time Vol. 2: Spinor and Twistor Methods in Space-time Geometry*. Cambridge University Press.

Robinson, I. 1961, Null Electromagnetic Fields, *J. Math. Phys.* 2, 290-291.

KILLING SPINORS ON SPHERES AND PROJECTIVE SPACES

S.GUTT

Chercheur qualifié au F.N.R.S.

The general results presented in my talk about Killing spinors on a manifold were obtained in collaboration with M.Cahen, L.Lemaire and P.Spindel; they have appeared in [1]. A Killing spinor is a spinor field ψ, defined on a Riemannian manifold which admits a spin structure, that satisfies a differential equation $\nabla\psi = \lambda\gamma\psi$ where λ is a constant. I shall just recall in these notes the definition involved and mention the explicit construction of Killing spinors on spheres and projective spaces; it leads to the triviality of the full spinor bundle on those spaces.

1. Definitions and notations

Let V be the real Euclidean space of dimension n with scalar product denoted $< , >$; let $\{ e_a ;\ a = 1,\ldots,n \}$ be an orthonormal basis of V and let C_n be the Clifford algebra of V, i.e. $C_n = \mathcal{E}(V)/I$ where $\mathcal{E}(V)$ is the tensor algebra and I is the ideal generated by $x \otimes y + y \otimes x + 2 \langle x,y \rangle 1$. We have $C_n = C_n^+ \oplus C_n^-$ where C_n^+ (resp. C_n^-) is the image of the space of tensors on V of even (resp. odd) degree.

If V is even dimensional, n = 2m, then $C_{2m}^{\mathbb{C}}$ (the complexification of C_{2m}) is isomorphic to the algebra of all linear endomorphisms of the exterior algebra ΛW of an isotropic subspace W of $V^{\mathbb{C}}_n$. If W is generated by $\{ f_k = e_{2k-1} + ie_{2k} ;\ 1 \leq k \leq m \}$ a realization $\tilde{\rho}$ of the Clifford multiplication is given by

$$\overset{2m}{\tilde{\rho}}(e_{2k-1}).\alpha = f_k \wedge \alpha - i(f_k^*)\alpha \qquad \forall \alpha \in \Lambda W$$

$$\overset{2m}{\tilde{\rho}}(e_{2k}).\alpha = -\sqrt{-1}(f_k \wedge \alpha + i(f_k^*)\alpha)$$

where $i(f_k^*)f_\ell = \delta_{k\ell}$

The space $S_{2m} = \Lambda W$ is called the space of spinors for V; it has complex dimension 2^m and $S_{2m} = S_{2m}^+ \oplus S_{2m}^-$ where S_{2m}^+ (resp. S_{2m}^-) is the space of even (resp. odd) forms on W; clearly $C^+S^\pm \subset S^\pm$ and $C^-S^\pm \subset S^\mp$.

If V is odd dimensional $n = 2m+1$, then $C^{\mathbb{C}}_{2m+1}$ is isomorphic to $\mathrm{End}(S_{2m+1}) \oplus \mathrm{End}(S'_{2m+1})$ where S_{2m+1} and S'_{2m+1} have complex dimension 2^m. This results from the isomorphism $C_n \simeq C^+_{n+1}$ obtained by extending the map α: $V \to C^+_{n+1} : \alpha(e_i) = e'_i e'_{n+1}$. Here $\{e_a;\ a = 1 \ldots n\}$ is an orthonormal basis of $V = \mathbb{R}^n$ and $\{e'_a;\ a = 1 \ldots n+1\}$ is an orthonormal basis of $V' = \mathbb{R}^{n+1}$.

The space S_{2m+1} is called the <u>space of spinors</u> for V; it is isomorphic to S^+_{2m+2}.

For all n, the space of spinors S_n carries an irreducible representation, denoted $\overset{n}{\tilde{\rho}}$, of the Clifford algebra C_n; in the odd dimensional case $\overset{n}{\tilde{\rho}}(e_a) = \overset{n+1}{\tilde{\rho}}(e_a)\, \overset{n+1}{\tilde{\rho}}(e_{n+1})\big|_{S^+_{n+1}}$.

The <u>Spin group</u> $\mathrm{Spin}(n)$ is the set of elements $x \in C^+_n$ which are invertible in C_n, which preserve $V = \mathbb{R}^n$ $(\subset C_n)$ (i.e. $xyx^{-1} \in V$ for all $y \in V$) and which have norm 1 (i.e. $x^\tau x = 1$ where τ is the unique antiautomorphism of C_n extending $I_{|V}$). $\mathrm{Spin}(n)$ has a natural representation on S_n, namely $\overset{n}{\tilde{\rho}}|_{\mathrm{Spin}(n)}$.

The homomorphism $\overset{n}{\theta} : \mathrm{Spin}(n) \to G\ell(V) : x \to [y \to xyx^{-1}]$ has for image $SO(n)$; in fact $\mathrm{Spin}(n)$ is the universal covering of $SO(n)$ and $\overset{n}{\theta}$ is the covering homomorphism. The differential $\overset{n}{\theta}_*$ at the identity

$$\overset{n}{\theta}_* : \mathrm{spin}(n) \to so(n)$$

is an isomorphism of Lie algebras, which can be written

$$\overset{n}{\theta}{}^{-1}_*(\Lambda^{ab} E_{ab}) = -\frac{1}{4}\Lambda^{ab} e_a e_b$$

E_{ab} is the $n \times n$ matrix with a 1 at the intersection of the a-th row and b^{th} column and zero elsewhere; $\Lambda^{ab} = -\Lambda^{ba}$.

Let (M,g) be an oriented smooth n-dimensional manifold with a Riemannian metric g. Let B denote the bundle of oriented orthonormal frames on M and let $p : B \to M$ be the canonical projection; B is a principal bundle with structure group $SO(n)$; the Levi Civita connection is a 1-form ω on B with values in $\mathfrak{so}(n)$.

A <u>Spin structure</u> on (M,g) is a principal bundle $\tilde{B}$ on M with structure group $\mathrm{Spin}(n)$ and projection $\tilde{p} : \tilde{B} \to M$ and a homomorphism $\varphi : \tilde{B} \to B$ such that i) $p \circ \varphi = \tilde{p}$ and ii) $\varphi(\xi . g) = \varphi(\xi) . \overset{n}{\theta}(g) \quad \forall \xi \in \tilde{B}, \forall g \in \mathrm{Spin}(n)$.

The manifold (M,g) admits a spin structure if and only if the second Stiefel-Whitney class $w_2(M)$ vanishes; in this case, the inequivalent spin structures on M are parametrized by $H^1(M,\mathbb{Z}_2)$ [2]. We assumes from now on that (M,g) admits a chosen spin structure $(\tilde{B},\varphi)$.

The <u>spinor bundle</u> $\mathcal{S}$ on M is the associated bundle $\mathcal{S} = \tilde{B} \times_{\mathrm{Spin}(n)} S_n$ where $\mathrm{Spin}(n)$ acts on S_n by $\overset{n}{\tilde{\rho}}$.

A <u>spinor field</u> ψ on M is a section ψ of the spinor bundle $\mathcal{S}$; equivalently it is a function $\tilde{\psi} : \tilde{B} \to S_n$ so that

$$\tilde{\psi}(\xi . g) = \tilde{\rho}(g^{-1})\tilde{\psi}(\xi) \quad \text{for all } g \in \mathrm{Spin}(n),\ \xi \in \tilde{B}$$

[The link is : $\psi(\tilde{p}(\xi)) = [\xi, \tilde{\psi}(\xi)]$ where $[\xi, v]$ is the equivalence class of the pair (ξ, v) in $\tilde{B} \times_{\mathrm{Spin}(n)} S_n$.]

The <u>covariant derivative</u> $\nabla\psi$ of a spinor field is associated to the connection $\tilde{\omega}$ on $\tilde{B}$ which is the pull back of the Levi Civita connection on B; explicitly

$$\tilde{\omega} = \theta_*^{-1}(\varphi^*\omega)$$

If X is a vector field on M

$$\widetilde{\nabla_X \psi} = \bar{X}\,\tilde{\psi}$$

where $\bar{X}$ is the unique horizontal lift of X on $\tilde{B}$ (i.e. $\tilde{\omega}(\bar{X}) = 0$ and $\tilde{p}_* \bar{X} = X$).

Consider the bundle $\mathcal{E} = T^*M \otimes \mathcal{S} \otimes \mathcal{S}^*$ which is the associated vector bundle $\tilde{B} \times_{\mathrm{Spin}(n)} (\mathbb{R}^{n*} \otimes S_n \otimes S_n^*)$ for the representation $\bar{\rho}$ of $\mathrm{Spin}(n)$ given by $(\overset{n}{\rho} \circ \overset{n}{\theta}) \otimes \overset{n}{\tilde{\rho}} \otimes \overset{n}{\tilde{\rho}}{}^*$. [$\overset{n}{\rho}$ denotes the usual representation of $SO(n)$ on $\mathbb{R}^n$; σ^* is the contragredient representation on R^* of the representation σ on R

i.e. $\langle \sigma^*(g)w^*, v\rangle = \langle w^*, \sigma(g^{-1})v\rangle]$.

The element γ is a section (of a trivial subbundle) of $\mathcal{E}$ whose associated function $\tilde{\gamma} : \tilde{B} \to \mathbb{R}^{*n} \otimes S_n \otimes S_n^*$ is the constant function

$$\tilde{\gamma}(\xi) = \sum_{k=1}^{m} e^k \otimes \gamma_k$$

$S_n \otimes S_n^*$ is identified with $\mathrm{End}(S_n)$; $\{ e_a ;\ 1 \leq a \leq n\}$ is an orthonormal basis of $\mathbb{R}^n$ and $\{e_*^a;\ a \leq n\}$ is its dual basis; $\gamma_k = \overset{n}{\tilde{\rho}}(e_k)$. This function has the correct equivariance property and thus defines a section. Indeed

$$\tilde{\gamma}(\xi . g) = \sum_{k=1}^{n} e_*^k \otimes \gamma_k$$

$$\bar{\rho}(g^{-1})\tilde{\gamma}(\xi) = \sum_{k=1}^{n} (\overset{n}{\rho} \circ \overset{n}{\theta})(g^{-1})(e_*^k) \otimes \overset{n}{\tilde{\rho}}(g^{-1}) \circ \gamma_k \circ \overset{n}{\tilde{\rho}}(g)$$

$$= \sum_{k=1}^{n} ((e_*^k) \circ (\overset{n}{\rho} \circ \overset{n}{\theta})(g)) \otimes \overset{n}{\tilde{\rho}}(\overset{n}{\theta}(g^{-1}).e_k)$$

$$= \sum_{k,\ell',\ell=1}^{n} e_*^{\ell}((\overset{n}{\rho} \circ \overset{n}{\theta})(g))^k{}_{\ell} \otimes (\overset{n}{\rho} \circ \overset{n}{\theta}(g^{-1}))^{\ell'}{}_k \gamma_{\ell'} = \sum_{\ell=1}^{n} e_*^{\ell} \otimes \gamma_{\ell}$$

A Killing spinor on (M,g) is a spinor field ψ so that $\nabla\psi = \lambda\gamma\psi$. Equivalently it is a function $\tilde{\psi} : \tilde{B} \to S_m$ so that

$$(*) \begin{cases} \tilde{\psi}(\xi . g) = \tilde{\rho}(g^{-1})\tilde{\psi}(\xi) \\ (\bar{X}\ \tilde{\psi})(\xi) = \lambda \sum_{k=1}^{n} X^k(\xi)\gamma_k \tilde{\psi}(\xi) \end{cases}$$

Here X denotes a vector field on M; $\bar{X}$ is its horizontal lift and $X^k(\xi)(k \leq n)$ are the components of X in the orthonormal frame $\varphi(\xi)$.

One can show that a manifold (M,g), admitting a spin structure and carrying a non zero Killing spinor, is an Einstein manifold with scalar curvature $\rho = 4n(n-1)\lambda^2$. $(**)$.

2. Killing spinors on spheres and projective spaces

We recall here the construction of spin structures on the sphere S^n and on the projective space $\mathbb{P}_{4n-1}$; similar constructions have been considered by Dabrowski and Trautman [3]. The Killing spinor fields are explicity described here under (see also [5])

a. Spin structure on spheres

Consider for M the sphere $S^n \subset \mathbb{R}^{n+1}$ with its usual metric.

The bundle B of its oriented orthonormal frames is isomorphic to $SO(n+1)$, with projection $p : SO(n+1) \to S^n$ given by $p(A) = A.e_{n+1}$ where $\{e_1 \dots e_{n+1}\}$ is the canonical basis of $\mathbb{R}^{n+1}$; A is identified with the frame at $A.e_{n+1}$ given by $(Ae_1, \dots, Ae_n)$ and $g \in SO(n)$ acts on $B = SO(n+1)$ by $A.g = A\begin{pmatrix} g & 0 \\ 0 & 1 \end{pmatrix}$.

A spin structure exists and is unique on S^n. It can be described as follows : $\tilde{B} = \mathrm{Spin}(n+1)$; $\tilde{p} : \tilde{B} \to S^n : A \to \overset{n+1}{\theta}(A).e_{n+1}$; $\tilde{\varphi} : \tilde{B} \to B$ coincides with the covering homomorphism $\overset{n+1}{\theta}$; $\mathrm{Spin}(n)$ acts on $\tilde{B} = \mathrm{Spin}(n+1)$ by right multiplication with the obvious inclusion of $\mathrm{Spin}(n)$ $(\subset C_n)$ in $\mathrm{Spin}(n+1)$ $(\subset C_{n+1})$.

Observe that $\tilde{B}$ is a principal bundle on S^n satisfying $p \circ \varphi = \tilde{p}$, $\varphi(\xi g) = \varphi(\xi)\overset{n}{\theta}(g)$ and its group structure is $\overset{n+1}{\theta}{}^{-1}(SO(n))$, which is isomorphic to $\mathrm{Spin}(n)$; indeed this group is a double cover of $SO(n)$ and is connected as $\exp te_1e_2$ lies in $\overset{n+1}{\theta}{}^{-1}(SO(n))$ and connects the two elements of the kernel of $\overset{n+1}{\theta}$.

b. Killing spinors on spheres

<u>Proposition</u>. <u>On the sphere</u> S^n $(n > 2)$ <u>a Killing spinor is defined by a function</u> $\tilde{\psi} = \mathrm{Spin}(n+1) \to S_n$ <u>given by</u>

$$\tilde{\psi}^+(g) = \tau^+(g^{-1})\psi_e^+ \quad \text{or} \quad \tilde{\psi}^-(g) = \tau^-(g^{-1})\psi_e^-$$

<u>where</u> $\psi_e^{\pm}$ <u>is any element of</u> S_n; $\tau^{\pm}$ <u>is the representation of</u> $\mathrm{Spin}(n+1)$ <u>on</u> S_n <u>whose differential is given by</u>

$$\tau_*^{\pm}\Big(\sum_{a,b\leq n+1} \Lambda^{ab} e_a e_b\Big) = \sum_{i,j\leq n} \Lambda^{ij} \overset{n}{\gamma}_i \overset{n}{\gamma}_j \pm 2 \sum_{i\leq n} \Lambda^{in+1} \overset{n}{\gamma}_i$$

It satisfies

$$\nabla\psi^{\pm} = \pm\frac{1}{2}\gamma\psi^{\pm}$$

Proof. Recall the description of the Levi Civita connection on a riemannian symmetric space [4]. As above $S^n = SO(n+1)/SO(n)$

Let $\sigma : SO(n+1) \to SO(n+1) : A \to \begin{pmatrix} I_n & 0 \\ 0 & -1 \end{pmatrix} A \begin{pmatrix} I_n & 0 \\ 0 & -1 \end{pmatrix}$ be the involutive automorphism induced by the symmetry at the point e_{n+1} and let $so(n+1) = k + p$ ($\sigma_*|_k = I$, $\sigma_*|_p = -I$) be the corresponding decomposition of the Lie algebra $so(n+1)$.

Clearly $k = \left\{ \begin{pmatrix} U & 0 \\ 0 & 1 \end{pmatrix}, U \in so(n) \right\}$ and p is identified with $S^n_{e_{n+1}}$.

Similarly define $\tilde{\sigma}_* : Spin(n+1) \to Spin(n+1)$ by $\tilde{\sigma}_* = \theta^{-1}_* \circ \sigma_* \circ \theta'$ so that $spin(n+1) = k' \oplus p'$ where $k' = \left\{ \sum_{i,j \leq n} \Lambda^{ij} e_i e_j \right\} \simeq spin(n)$.

The Levi Civita connection ω on $B = SO(n+1)$ is given by

$$\omega_A(A_* U) = U_k \qquad \begin{matrix} A \in SO(n+1) \\ U \; so(n+1) \end{matrix}$$

where A_* denotes the differential of the left translation by A and U_k is the component of U in k in the above decomposition.

Thus the spin connection on $\tilde{B} = Spin(n+1)$ is given by

$$\tilde{\omega}_{A'}(A'_* U') = U'_{k'}, \qquad A' \in Spin(n+1),\ U' \in spin(n+1).$$

Let $X \subset TS^n$ be a vector field; it can be written $X_{\theta(g).e} = \sum_{\substack{k \leq n \\ \ell \leq n+1}} X^k_{(g)} . (\theta(g))^{\ell}_{k} e_{\ell}$ when viewed as a vector in $\mathbb{R}^{n+1}$. Its horizontal lift $\bar{X}$ on $\tilde{B} = Spin(n+1)$ at a point g is given by

$$\bar{X}(g) = \frac{d}{dt}(g.\exp - \frac{t}{2} X^k \overset{n+1}{e_k} \overset{n+1}{e_{n+1}})\Big|_{t=0}$$

so that the equation (*) gives for a Killing spinor $\tilde{\psi}$:

$$\frac{d}{dt}\tilde{\psi}(g.\exp - \frac{t}{2} X^k \; e_k \; e_{n+1})\Big|_{t=0} = \lambda X^k \overset{n}{\gamma_k} \tilde{\psi}(g)$$

$$\tilde{\psi}(g.\exp \sum_{a,b \leq n} \Lambda^{ab} e_a e_b) = \exp(-\sum \Lambda^{ab} \overset{n}{\gamma_a} \overset{n}{\gamma_b}) . \tilde{\psi}(g)$$

This implies, as $\lambda = \pm\frac{1}{2}$ (by (**)) that

$$\frac{d}{dt}\tilde{\Psi}(g.\exp tU)\Big|_{t=0} = \tau_*^{\pm}(-U)\tilde{\Psi}(g) \qquad \forall\, g \in \mathrm{Spin}(n+1)$$
$$\forall\, U \in \mathrm{spin}(n+1)$$

hence the result.

c. Spin structure on projective spaces

Consider the projective space $P_n = S^n/\{Id,\sigma\}$ where σ is the antipodal map ($\sigma : S^n(\subset \mathbb{R}^{n+1}) \to S^n : x \to -x$), and denote by π the projection $\pi : S^n \to P_n$. An orthonormal basis of P_n is the image under π of a pair of orthonormal basis on S^n related by σ_* ; so P_n is orientable if and only if n is odd (so that $-Id$ in $\mathbb{R}^{n+1}$ is in $SO(n+1)$); in this case the bundle B' of oriented orthonormal frames on P_n is given by $SO(n+1)/\{\pm Id\}$ and the following diagram commutes :

$$\begin{array}{ccc} S^{2m-1} & \xleftarrow{\ p\ } & B = SO(2m) \\ \downarrow \pi & & \downarrow \\ P_{2m-1} & \xleftarrow{\ p'\ } & B' = SO(2m)/\{\pm Id.\} \end{array}$$

If P_{2m-1} admits a spin structure $(\tilde{B}',\varphi')$ then the pull back $\pi^{-1}\tilde{B}'$ of $\tilde{B}'$ by π gives a spin structure on S^{2m-1} (thus $\mathrm{Spin}(2m)$) and naturally double covers $\tilde{B}'$.

Indeed $\pi^{-1}\tilde{B}' = \{(x,b),\ x \in S^{2m-1},\ b \in \tilde{B}'|_{\pi(x)}\}$ and one defines $\varphi : \pi^{-1}\tilde{B}' \to B$: $(x,b) \to A$ where $\{A,-A\} = \varphi'(b)$ and $Ae_{n+1} = x$.
Thus $(\pi^{-1}\tilde{B}',\varphi)$ is a spin structure on S^{2m-1}. Furthermore $\tilde{\pi} : \pi^{-1}\tilde{B}' \to \tilde{B}'$ $(x,b) \to b$ is a double covering.

Hence we have the commutativity

$$\begin{array}{ccc} SO(2m) & \xleftarrow{\ \theta^{2m}\ } & \tilde{B} = \mathrm{Spin}(2m) \\ \downarrow & & \downarrow \tilde{\pi} \\ SO(2m)/(\pm Id) & \xleftarrow{\ \varphi'\ } & \tilde{B}' \end{array}$$

which shows that $\tilde{B}'$ must be a quotient of $\mathrm{Spin}(2m)$ by a central subgroup isomorphic to $\mathbb{Z}_2$. But the center Z of $\mathrm{Spin}(2m)$ which is $\theta^{2m\,-1}\{\text{center of } SO(2m)\}$

is given by $\{I, -I, \tilde{\gamma} = e_1e_2 \ldots e_{2m-1}e_{2m}, -\tilde{\gamma}\}$; so $Z = Z_4$ if m is odd and $Z = \mathbb{Z}_2 \times Z_2$ if m is even.

The $\mathbb{Z}_2$ subgroup used to make the quotient is not $\{\pm \mathrm{Id}\}$ because the fibres

$$\tilde{B}'|_{\pi(e)} = \varphi'^{-1} p'^{-1} \pi(e) = \tilde{\pi}(\theta_{2m}^{-1}\{\begin{pmatrix} A & 0 \\ 0 & -1 \end{pmatrix} \cup \begin{pmatrix} -A & 0 \\ 0 & -1 \end{pmatrix} \mid A \in SO(2m-1)\})$$
$$= \tilde{\pi}(\{A'\} \cdot \cup \{\tilde{\gamma} A'\} \mid A' \in \mathrm{Spin}(2m-1))$$

must be isomorphic to spin(2m-1).

Hence P_n admits a spin structure only if $n = 4m - 1$; it then admits two inequivalent spin structures $\tilde{B}^{\epsilon}$, where $\epsilon = \pm 1$, given by

$$\begin{array}{c} \tilde{B}^{\epsilon} = \mathrm{Spin}(4m)/\{I, \epsilon\tilde{\gamma}\} \\ \downarrow \\ P_{4m-1} \end{array}$$

Let us denote $\tilde{\pi}_{\epsilon} : \mathrm{Spin}(4m) \to \tilde{B}^{\epsilon}$.

d. Killing spinors on projective spaces

Proposition. A spinor ψ is a Killing spinor $(\nabla\psi = \lambda\gamma\psi)$ on P_{4m-1} with spin structure $\tilde{B}^{\epsilon}$ $(\epsilon = +1$ or $\epsilon = -1)$ if and only if (i) $\lambda = \epsilon(-1)^m \frac{1}{2}$ and (ii) ψ is given by a function $\tilde{\psi} : \tilde{B}^{\epsilon} \to S_{4m-1}$ such that $\tilde{\pi}_{\epsilon}^{*}\tilde{\psi} : \mathrm{Spin}(4m-1) \to S_{4m-1} \simeq S^{+}_{4m}$ satisfies $\tilde{\pi}_{\epsilon}^{*}\tilde{\psi}(A) = \tau^{\epsilon'}(A^{-1})\psi_e$; where $\psi_e \in S_{4m-1}$ and $\epsilon' = \epsilon(-1)^m$.

Proof. It is clear that $\tilde{\pi}_{\epsilon}^{*}\tilde{\psi}$ must define a Killing spinor ψ' on S^{4m-1} with $\nabla\psi' = \lambda\gamma\psi'$ with $\lambda = \pm\frac{1}{2}$ so that $\tilde{\pi}_{\epsilon}^{*}\tilde{\psi} = \tau^{\pm}(A^{-1})\psi_e$ where $\psi_e \in S_{4m-1} \simeq S^{+}_{4m}$ and that $(\tilde{\pi}_{\epsilon}^{*}\tilde{\psi})(\epsilon\tilde{\gamma}) = (\tilde{\pi}_{\epsilon}^{*}\tilde{\psi})(\mathrm{Id})$ so one must have

$$\tau^{\pm}(\epsilon\tilde{\gamma})\psi_e = \psi_e$$

We have

$$\tau^{\pm}(\tilde{\gamma}) = \tau^{\pm}(\exp \frac{\pi}{2}(e_1e_2 + \ldots + e_{4m-1}e_{4m}))$$
$$= \exp \frac{\pi}{4} \tau^{+}_{*}(e_1e_2 - e_2e_1 + \ldots + e_{4m-1}e_{4m} - e_{4m}e_{4m-1})$$
$$= \exp \frac{\pi}{2}(\gamma_1^{4m-1}\gamma_2^{4m-1} + \ldots + \gamma_{4m-3}^{4m-1}\gamma_{4m-2}^{4m-1} \pm \gamma_{4m-1}^{4m-1})$$

$$= \exp \frac{\pi}{2}\left(\gamma_1^{4m}\gamma_2^{4m}\Big|_{S^+_{4m}} + \ldots \pm \gamma_{4m-1}^{4m}\gamma_{4m}^{4m}\Big|_{S^+_{4m}}\right)$$

$$= \pm\, (\gamma_1^{4m} \ldots \gamma_{4m}^{4m})\Big|_{S^+_{4m}}$$

Observe that

$$(\gamma_{2k-1}^{4m}\gamma_{2k}^{4m})(\alpha) = \left(f_k \wedge - i(f_k^*)\right)\circ\left(- \sqrt{-1}(f_k \wedge + i(f_k^*))\right)\alpha$$

$$= \sqrt{-1}\,\alpha \quad \text{if } \alpha = f_{i_1}\wedge \ldots \wedge f_{i_\ell} \quad \text{with } k \notin \{i_1 \ldots i_\ell\}$$

$$= -\sqrt{-1}\,\alpha \quad \text{if } \alpha = f_{i_1}\wedge \ldots \; f_{i_\ell} \quad \text{with } k \in \{i_1 \ldots i_\ell\}$$

so that $(\gamma_1^{4m} \ldots \gamma_{4m}^{4m})(f_{i_1}\wedge \ldots \wedge f_{i_\ell}) = (\sqrt{-1})^{2m}(-1)^\ell$

and $\tau^{\pm}(\tilde{\gamma}) = \pm\,(-1)^m \mathrm{Id}$.

Therefore we must have, for a non zero Killing spinor ψ that

$$\pm\,\epsilon\,(-1)^m = 1.$$

The reverse is clear.

3. Triviality of spinor bundles

The results of § 2 prove, as arose from discussions with F.Burstall, J.Rawnsley and A.Trautman, that the spinor bundle on S^m or P_{4m-1} are trivial. Indeed, a nonzero Killing spinor does not vanish at any point and the Killing spinors on those spaces give a global trivialization of the spinor bundles.

Remark that those results can been seen directly as follows :

Proposition. *Let V be a vector space carrying a representation σ of $\mathrm{Spin}(m)$ induced by a representation σ of the Clifford algebra C_n and let $\mathcal{E}$ be the associated vector bundle on the sphere S^n : $\mathcal{E} = \mathrm{Spin}(n+1) \times_{\mathrm{Spin}(n)} V$; then $\mathcal{E}$ is a trivial bundle.*

A global trivialization of $\mathcal{E}$ is given by

$$\mathcal{E} \to S^n \times V \;:\; [A,v] \to (\tilde{p}(A), \tau^{\epsilon'}(A)v) \qquad A \in \mathrm{Spin}(n+1),\ v \in V,\ \epsilon' = \pm 1$$

where $\tau^{\epsilon'}$ is the representation of $\mathrm{Spin}(n+1)$ on V whose differential is :

$$\tau_*^{\epsilon'}\Big(\sum_{a,b=1}^{n+1}\Lambda^{ab}e_a e_b\Big) = \sum_{a,b=1}^{n}\sigma(e_a)\sigma(e_b) + 2\epsilon'\sum_{a=1}^{n}\Lambda^{an+1}\sigma(e_a)$$

Similarly, if $n = 4m - 1$ let $\tilde{\mathcal{E}}^{\epsilon}$ be the vector bundle on P_{4m-1} associated to V and to the spin structure $\tilde{B}^{\epsilon}$; $\tilde{\mathcal{E}}^{\epsilon} = \mathrm{Spin}(4m)/\{I, \epsilon e_1 \dots e_{4m}\} \times_{\mathrm{Spin}(4m-1)} V$. If $\sigma(e_1 \dots e_{4m-1}) = \epsilon''\,\mathrm{Id}$, the bundle $\tilde{\mathcal{E}}^{\epsilon}$ admits a global trivialization

$$\tilde{\mathcal{E}}^{\epsilon} \to P_{4m-1} \times V : [A\{I, \epsilon\tilde{\gamma}\}, v] \to (\pi.\tilde{p}(A), \tau^{\epsilon'}(A)v)$$

for $\epsilon' = \epsilon\epsilon''$.

Acknowledgment : This work has been done in collaboration with Michel Cahen whom I thank wholeheartedly

Bibliography

1 M.Cahen, S.Gutt, L.Lemaire, P.Spindel : Bull.Soc.Math.de Belgique, volume en l'honneur de G.Hirsch (sous presse)

2 M.Atiyah, R.Bott, A.Shapiro: Topology Supp 1 to vol 3 (1964) p.3-38.

3 L.Dalrowski, A.Trautman: J.Math.Phys. 27(8), 1986 p.2022-2028.

4 S.Helgason : Differential geometry, Lie groups and symmetric spaces. Academic Press.

5 P.Van Niewenhuizen : Relativity and topology II. Les Houclos (1983) North Holland pp.825-932.

SPINOR STRUCTURES ON HOMOGENEOUS RIEMANNIAN SPACES

Ludwik Dabrowski*
Scuola Internazionale Superiore di Studi Avanzati, Trieste, Italy

and

Andrzej Trautman**
Programs in Mathematics, University of Texas at Dallas, Richardson, Texas, U.S.A.

Summary

It is shown how pin structures on a homogeneous Riemannian space M can be constructed as bundles associated with the universal covering space bundle $\pi_1(G) \longrightarrow \tilde{G} \longrightarrow G$ of the Lie group G of isometries acting effectively on M.

* Permanent address: Instytut Fizyki Teoretycznej, Uniwersytet Wrocławski, ul. Cybulskiego 36, 50-205 Wrocław, Poland

** Permanent address: Instytut Fizyki Teoretycznej, Uniwersytet Warszawski, ul. Hoża 69, 00-681 Warszawa, Poland

In a previous paper [1], referred to in the sequel as DT, we gave an explicit construction of spinor structures on projective spaces. We now show how the method outlined there can be used to find the spinor structures on any homogeneous Riemannian space. For simplicity, we restrict ourselves to properly Riemannian spaces which are smooth, connected and compact, but not necessarily orientable. We follow, with slight modifications, the terminology and notation of DT.

Let M be a connected and compact Riemannian n-manifold admitting a Lie group G of isometries acting transitively on M. This group may be assumed to be connected: if it is not, then one can replace it by the connected component of its neutral element. Moreover, the action of G on M may be taken to be effective: if it is not, then one can replace G by the quotient group G/N, where N is the kernel of inefficiency of the action. (Note that in DT we found it convenient to consider an ineffective action of U(n+1, K) on KP_n; K = $\mathbb{R}$, $\mathbb{C}$, $\mathbb{H}$). If H is the stability group at a point of M, then M can be identified with the quotient G/H. The principal H-bundle

$$H \to G \to G/H = M$$

is a restriction of the bundle of all orthonormal frames of M to the subgroup H of the full orthogonal group O(n). Let, as in DT, $Pin_+(n)$ and $Pin_-(n)$ be the groups occuring in the two inequivalent exact sequences

$$1 \to \mathbb{Z}_2 \to Pin_\pm(n) \to O(n) \to 1$$

which extend the sequence

$$1 \to \mathbb{Z}_2 \to \mathrm{Spin}(n) \xrightarrow{\rho} SO(n) \to 1$$

in such a way that $\rho = \rho_\pm | \mathrm{Spin}(n)$. Recall that

$$\rho_\pm(s)x = \alpha(s)xs^{-1}$$

where $s \in \mathrm{Pin}_\pm(n), x \in \mathbb{R}^n$ and α is the main automorphism of the Clifford algebra of $\mathbb{R}^n$ equipped with the quadratic form $\pm\phi(x)$,

$$\phi(x) = x_1^2 + \ldots + x_n^2 \; .$$

We denote by $\hat{H}_\pm$ the subgroup of $\mathrm{Pin}_\pm(n)$ that covers H,

$$\hat{H}_\pm = \rho_\pm^{-1}(H) \; .$$

For simplicity, we often omit the subscripts $\pm$ and write Pin, ρ and $\hat{H}$ instead of Pin_+, ρ_+ and $\hat{H}_+$ or Pin_-, ρ_- and $\hat{H}_-$, respectively.

A pin structure on the homogeneous Riemannian n-manifold M is given by a principal $\hat{H}$-bundle $\hat{G} \to M$ together with a morphism $\eta: \hat{G} \to G$ of principal bundles such that there is a commutative diagram

(1)
$$\begin{array}{ccccc} \hat{G} \times \hat{H} & \longrightarrow & \hat{G} & \searrow & \\ {\scriptstyle \eta \times \rho} \downarrow & & {\scriptstyle \eta} \downarrow & & M \\ G \times H & \longrightarrow & G & \nearrow & \end{array}$$

where the horizontal arrows denote the action maps. Note that $\mathbb{Z}_2$ is a subgroup of $\hat{H}$ and there is a principal bundle

(2) $$\mathbb{Z}_2 \longrightarrow \hat{G} \xrightarrow{\eta} G \;.$$

If M is orientable, then H is a subgroup of SO(n), its cover $\hat{H}$ is a subgroup of Spin(n) and our definition describes a spin structure on M.

The universal covering group $\tilde{G}$ of the group G is the total space of the principal bundle

(3) $$\pi_1(G) \longrightarrow \tilde{G} \longrightarrow G \;.$$

The subgroup of $\tilde{G}$ that covers H is denoted $\tilde{H}$. Given a homomorphism of groups,

(4) $$h : \pi_1(G) \longrightarrow \mathbb{Z}_2$$

one can construct the principal $\mathbb{Z}_2$-bundle

(5) $$\mathbb{Z}_2 \longrightarrow G_h \xrightarrow{\sigma} G$$

associated with (3) by means of (4). The bundle (5) is trivial, $G_h = G \times \mathbb{Z}_2$, if, and only if, h is a trivial (constant) homomorphism. Otherwise, G_h is connected and $\pi_1(G_h)$ = kerh.

The manifold G_h has a natural structure of a Lie group which makes σ into a homomorphism and the sequence (5) to be exact. The subgroup $H_h = \sigma^{-1}(H)$ acts on G_h by right translations and defines a principal H_h-bundle $G_h \to M$ so that there is a commutative diagram

$$\begin{array}{ccccc} G_h \times H_h & \longrightarrow & G_h & \searrow & \\ \sigma\times\sigma \downarrow & & \sigma \downarrow & & M \\ G \times H & \longrightarrow & G & \nearrow & \end{array} \qquad (6)$$

resembling (1). We can now formulate our main

Theorem. The diagram (6) defines a pin structure on the homogeneous Riemannian manifold M if, and only if, the Lie groups H_h and $\hat{H}$ are isomorphic as coverings of H. Conversely, with any structure (1) on M one can associate a homomorphism (4) and an isomorphism

$$i : \hat{H} \longrightarrow H_h \qquad (7)$$

such that the diagrams (1) and (6) are isomorphic to each other.

To prove the theorem, assume h to be such that there is an isomorphism (7). Putting $\hat{G} = G_h$, $\eta = \sigma$, and defining the action of $\hat{H}$ on $\hat{G}$ by $\hat{G} \times \hat{H} \ni (a,b) \mapsto a\, i(b) \in \hat{G}$, one obtains a pin structure (1). To prove the converse, consider a pin structure (1). If η is trivial, then $\hat{G} = G \times \mathbb{Z}_2$, $\hat{H} = H \times \mathbb{Z}_2$ and one takes h to be the constant homomorphism. If η is non-trivial, then $\hat{G}$ is connected. The projection η then induces a monomorphism of fundamental groups, $\eta_* : \pi_1(\hat{G}) \to \pi_1(G)$, and there is a principal bundle

$$(8) \qquad \pi_1(\hat{G}) \longrightarrow \tilde{G} \longrightarrow \hat{G} \; .$$

Since η_* is not surjective, one can define h by requiring that it be surjective with kernel equal to the image of η_*. In other words, h is defined by the exact sequence

$$(9) \qquad 1 \longrightarrow \pi_1(\hat{G}) \xrightarrow{\eta_*} \pi_1(G) \xrightarrow{h} \mathbb{Z}_2 \longrightarrow 1 \, .$$

Since $\hat{G}$ and G_h may be identified with $\tilde{G}/\pi_1(G)$ and $(\tilde{G} \times \pi_1(G)/\pi_1(\hat{G}))/\pi_1(G)$, respectively, it is clear that the bundles (2) and (5) are isomorphic and so are the diagrams (1) and (6).

To illustrate the theorem, consider the following simple examples.

Example 1. Let M be the circle $S_1 = U(1)$ acted upon by $G = U(1)$. In this case $\tilde{G} = \mathbb{R}$, $\pi_1(G) = \mathbb{Z}$, H is trivial, $\hat{H} = \mathbb{Z}_2$. The constant homomorphism $\mathbb{Z} \to \mathbb{Z}_2$ yields the trivial spin structure, whereas the homomorphism defined by $1 \to 1 \bmod 2$ corresponds to the other spin structure

$$U(1) \xrightarrow{\text{square}} U(1) \xrightarrow{\text{id}} U(1) .$$

Example 2. Let M = $\mathbb{R}P_{2k}$ (k = 1,2,...) be a non-orientable real projective space with its standard Riemannian metric. The group G = SO(2k+1) acts transitively and effectively on M. One has $\pi_1(G) = \mathbb{Z}_2$ and $\tilde{G}$ = Spin(2k+1). Identify $\mathbb{R}P_{2k}$ with the set of all pairs (x, -x), where $x \in S_{2k} \subset \mathbb{R}^{2k+1}$. Let $(e_1,..., e_{2k+1})$ be the standard orthonormal frame in $\mathbb{R}^{2k+1}$. The stability group H of the pair $(e_{2k+1}, -e_{2k+1})$ is isomorphic to O(2k) embedded in SO(2k+1) as follows: if $a \in O(2k)$ and $x = y + ze_{2k+1}$, where $y = y^1e_1 + ... + y^{2k} e_{2k}$, with $y^1,...,y^{2k}$, $z \in \mathbb{R}$, then the image of a in SO(2k+1) maps x into $x' = ay + (\det a)\, z\, e_{2k+1}$. The group $\tilde{H} \subset$ Spin (2k+1) consists of all elements that are products of either an even number of unit vectors belonging to $\mathbb{R}^{2k}$ or an odd number of such vectors by e_{2k+1}. These elements of $\tilde{H}$ are called even and odd, respectively. The set of all even elements of $\tilde{H}$ is a subgroup isomorphic to Spin(2k). In this case, the trivial homomorphism $\mathbb{Z}_2 \rightarrow \mathbb{Z}_2$ is of no interest; let us consider the only other possibility, h = id, so that $G_h = \tilde{G}$ and $H_h = \tilde{H}$. Depending on whether k is even or odd, we take $\hat{H}$ to be $Pin_+(2k)$ or $Pin_-(2k)$ and define, in each case, two inequivalent isomorphisms $i_a : \hat{H} \rightarrow \tilde{H}$ (a = 1,2),

$$i_a(s) = \begin{cases} s & \text{if s is even,} \\ (-1)^a s\varepsilon & \text{if s is odd,} \end{cases}$$

where $\varepsilon = e_1 \dots e_{2k+1}$. Since $\varepsilon^2 = (-1)^k e_1^2 \dots e_{2k+1}^2$, our definition of $\hat{H}$ guarantees that $\varepsilon^2 = 1$ for all k; this is important to ensure that i_a is an isomorphism. The two pairs (h, i_a), a = 1,2, define the two inequivalent pin structures that are known to exist on even-dimensional real projective spaces.

Example 3. Let M be a Lie group G acting on itself by left translations and equipped with a left-invariant Riemannian metric. In this case H is trivial and $\tilde{H} = \pi_1(G)$. There is always at least one spin structure on G, corresponding to the trivial homomorphism h. The orthogonal groups have another spin structure, corresponding to h = id., but the groups $SU(2k+1)/\mathbb{Z}_{2k+1}$ and Sp(n) have not. The n-dimensional torus has 2^n distinct spin structures.

Acknowledgements

We thank P. Gilkey, H. and Y. Kerbrat for useful discussions. One of us (A.T.) was supported by the National Science Foundation under Grant No. PHY 8513522 and by the Polish Research Program CPBP 01.03.

References

1. L. Dąbrowski and A. Trautman, Spinor Structures on Spheres and Projective Spaces. J.Math.Phys. 27 (1986) 2022-2028.

CLASSICAL STRINGS AND MINIMAL SURFACES

H. Urbantke

Institut für Theoretische Physik

Universität Wien

Abstract

Real Lorentzian forms of some complex or complexified Euclidean minimal surfaces are obtained as an application of H.A. Schwarz' solution to the initial value problem or a search for surfaces admitting a group of Poincaré transformations.

1. Introduction

In this paper I shall not derive anything new but just discuss some old ($\geq$ 100 years) results on minimal surfaces in the guise they take when everything is translated to the situation where the embedding space is pseudo-Euclidean of Lorentzian signature. They seem to be of more interest here just because they can take place in the real domain here whereas in the Euclidean case one has to consider analytic continuations (which are known to exist). The existence of null coordinates on time-like two-surfaces of Minkowski space needs no deeper result. With the help of these one gets a real version of the H.A. Schwarz solution to the initial 'value' problem of the classical relativistic string - namely to pass its world sheet through a given initial strip. (In the Euclidean case, the classical problem is that of Plateau - namely to pass a minimal surface through one or more closed boundary curves; this is much more difficult.) What I shall do is to give a short derivation of Schwarz' formula and apply it in some simple situations.

2. The H.A. Schwarz Formula

Let M^D be the D-dimensional Minkowski space with its standard metric $dS^2 = G_{AB}\, dX^A \otimes dX^B$, $G_{AB} = \mathrm{diag}(1,\ldots,1,-1)$ when referred to Cartesian orthonormal coordinates. Let $\iota\colon M^d \to M^D$ be an imbedded d-dimensional submanifold given locally by $X^A(x^a)$; then we have an induced metric $ds^2 = i^* dS^2 = g_{ab} dx^a \otimes dx^b$, $g_{ab}(x) = G_{AB}(\iota(x))\ \partial X^A/\partial x^a\ \partial X^B/\partial x^b$, and the volume on M^d is given by

$$\int |\det g_{ab}|^{1/2}\, d^d x\ .$$

The imbedding is extremal (or, rather, stationary) with respect to this volume function if the imbedding functions $X^A(x^a)$ satisfy $\Delta_g X^A = 0$ which formally looks like Laplace's equation w.r. to g_{ab} but is highly non-linear because g_{ab} depends on ι. (Geometrically it tells us that the so-called mean curvature normal to our surface vanishes.) Using the Hodge star operation w.r. to g_{ab}, it can also be written $d{*}dX^A\big|_{M^d} = 0$, i.e. $*dX^A$ is a closed (d-1)-form on M^d; we also note the identity

$$*dX^A\Big|_{M^d} = \frac{1}{(d-1)!}\, t^{AA_2\ldots A_d}\, dX_{A_2} \wedge \ldots \wedge dX_{A_d}\Big|_{M^d} ,$$

where $t^{A_1\ldots A_d}$ is the normalized tangent d-vector on $\iota(M^d)$, i.e. the image of the contravariant volume tensor under ι.

Specializing to the case d = 2, we consider the time-like world sheet of a string in M^D. Then at each point of it we have two independent null directions, and these integrate to give two systems of null lines on the sheet. If $x^{\pm}$ are any parameters on these lines, one can parametrize the sheet as $X^A(x^+,x^-)$ and linearize $\Delta_g X^A = 0$ in this way: it simply becomes $\partial^2 X^A/\partial x^+\partial x^- = 0$, so that $X^A(x^+,x^-) = X^A_+(x^+) + X^A_-(x^-)$, where we have the only condition that $u \to X^A_{\pm}(u)$ are null curves, $g_{AB}X^{A'}_{\pm}X^{B'}_{\pm} = 0$, which generate the sheet.

It is at this point where spinors could enter the subject. Now, some time ago, I stepped onto the relation between null curves in complexified Minkowski space and holomorphic curves in projective twistor space in a completely different context [2]; in the present one it was discussed by Budinich [3], by Shaw [4] (who tries to take care of the reality conditions) and by Hughston and Shaw [5] for D = 6. The existence of a representation of null curves in arbitrary dimensions without integral signs was discussed by Pinl [6]. Here I turn to a different aspect of classical strings, the solution of the initial value problem.

Now with the null coordinate representation we may check easily that $*dX^A = dX^A_+ - dX^A_-$ on the sheet, indeed an exact differential. Together with $dX^A = dX^A_+ + dX^A_-$, this may be solved for $X^A_{\pm}$, considered as functions on the sheet:

$$X^A_{\pm} = \frac{1}{2}\,(X^A \pm \int *dX^A) ,$$

and this relation may now be used to solve the initial value problem as follows. Since $X^A_\pm$ are constant along corresponding null lines, their values can be taken from the value of $\frac{1}{2}(X^A \pm \int *dX^A)$ at the intersection of such a null line with a given non-null initial curve, because the line integral $\int *dX^A$ may be evaluated along it due to exactness and because, according to the formula above, the integrand $*dX^A|_{M^2} = t^{AB}\, dX_B|_{M^2}$ is given there by the initial data. Explicitly, if $X^A(x^0{=}0,x^1{=}\sigma) = Y^A(\sigma)$ and $X^A{}_{,0}(x^0{=}0,x^1{=}\sigma) = Z^A(\sigma)$ are the data, so that $t^{AB} = (Y'^A Z^B - Y'^B Z^A)((Y'^A Z_A)^2 - Y'^A Y'_A Z^B Z_B)^{-1/2}$, define (H.A. Schwarz, 1874)

$$X^A_\pm(\sigma) = \frac{1}{2}\,(Y^A(\sigma) \pm \int_{\sigma_0}^{\sigma} t^{AB}\, Y'_B\, d\sigma) \; ;$$

then $\sigma \to X^A_\pm(\sigma)$ are null curves that generate the sheet according to $X^A(x^+,x^-) = X^A_+(x^+) + X^A_-(x^-)$. It is evident that only the initial strip, i.e. the unparametrized initial curve and the (time-like!) tangent plane elements (contact elements) spanned by Y', Z are relevant and that these 'geometric initial values' are taken: $X^A(\sigma,\sigma) = Y^A(\sigma)$, $X'^{[A}_+(\sigma,\sigma)X'^{B]}_-(\sigma,\sigma) \propto Y'^{[A}(\sigma)Z^{B]}(\sigma)$. Note that the formula does not take us off a characteristic strip, i.e. one where the curve is null!

3. Some Applications

For a more concise notation, we introduce an origin and an orthonormal frame in M^D: $E_1, E_2, \ldots, E_D$, and write $Y := Y^A E_A$ etc., $t^{AB} Y'_B E_A =: t \llcorner Y'$ when convenient.

A circular string

We consider the initial position as given by the circle

$$Y(\sigma) = a(\cos\sigma E_1 + \sin\sigma E_2) \;, \qquad 0 \le \sigma < 2\pi \;,$$

and distribute the initial velocities $Z(\sigma)$ as

$$Z(\sigma) = (1 - v^2)^{-1/2}(E_D + v\cos\sigma E_1 + v\sin\sigma E_2)$$

so that

$$t \llcorner Y' = a\,Z$$

and

$$X_\pm(\sigma) = \frac{a}{2}\,(\cos\sigma E_1 + \sin\sigma E_2) \pm \frac{a}{2}\,(1 - v^2)^{-1/2}(\sigma E_D + v\sin\sigma E_1 - v\cos\sigma E_2)\ .$$

Putting $v = \sin\alpha$, the world sheet becomes

$$X(x^+,x^-) = \frac{a}{2\cos\alpha}\,[(x^+ - x^-)E_D + (\cos(x^+ - \alpha) + \cos(x^- + \alpha))E_1 + $$
$$+ (\sin(x^+ - \alpha) + \sin(x^- + \alpha))E_2]\ .$$

Eliminating x^+, x^- between X^1, X^2, $X^D = T$ and writing $R^2 = (X^1)^2 + (X^2)^2$, we get

$$\frac{R}{a}\cos\alpha = \cos\left(\frac{T}{a}\cos\alpha - \alpha\right)\ .$$

This represents, in the 3-space spanned by E_1, E_2, E_D, a surface obtained by rotating a cosine curve around E_D. In particular, we see the circular string contracting towards a point if we go either to the future or the past. One might ask whether the occurrence of such a singularity on the world sheet of a closed string is due to the assumed symmetry or is more general.

A 'uniformly accelerated' string

Consider the strip given by $Y(\tau) = a(\mathrm{sh}\,\tau E_D + \mathrm{ch}\,\tau E_2)$, $Z(\tau) = E_1$, which is not an initial strip in the causal sense, but is made up of timelike tangent plane elements. Thus our formula may be applied and we get $t \llcorner Y' = -\,aE_1$, and thus

$$X(x^+,x^-) = \frac{a}{2}\,[(\mathrm{sh}\,x^+ + \mathrm{sh}\,x^-)E_D + (\mathrm{ch}\,x^+ + \mathrm{ch}\,x^-)E_2 - (x^+ - x^-)E_1]\ .$$

Eliminating, we get the surface

$$\frac{1}{a}\sqrt{(X^2)^2 - T^2} = \mathrm{ch}(X^1/a)$$

in 1,2,D-space, representing the second type of Lorentzian catenoid (the first being the sheet above). The curves cut out by X^1 = const. all represent world lines of uniformly accelerating particles.

Ruled sheets

In the real Euclidean case it is known (E.C. Catalan, 1842) that the only class of minimal surfaces containing a ruling by a one-parameter family of straight lines are the helicoids. In the complex case, S. Lie (1879) discovered another type, completing the list of possibilities. In real Minkowski space, there are more classes of real forms of these than in real Euclidean space, and some of them have a physical taste. To get these real forms, it is enough to give a classification of the 'initial' strips under the Poincaré group when the 'initial' curve is a straight line - its axis a (the quotes refer to the possibility of a timelike line). First one has to distinguish between the case of strips where all contact elements are parallel along the axis - which simply lead to planar sheets* - and the case where the plane of the contact element rotates as the point of contact progresses along the axis. If the axis is not isolated but belongs to the ruling, as we of course shall assume, it follows that this rotation is correlated projectively with the progression along the line, and it is enough to classify strips of this type. It turns out that light-like axes do not lead to solutions here**, so we must distinguish just the timelike and spacelike cases. Next, upon rotation around the axis, the contact plane sweeps out a 3-plane π containing the axis a, within which the whole sheet will remain. This 3-plane may be timelike (and must be so for a timelike axis), spacelike, or lightlike, corresponding to the induced metric in it being indefinite, definite, or semidefinite, respectively. Finally one has to distinguish the limiting position t_∞ of the plane of the contact element as the contact point moves to infinity as being timelike, spacelike, or lightlike: for a and therefore π timelike, t_∞ must be timelike; for a spacelike, π timelike, all three cases for t_∞ may occur; for a spacelike,

* and cylinders with lightlike generators, if the axis is null.

** This and the preceding footnote illustrate the peculiarities of characteristic strips.

π lightlike, t_∞ may be spacelike or lightlike; and for a and π spacelike, t_∞ must be spacelike (proper Euclidean case!) – so there will be no causal interpretation for the latter three cases, and we shall not write down the corresponding solutions explicitly.

A string of dust

If a, π, t_∞ are timelike, we may take a as the T-axis, t_∞ as the (T,X^1)-plane, π as the (T,X^1,X^2)-space and the point of contact of the (T,X^2)-plane on a as the origin. Then the contact correlation between points on a and planes through a may be specified by

$$Y(\sigma) = p\,\sigma E_D\,, \qquad Z(\sigma) = \sigma E_1 + E_2 \qquad -\infty < \sigma < \infty\,.$$

Schwarz' formula then yields

$$X(x^+,x^-) = \frac{p}{2}\left[(x^+ + x^-)E_D + \left(\sqrt{1+(x^+)^2} - \sqrt{1+(x^-)^2}\right)E_1 + (\operatorname{ar\,sh} x^+ - \operatorname{ar\,sh} x^-)E_2\right].$$

Eliminating x^+, x^- we obtain

$$X^2 = p \operatorname{ar\,th}(X^1/T)\,.$$

This is a Lorentzian helicoid, arising from a simultaneous Lorentz rotation of the T axis in the (T,X^1)-plane and a translation in the X^2-direction by an amount proportional to the (hyperbolic) angle of rotation. At infinite distance along the X^2-axis, the straight world lines of the sheet become lightlike; so it may be interpreted as the history of an infinite string of dust.

A rigidly rotating straight string

If a and t_∞ are spacelike while π is timelike, we may take a as the X^1-axis, t_∞ as the (X^1,X^2)-plane, π as (X^1,X^2,T)-space and the point of contact of the (T,X^1)-plane on a as the origin. Then the contact correlation may be specified by assigning some value p to the X^1-coordinate of of the contact point of a lightlike contact element:

$$Y(\sigma) = p\,\sigma E_1\ , \qquad Z(\sigma) = E_D + \sigma E_2\ ; \qquad |\sigma| < 1$$

(this parameter range covers only the causal part of the strip). A similar integration and elimination gives the helicoid

$$T = p \text{ arc tg } (X^2/X^1)\ ,$$

arising from a rotation of the X^1-axis in the (X^1,X^2)-plane together with a translation along the T axis by an amount proportional to the angle of rotation. The sheet thus may be considered as the history of a finite rigidly rotating straight string (as judged from the initial inertial frame).

Two types of semi-infinite strings

If a is spacelike, t_∞ and π timelike, we take a to be the X^1-axis, t_∞ the (T,X^1)-plane, π as (X^1,X^2,T)-space, the point of contact of the (X^1,X^2)-plane as the origin and the point of contact of a lightlike element at distance p from it. Then the contact correlation is specified by

$$Y(\sigma) = p\,\sigma E_1\ , \qquad Z(\sigma) = E_2 + \sigma E_D\ , \qquad \sigma > 1\ ,$$

where the parameter range covers one of the two causal parts of the strip. There results the Lorentzian helicoid

$$X^2 = p \text{ ar th } (T/X^1)\ ,$$

which geometrically arises from a Lorentz rotation of the X^1-axis in the (T,X^1)-plane together with a translation along the X^2-direction as before. Thus this string remains straight for a uniformly accelerated system of observers.

The ruled sheets considered so far can all be obtained directly as Minkowskian real forms of the complexification of a Euclidean helicoid. There is no real Euclidean analog of the complexification (found by S. Lie) of the following sheet, however.

If a is spacelike, π timelike and t_∞ lightlike, we take a = X^1-axis, $\pi = (X^1, X^2, T)$-space, point of contact of the second lightlike plane element of the strip = origin. The correlation may be given as

$$Y = \sigma E_1, \qquad Z = (\sigma+1)E_D + (\sigma-1)E_2, \qquad 0 < \sigma < \infty .$$

If, after integration, σ is replaced by σ^2, the parametric form of the sheet becomes rational

$$X(x^+,x^-) = \frac{1}{2}\Big[\Big(\frac{(x^-)^3 - (x^+)^3}{3} + x^- - x^+\Big)E_D + \Big(\frac{(x^-)^3 - (x^+)^3}{3} + x^+ - x^-\Big)E_2 + \Big((x^+)^2 + (x^-)^2\Big)E_1\Big]$$

and eliminating $x^\pm$ gives the ('Cayley type') cubic equation

$$6(X^2 + T) + 6X^1(X^2 - T) - (X^2 - T)^3 = 0 .$$

The lines of the ruling are obtained by varying $x^\pm$ such that $x^+ - x^-$ remains fixed. It is not immediate now how they arise geometrically because a group of 'null rotations' (singular Lorentz transformations) is involved now. There seems to be no appealing physical description of this situation.

All the examples considered here admit some one-parameter subgroup of the Poincaré group in M^D as a group of self-transformations and one would have encountered them in a systematic search for string histories having this property. This program has not been carried out yet; we may just mention, in addition to the Lorentzian catenoids described above (which admit one-parameter groups of spacelike or timelike rotations) the following real pseudoeuclidean analog of Geiser's minimal surface: In a parametric representation, it is given by

$$X(u,v) = u\,v\,E_1 + \frac{u}{2}(E_2 - E_D) + \frac{1}{2}(Ku^3 - uv^2)(E_2 + E_D) ,$$

or, upon elimination, by

$$(X^1)^2 + (X^2)^2 - T^2 = K(X^2 - T)^4 .$$

This equation makes it obvious that the surface, which is timelike for $K < 0$ away from the origin, admits the null rotations

$$X^2 - T \to X^2 - T , \qquad X^1 \to X^1 + \varepsilon(X^2 - T) ,$$

$$X^2 + T \to X^2 + T - 2\varepsilon X^1 - \varepsilon^2(X^2 - T) .$$

(If one here adds the increments 2ε, ε^2, $-2\varepsilon^3/3$, one gets the invariance group of the Cayley surface.) Again, there seems to be no appealing physical description.

References

[1] As a reference for the old results on minimal surfaces in real or complex Euclidean 3-space I used
K. Strubecker, Differentialgeometrie III. Sammlung Göschen, Bd. 1180, Berlin 1969.

[2] H.K. Urbantke, Rep. Math. Phys. 21, 111 (1985).

[3] P. Budinich, Null Vectors, Spinors and Strings, SISSA preprint 10/86/E.P., Trieste.

[4] W.T. Shaw, Class. Qauntum Grav. 2, L113 (1985).

[5] L.P. Hughston, W.T. Shaw, Minimal Curves in 6 Dimensions. Preprint Univ. Oxford, 1986.

[6] M. Pinl, Monatsh. f. Math. u. Phys. 44, 1 (1935).

REPRESENTING SPINORS WITH DIFFERENTIAL FORMS

I M Benn, R W Tucker

Dec. 1986

ABSTRACT. The notion of a spinor field is approached in terms of a section of a bundle of irreducible representation modules for the Clifford algebra associated with a pseudo-Riemannian manifold. The existence of such a bundle is related to a lifting problem for transition maps of the Clifford algebra bundle. The notion of inequivalent spinor structures is reformulated in terms of inequivalent bundles of irreducible representation modules. Such a description of spinor fields de-emphasises the role played by orthonormal tangent frames that feature in the more traditional approach adopted by physicists. Using the vector space isomorphism between Clifford and exterior algebras we show how a bundle carrying an irreducible representation of the Clifford bundle can be constructed out of equivalence classes of certain differential forms, thus enabling spinor fields to be locally represented by local differential forms.

1. Introduction.

There are several equivalent definitions of a spinor structure (see Milnor[**1**] for example). The one most familiar to physicists (Milnor's first definition) is given in terms of a covering of the principal bundle of (oriented) orthonormal frames with a principal Spin bundle. One then defines spinor fields as sections of an associated bundle where the typical fibre carries the spin represenation of the Spin group. Due to the relation between the principal bundles this associated bundle inherits a connection from that in the orthonormal frame bundle, and one may also define a Clifford multiplication of sections by the othonormal vectors from a section of the othonormal frame bundle. Hence one may define a Dirac operator on spinor fields. (A self-contained exposition of this approach can be found in Petry[**2**] .)

Although the above procedure gives an elegant prescription for defining spinor fields and Dirac operators it is not especially direct. Also the essential role of the orthonormal frames has had what we regard as the unfortunate consequence of making all too common in the physics literature statements such as "...thus the introduction of spinor fields requires not just the metric tensor but its 'square root' the vierbein....". The desire to relinquish the apparent restriction to orthonormal frames has motivated Ne'eman[**3**] to contemplate the infinite-dimensional representations of certain coverings of the general linear group. We would like to draw an analogy with the tangent bundle of a pseudo-Riemannian manifold. This bundle may be directly constructed out of the collection of all tangent vectors. The pseudo-Riemannian connection may be defined by the Koszul definition. Alternatively we could form the principal bundle of linear frames and regard the tangent bundle as being an associated bundle. A connection, in the sense of Ehresmann, in the frame bundle then gives a connection in the tangent bundle. The pseudo-Riemannian structure gives a reduction of the general linear frame bundle to the orthonormal frame bundle so we could also regard the tangent bundle as being associated with this principal bundle. Of course this is not to say that one is constrained to always choose orthonormal frames. The beauty of the Ehresmann approach to connections in principal fibre bundles is its generality, if one is only interested in pseudo-Riemannian connections then the Koszul approach is more direct.

Our approach to defining spinor fields will be to work directly with the vector bundles rather than going via the principal bundles. We shall focus on the Clifford algebra and its representations rather than the orthogonal and Spin groups, thus de-emphasising the role of the orthonormal frames. This approach introduces nothing new other than a change of emphasis. Nevertheless such a change of emphasis can often have a bearing on the formulation of a physical theory.

We shall define spinor fields as sections of a bundle of irreducible representation modules for the Clifford algebra of the cotangent space. We have found it convenient to form such a bundle of modules out of equivalence classes of certain differential forms. This forms the subject of the final section.

2. Algebra Bundles and Their Irreducible Representation Modules.

Let A be a central simple algebra over a field F. Then, by Wedderburn's theorem,

$$A \simeq M_r(F) \otimes D(F) \equiv M_r(D) \qquad 2.1$$

where D is a central division algebra over F and $M_r(F)$ is a total matrix algebra, consisting of all order r matrices with entries in F. Thus A is isomorphic to the algebra of order r matrices with entries in D. Any irreducible representation of a simple algebra is equivalent to that carried by any minimal left ideal. Thus irreducible representations of A are carried by right D-modules of D-dimension r, isomorphic to D^r.

Let $\mathcal{A}(M)$ be a bundle of algebras over a paracompact manifold M, such that if $\mathcal{A}_x$ is the fibre above x then $\mathcal{A}_x \simeq A$. If $\{U_\alpha\}$ is an open covering of M then for each U_α there is a trivialising map $\Phi_\alpha : \pi_{\mathcal{A}}^{-1}(U_\alpha) \longrightarrow A$. If $U_{\alpha\beta} \equiv U_\alpha \cap U_\beta$ and $x \in U_{\alpha\beta}$, $u \in \pi_{\mathcal{A}}^{-1}(x)$ then

$$\Phi_\alpha(u) = f_{\alpha\beta}(x).\Phi_\beta(u) \qquad 2.2$$

for $f_{\alpha\beta}(x) \in Aut\, A$: the maps $f_{\alpha\beta} : U_{\alpha\beta} \longrightarrow Aut\, A$ being the transition maps of $\mathcal{A}(M)$. Since A is central simple every automorphism is inner. Thus

$$f_{\alpha\beta}(x).\Phi_\beta(u) = \tilde{f}_{\alpha\beta}(x)\Phi_\beta(u)\tilde{f}_{\alpha\beta}(x)^{-1} \qquad 2.3$$

for some $\tilde{f}_{\alpha\beta}(x) \in A^*$, the group of invertible elements from A. Equivalently

$$f_{\alpha\beta}(x).\Phi_\beta(u) = \Phi_\beta(a_{\alpha\beta}ua_{\alpha\beta}{}^{-1}) \quad 2.4$$

for some $a_{\alpha\beta} \in \mathcal{A}_x$.

Let $\mathcal{I}(M)$ be a bundle of right D-modules such that the fibre of $\mathcal{I}(M)$ above $x \in M$ carries an irreducible representation of $\mathcal{A}_x$. For $\{U_\alpha\}$ an open covering of M there is a trivialising map $\phi_\alpha : \pi_{\mathcal{I}}^{-1}(U_\alpha) \longrightarrow D^r$ for each U_α. For $u \in \pi_{\mathcal{I}}^{-1}(x)$

$$\phi_\alpha(u) = \tilde{f}_{\alpha\beta}(x)\phi_\beta(u) \quad 2.5$$

for transition maps $\tilde{f}_{\alpha\beta}(x)$ in A^*. Equivalently

$$\phi_\alpha(u) = \phi_\beta(a_{\alpha\beta}u) \quad 2.6$$

for some $a_{\alpha\beta} \in \mathcal{A}_x$. Our requirement that the transition maps of $\mathcal{I}(M)$ be in A^* is equivalent to requiring that a right action of D be defined on the bundle. For if $u \in \pi_{\mathcal{I}}^{-1}(x)$, for $x \in U_\alpha$, then for $q \in D$ we define uq to be the unique element of $\pi_{\mathcal{I}}^{-1}(x)$ such that

$$\phi_\alpha(uq) = \phi_\alpha(u)q \ . \quad 2.7$$

If $U_{\alpha\beta} \neq \emptyset$ then this only makes sense if

$$\tilde{f}_{\beta\alpha}(x)(\phi_\alpha(u)q) = (\tilde{f}_{\beta\alpha}(x)\phi_\alpha(u))q \ :$$

that is, if the transition maps are D-linear. We could consider the more general case of a bundle of irreducible modules, each fibre carrying an irreducible representation of $\mathcal{A}_x$, but without this D-linear structure. Each fibre would have a D-linear structure but the transition maps would lie in $Gl(rs, F)$ rather than $Gl(r, D)$ where $Dim_F D = s$. *Unless we specifically state otherwise we shall always mean our bundles of irreducible A-modules to be those admitting a D-linear structure, the transition maps lying in* A^*. We will consider the more general case in section 7.

Definition: Two bundles of irreducible modules, $\mathcal{I}(M)$ and $\mathcal{I}'(M)$, for the algebra bundle $\mathcal{A}(M)$ are equivalent iff there is a bundle isomorphism $\rho : \mathcal{I}(M) \longrightarrow \mathcal{I}'(M)$ such that $\rho(a_x u) = a_x\rho(u) \ \forall a_x \in \mathcal{A}_x, \ \forall u \in \pi_{\mathcal{I}}^{-1}(x)$.

2.8

3. Čech cohomology and the characteristic class of a bundle of algebras.

(The following should not be regarded as an obstacle to reading the subsequent sections: although this section will put them in proper perspective.)

Let $\mathcal{U} = \{U_\alpha\}$ be a good open cover of a paracompact manifold M. For a field F let $\mathcal{F}^*$ be the space of all non-zero F-valued germs of functions on M. Under multiplication[1] $\mathcal{F}^*$ forms an Abelian group (the function germs are non-vanishing). Let $\Phi^p(\mathcal{U},\mathcal{F}^*)$ be the set of all mappings of an ordered set of p open sets from $\mathcal{U}$ (whose intersection is non-empty) into the set of F^*-valued functions on their intersection. Under multiplication elements of $\Phi^p(\mathcal{U},\mathcal{F}^*)$ form an Abelian group. Elements of $\Phi^p(\mathcal{U},\mathcal{F}^*)$ will be called p-cochains. If $\phi \in \Phi^p(\mathcal{U},\mathcal{F}^*)$ then the image in $\mathcal{F}^*$ of $(U_{\alpha_1}, U_{\alpha_2}, ..., U_{\alpha_p})$ under ϕ will be denoted $\phi_{\alpha_1\alpha_2....\alpha_p}$. If $x \in U_{\alpha_1} \cap U_{\alpha_2} \cap .... \cap U_{\alpha_p}$ then the element in F^* obtained by evaluating an element of $\mathcal{F}^*$ on x will be denoted $\phi_{\alpha_1\alpha_2....\alpha_p}(x)$. The coboundary operator

$$\delta : \Phi^p(\mathcal{U},\mathcal{F}^*) \to \Phi^{p+1}(\mathcal{U},\mathcal{F}^*) \qquad 3.1$$

is defined by

$$(\delta\phi)_{\alpha_0\alpha_1...\alpha_{p+1}} = \Pi_{i=0}^{p+1}\, \phi^{(-1)^i}_{\alpha_0\alpha_1...\hat{\alpha}_i...\alpha_{p+1}} \qquad 3.2$$

where $\hat{\alpha}_i$ denotes the omision of that argument. A p-cochain ϕ such that $\delta\phi = 0$ is called a p-cocycle. A p-cochain ϕ such that $\phi = \delta\psi$, for some $(p-1)$-cochain ψ, is called a coboundary. The quotient of the group of p-cocycles modulo the group of p-coboundaries is the p'th Čech cohomology group, with values in $\mathcal{F}^*$, $\check{H}^p(M,\mathcal{F}^*)$. The crucial point is that these cohomology groups do not depend (up to isomorphism) on the open cover chosen.

We now consider the bundle of central simple algebras $\mathcal{A}(M)$. The transition maps

$$f_{\alpha\beta} : U_{\alpha\beta} \to Aut(A) \qquad 3.3$$

satisfy

$$f_{\alpha\beta} f_{\beta\gamma} = f_{\alpha\gamma} \quad \text{on } U_\alpha \cap U_\beta \cap U_\gamma . \qquad 3.4$$

This coherence condition may be rephrased in terms of the 1-cocycle f:

$$(\delta f)_{\alpha\beta\gamma} = 1 \quad \text{on } U_\alpha \cap U_\beta \cap U_\gamma . \qquad 3.5$$

Since all automorphisms of a central simple algebra are inner we have a surjective group homomorphism

$$\chi : A^* \to Aut\, A$$

[1] It is here more natural to use multiplicative notation throughout rather than the additive notation usually used.

given by the action

$$\chi(s).a = sas^{-1} \quad \forall a \in A \,. \tag{3.6}$$

As the centre of A is F the invertible elements in the field, F^*, constitute the kernel of χ and we have the exact sequence of groups:

$$1 \to F^* \to A^* \to Aut\, A \to 1\,. \tag{3.7}$$

A set of *lifts* of the transition maps of $\mathcal{A}(M)$ is a set of elements in A^*, $\{\tilde{f}_{\alpha\beta}\}$ such that $\chi(\tilde{f}_{\alpha\beta}) = f_{\alpha\beta}$ on all intersections. We may use such a set of lifts to construct a Čech 2-cocycle:

$$\phi^{\tilde{f}}_{\alpha_1\alpha_2\alpha_3} = \tilde{f}_{\alpha_1\alpha_2}\tilde{f}_{\alpha_2\alpha_3}\tilde{f}^{-1}_{\alpha_1\alpha_3}\,. \tag{3.8}$$

Note $\chi(\phi^{\tilde{f}}_{\alpha_1\alpha_2\alpha_3}) = f_{\alpha_1\alpha_2}f_{\alpha_2\alpha_3}f^{-1}_{\alpha_1\alpha_3} = 1$ since the $\{f_{\alpha\beta}\}$ are transition maps for $\mathcal{A}(M)$. Hence $\phi^{\tilde{f}}_{\alpha_1\alpha_2\alpha_3} \in \mathcal{F}^*$, the kernel of χ. That $\phi^{\tilde{f}}$ is a 2-cocycle follows by computing

$$\begin{aligned}(\delta\phi^{\tilde{f}})_{\alpha_1\alpha_2\alpha_3\alpha_4} &= \phi^{\tilde{f}}_{\alpha_2\alpha_3\alpha_4}\phi^{\tilde{f}}_{\alpha_1\alpha_3\alpha_4}{}^{-1}\phi^{\tilde{f}}_{\alpha_1\alpha_2\alpha_4}\phi^{\tilde{f}}_{\alpha_1\alpha_2\alpha_3}{}^{-1} \\ &= \tilde{f}_{\alpha_2\alpha_3}\tilde{f}_{\alpha_3\alpha_4}\tilde{f}^{-1}_{\alpha_2\alpha_4}[\tilde{f}_{\alpha_1\alpha_2}\tilde{f}_{\alpha_2\alpha_4}\tilde{f}^{-1}_{\alpha_1\alpha_4}][\tilde{f}_{\alpha_1\alpha_4}\tilde{f}^{-1}_{\alpha_3\alpha_4}\tilde{f}^{-1}_{\alpha_1\alpha_3}]\tilde{f}_{\alpha_1\alpha_3}\tilde{f}^{-1}_{\alpha_2\alpha_3}\tilde{f}^{-1}_{\alpha_1\alpha_2}\end{aligned}$$

where we have commuted $\phi^{\tilde{f}}{}^{-1}_{\alpha_1\alpha_3\alpha_4}$ with $\phi^{\tilde{f}}_{\alpha_1\alpha_2\alpha_4}$ in the above. Finally commuting $\phi^{\tilde{f}}_{\alpha_2\alpha_3\alpha_4}$ to a position in front of $\tilde{f}^{-1}_{\alpha_1\alpha_2}$ we verify

$$\delta\phi^{\tilde{f}} = id_{(3)}\,. \tag{3.9}$$

To see that the cohomology class of $\phi^{\tilde{f}}$ does not in fact depend on the choice of lifts consider new lifts:

$$\tilde{g}_{\alpha_1\alpha_2} = \tilde{f}_{\alpha_1\alpha_2}\lambda_{\alpha_1\alpha_2} \quad \forall \alpha_1, \alpha_2 \tag{3.10}$$

where $\lambda_{\alpha_1\alpha_2} \in \mathcal{F}^*$. The 2-cochain $\phi^{\tilde{g}}$ associated with these lifts is defined by

$$\begin{aligned}\phi^{\tilde{g}}_{\alpha_1\alpha_2\alpha_3} &= \tilde{g}_{\alpha_1\alpha_2}\tilde{g}_{\alpha_2\alpha_3}\tilde{g}^{-1}_{\alpha_1\alpha_3} \\ &= \phi^{\tilde{f}}_{\alpha_1\alpha_2\alpha_3}\lambda_{\alpha_1\alpha_2}\lambda_{\alpha_2\alpha_3}\lambda^{-1}_{\alpha_1\alpha_3}\,.\end{aligned}$$

But $(\delta\lambda)_{\alpha_1\alpha_2\alpha_3} = \lambda_{\alpha_2\alpha_3}\lambda^{-1}_{\alpha_1\alpha_3}\lambda_{\alpha_1\alpha_2}$ hence

$$\phi^{\tilde{g}}_{\alpha_1\alpha_2\alpha_3} = (\phi^{\tilde{f}}\delta\lambda)_{\alpha_1\alpha_2\alpha_3} \quad \forall \alpha_1, \alpha_2, \alpha_3. \tag{3.11}$$

Since they differ by a coboundary the 2-cocycles $\phi^{\tilde{f}}$ and $\phi^{\tilde{g}}$ lie in the same equivalence class of $\check{H}^2(M, \mathcal{F}^*)$. Thus any set of lifts may be used to define a characteristic class $w(\mathcal{A}) \in \check{H}^2(M, \mathcal{F}^*)$ of the algebra bundle $\mathcal{A}(M)$.

Now if we can *coherently* lift the transition maps of $\mathcal{A}(M)$ to a set of lifts $\{\tilde{f}_{\alpha_1\alpha_2}\}$, so that

$$\tilde{f}_{\alpha_1\alpha_2}\tilde{f}_{\alpha_2\alpha_3} = \tilde{f}_{\alpha_1\alpha_3} \quad \forall \alpha_1, \alpha_2, \alpha_3 \,,$$

then $\phi^{\tilde{f}}_{\alpha_1\alpha_2\alpha_3} = 1$ and $w(\mathcal{A}) = [id_{(2)}]$. Conversely if the characteristic class $w(\mathcal{A})$ is trivial then for any lifts $\phi^{\tilde{f}} = \delta\lambda$ for some 1-cochain λ. Although the lifts $\tilde{f}$ are not necessarily coherent we may choose new lifts

$$\tilde{g}_{\alpha_1\alpha_2} = \tilde{f}_{\alpha_1\alpha_2}\lambda^{-1}_{\alpha_1\alpha_2} \quad \forall \alpha_1, \alpha_2 \ . \qquad 3.12$$

These lifts are coherent since

$$\begin{aligned}\phi^{\tilde{g}}_{\alpha_1\alpha_2\alpha_3} &= \phi^{\tilde{f}}_{\alpha_1\alpha_2\alpha_3}\lambda^{-1}_{\alpha_1\alpha_2}\lambda^{-1}_{\alpha_2\alpha_3}\lambda_{\alpha_1\alpha_3} \\ &= (\delta\lambda)_{\alpha_1\alpha_2\alpha_3}\lambda^{-1}_{\alpha_1\alpha_2}\lambda^{-1}_{\alpha_2\alpha_3}\lambda_{\alpha_1\alpha_3} \\ &= \lambda_{\alpha_2\alpha_3}\lambda^{-1}_{\alpha_1\alpha_3}\lambda_{\alpha_1\alpha_2}\lambda^{-1}_{\alpha_1\alpha_2}\lambda^{-1}_{\alpha_2\alpha_3}\lambda_{\alpha_1\alpha_3} \\ &= 1 \ .\end{aligned}$$

Thus the incoherencies in arbitrary lifts can always be removed into a Čech coboundary if $w(\mathcal{A})$ is trivial.

If the class in $\check{\mathrm{H}}^2(M, \mathcal{F}^*)$, determined by any set of lifts, is trivial then the coherent lifts can be classified by an element of $\check{\mathrm{H}}^1(M, \mathcal{F}^*)$. If $\tilde{f}$ and $\tilde{g}$ are two sets of coherent lifts then a 1-cocycle is defined by

$$\psi^{\tilde{f}\tilde{g}}_{\alpha_1\alpha_2} = \tilde{f}_{\alpha_1\alpha_2}\tilde{g}^{-1}_{\alpha_1\alpha_2} \ . \qquad 3.13$$

Since $\{\tilde{f}_{\alpha\beta}\}$ and $\{\tilde{g}_{\alpha\beta}\}$ are both lifts of the transition maps $\{f_{\alpha\beta}\}$ then $\chi(\psi^{\tilde{f}\tilde{g}}) = 1$ and so certainly $\psi^{\tilde{f}\tilde{g}}_{\alpha\beta} \in \mathcal{F}^*$. It is in fact a 1-cocycle since

$$\begin{aligned}(\delta\psi^{\tilde{f}\tilde{g}})_{\alpha_1\alpha_2\alpha_3} &= \psi^{\tilde{f}\tilde{g}}_{\alpha_2\alpha_3}{\psi^{\tilde{f}\tilde{g}}}^{-1}_{\alpha_1\alpha_3}\psi^{\tilde{f}\tilde{g}}_{\alpha_1\alpha_2} = \psi^{\tilde{f}\tilde{g}}_{\alpha_2\alpha_3}\psi^{\tilde{f}\tilde{g}}_{\alpha_1\alpha_2}{\psi^{\tilde{f}\tilde{g}}}^{-1}_{\alpha_1\alpha_3} \\ &= \tilde{f}_{\alpha_2\alpha_3}\tilde{g}_{\alpha_3\alpha_2}\tilde{f}_{\alpha_1\alpha_2}\tilde{g}_{\alpha_2\alpha_1}\tilde{g}_{\alpha_1\alpha_3}\tilde{f}_{\alpha_3\alpha_1} \\ &= \tilde{f}_{\alpha_1\alpha_2}\tilde{f}_{\alpha_2\alpha_3}\tilde{g}_{\alpha_3\alpha_2}\tilde{g}_{\alpha_2\alpha_3}\tilde{f}_{\alpha_3\alpha_1} \\ &= 1 \ .\end{aligned}$$

We have used repeatedly $\tilde{f}_{\alpha\beta} = \tilde{f}^{-1}_{\beta\alpha}$. Two sets of lifts, $\tilde{f}$ and $\tilde{g}$, are *equivalent* if $\psi^{\tilde{f}\tilde{g}}$ is cohomologous to the identity. That is, $\psi^{\tilde{f}\tilde{g}} = \delta\mu$ for some 0-cochain μ, or

$$\tilde{g}_{\alpha\beta} = \mu_\alpha\tilde{f}_{\alpha\beta}\mu^{-1}_\beta \ . \qquad 3.14$$

Inequivalent lifts define an element in $\check{H}^1(M, \mathcal{F}^*)$. In fact the number of inequivalent lifts is precisely the dimension of $\check{H}^1(M, \mathcal{F}^*)$. For if $\tilde{f}$ is one set of coherent lifts and λ is any 1-cocycle then we may define lifts by $\tilde{g} = \tilde{f}\lambda$. These are certainly lifts since λ is a cochain; they are coherent since λ is a cocycle. We then have $\psi^{\tilde{g}\tilde{f}} = \lambda$ and we obtain classes of lifts as λ ranges through the cohomology classes in $\check{H}^1(M, \mathcal{F}^*)$.

We now restrict to the case in which the field F is the real numbers $\mathcal{R}$. Let $|a|$ be the determinant of the $\mathcal{R}$-linear map from A to A given by $x \mapsto ax$. If A_1^* is the sub-group of A^* consisting of elements of determinant plus or minus one and $\chi_1 : A_1^* \to Aut\, A$ is the restriction of χ then we have $ker\, \chi_1 = \pm 1 \simeq Z_2$. We thus have coherent lifts of the transition maps of $\mathcal{A}(M)$ into A_1^* if and only if $\varpi(\mathcal{A}) \in \check{H}^2(M, Z_2)$ is trivial. As topological Abelian groups $\mathcal{R}^* \simeq \mathcal{R}^+ \times Z_2$ and $\check{H}^2(M, \mathcal{R}^*) \simeq \check{H}^2(M, \mathcal{R}^+) \times \check{H}^2(M, Z_2)$. Since $\check{H}^2(M, \mathcal{R}^+)$ is trivial the obstruction to coherently lifting into A_1^* is the same as that to coherently lifting into A^*.

4. Bundles of Irreducible Representation Modules as Equivalence Classes of Lifts.

In this section we will establish the following:

There exists a bundle of irreducible representation modules for an algebra bundle $\mathcal{A}(M)$ iff the transition maps of $\mathcal{A}(M)$ can be coherently lifted from $Aut\, A$ to A^*. The equivalence classes of coherent lifts are in one to one correspondence with the equivalence classes of such bundles of representation modules.

4.1

Assume firstly that transition maps $\{f_{\alpha\beta}\}$ for $\mathcal{A}(M)$, determined by the trivialising maps $\{\Phi_\alpha\}$, can be coherently lifted to $\{\tilde{f}_{\alpha\beta}\} \in A^*$. Then we can make a bundle of D-linear spaces in the standard way. We set $X_\alpha = U_\alpha \times D^r$, $X = \bigcup_\alpha X_\alpha$ and form the bundle $\mathcal{I}(M) = X/\varsigma$ where the equivalence relation ς is given by

$$(\alpha, x, m) \sim (\beta, y, n) \text{ iff } x = y \text{ and } n = \tilde{f}_{\beta\alpha}(x)m \qquad 4.2$$

for $x, y \in M$, $m, n \in D^r$. If $u \in \pi_{\mathcal{I}}^{-1}(U_\alpha)$ then it is represented by the triple (α, x, m). The trivialising map ϕ_α is such that $\phi_\alpha(u) = m$. The bundle $\mathcal{I}(M)$ that we have constructed carries a representation of $\mathcal{A}(M)$ by defining $a_x u$ such that

$$\phi_\alpha(a_x u) = \Phi_\alpha(a_x)\phi_\alpha(u) \quad \forall a_x \in \mathcal{A}_x \ \forall u \in \pi_{\mathcal{I}}^{-1}(x) \ \forall x \in M \ . \qquad 4.3$$

This is well defined, since the above expression is equal to that formed with the maps ϕ_β and Φ_β if and only if

$$\Phi_\beta(a_x) = f_{\beta\alpha}(x).\Phi_\alpha(a_x) = \tilde{f}_{\beta\alpha}(x)\Phi_\alpha(a_x)\tilde{f}_{\beta\alpha}^{-1} \ . \qquad 4.4$$

Thus given a set of coherent lifts of the transition maps of $\mathcal{A}(M)$ we can construct a bundle of irreducible modules.

Conversely now suppose that a bundle of irreducible modules for the algebra bundle $\mathcal{A}(M)$ exists. We pick an arbitrary set of trivialising maps for $\mathcal{A}(M)$, $\{\Phi_\alpha\}$, with transition maps $\{f_{\alpha\beta}\}$. For each U_α we may choose a trivialising map ϕ_α for $\pi_I^{-1}(U_\alpha)$ such that (4.3) is satisfied. If the resulting transition maps are $\{\tilde{f}_{\alpha\beta}\}$ then since the bundle $\mathcal{I}(M)$ carries a representation of $\mathcal{A}(M)$ they satisfy (4.4). That is, under the homomorphism χ we have $\chi(\tilde{f}_{\beta\alpha}(x)) = f_{\beta\alpha}(x)$ and the resulting transition maps for $\mathcal{I}(M)$ are coherent lifts of those for $\mathcal{A}(M)$. This prescription of using trivialising maps for $\mathcal{A}(M)$ to obtain trivialising maps for $\mathcal{I}(M)$ via (4.3) is not unique. Suppose that $\{\phi_\alpha\}$ and $\{\phi'_\alpha\}$ both satisfy (4.3) with $\phi'_\alpha(u) = \mu_\alpha(x)\phi_\alpha(u)$ for $\mu_\alpha(x) \in A^*$. Then we must have $\mu_\alpha(x)^{-1}\Phi_\alpha(a_x)\mu_\alpha(x) = \Phi_\alpha(a_x)$ for all $a_x \in \mathcal{A}_x$. That is, $\mu_\alpha(x) \in F^*$ since A is central simple. The transition maps are then related by $\tilde{f}'_{\alpha\beta}(x) = \tilde{f}_{\alpha\beta}(x)\mu_\alpha(x)\mu_\beta(x)^{-1}$ and are thus equivalent lifts of those of $\mathcal{A}(M)$. So if we have the bundle $\mathcal{I}(M)$ we may use (4.3) to obtain coherent lifts of the transition maps of $\mathcal{A}(M)$: any such lifts lying in one equivalence class.

Suppose now that we have two bundles of irreducible representation modules, $\mathcal{I}(M)$ and $\mathcal{I}'(M)$, with the bundle isomorphism $\rho : \mathcal{I}(M) \longrightarrow \mathcal{I}'(M)$ defining their equivalence. For some given set of local trivialisations of $\mathcal{A}(M)$ let $\{\phi_\alpha\}$ be trivialising maps for $\mathcal{I}(M)$ satisfying (4.3) and with resulting transition maps $\{\tilde{f}_{\alpha\beta}\}$. Then $\{\phi'_\alpha\}$ are local trivialising maps for $\mathcal{I}'(M)$ if $\phi'_\alpha \equiv \phi_\alpha \circ \rho^{-1}$. Further,

$$\phi'_\alpha(\rho(a_x u)) = \phi_\alpha(a_x u) = \Phi_\alpha(a_x)\phi_\alpha(u)$$

since ϕ_α satisfies (4.3). Using $\rho(a_x u) = a_x \rho(u)$ then gives

$$\phi'_\alpha(a_x \rho(u)) = \Phi_\alpha(a_x)\phi'_\alpha(\rho(u))$$

and we see that ϕ'_α also satisfies (4.3). It readily follows that the transition maps for $\mathcal{I}'(M)$ given by $\{\phi'_\alpha\}$ are the same as those for $\mathcal{I}(M)$ given by $\{\phi_\alpha\}$. Thus certainly equivalent bundles of irreducible representation modules for $\mathcal{A}(M)$ give rise to equivalent lifts.

It remains to establish a bundle isomorphism demonstrating the equivalence of bundles of representation modules which are associated with equivalent lifts of the transition functions of $\mathcal{A}(M)$. For given local trivialisations of $\mathcal{A}(M)$ suppose that ϕ_α and ϕ'_α are trivialising maps for $\mathcal{I}(M)$ and $\mathcal{I}'(M)$ respectively such that (4.3) is satisfied. If the resulting transition maps $\tilde{f}_{\alpha\beta}$ and $\tilde{f}'_{\alpha\beta}$ are equivalent then

$$\tilde{f}'_{\alpha\beta}(x) = \tilde{f}_{\alpha\beta}(x)\mu_\alpha(x)\mu_\beta(x)^{-1} \quad \text{for some } \mu_\alpha(x) \in F^*,\ \forall x \in U_{\alpha\beta}\ . \qquad 4.5$$

For each $u \in \pi_I^{-1}(x)$ there is a unique $u' \in \pi_{I'}^{-1}(x)$ such that $\phi_\alpha(u) = \phi'_\alpha(u')$. So we can define $\varsigma_\alpha : \pi_I^{-1}(U_\alpha) \longrightarrow \pi_{I'}^{-1}(U_\alpha)$ by requiring

$$\phi_\alpha(u) = \phi'_\alpha(\varsigma_\alpha(u))\ . \qquad 4.6$$

There is a unique $a_{x\alpha} \in \mathcal{A}_x$ such that

$$\Phi_\alpha(a_{x\alpha}) = \mu_\alpha(x) \ . \tag{4.7}$$

We now define

$$\begin{aligned} \rho_\alpha : \pi_I^{-1}(U_\alpha) &\longrightarrow \pi_{I'}^{-1}(U_\alpha) \\ u &\longrightarrow a_{x\alpha}\varsigma_\alpha(u) \ . \end{aligned} \tag{4.8}$$

These maps have been constructed such that if $u \in \pi_I^{-1}(U_{\alpha\beta})$ then $\rho_\alpha(u) = \rho_\beta(u)$. For

$$\begin{aligned} \phi'_\beta(\rho_\beta(u)) = \phi'_\beta(a_{x\beta}\varsigma_\beta(u)) &= \Phi_\beta(a_{x\beta})\phi'_\beta(\varsigma_\beta(u)) \quad \text{by (4.3)} \\ &= \Phi_\beta(a_{x\beta})\phi_\beta(u) \quad \text{by (4.5)} \\ &= \Phi_\beta(a_{x\beta})\tilde{f}_{\beta\alpha}(x)\phi_\alpha(u) = \Phi_\beta(a_{x\beta})\tilde{f}_{\beta\alpha}(x)\phi'_\alpha(\varsigma_\alpha(u)) \\ &= \Phi_\beta(a_{x\beta})\tilde{f}_{\beta\alpha}(x)\Phi_\alpha(a_{x\alpha})^{-1}\phi'_\alpha(a_{x\alpha}\varsigma_\alpha(u)) \\ &= \Phi_\beta(a_{x\beta})\tilde{f}_{\beta\alpha}(x)\Phi_\alpha(a_{x\alpha})^{-1}\tilde{f}'_{\alpha\beta}(x)\phi'_\beta(a_{x\alpha}\varsigma_\alpha(u)) \\ &= \Phi_\beta(a_{x\beta})\tilde{f}_{\beta\alpha}(x)\Phi_\alpha(a_{x\alpha})^{-1}\tilde{f}'_{\alpha\beta}(x)\phi'_\beta(\rho_\alpha(u)) \\ &= \mu_\beta(x)\tilde{f}_{\beta\alpha}(x)\mu_\alpha(x)^{-1}\tilde{f}'_{\alpha\beta}(x)\phi'_\beta(\rho_\alpha(u)) \quad \text{by (4.7).} \end{aligned}$$

Since $\mu_\alpha(x) \in F^*$ it follows from (4.5) that $\phi'_\beta(\rho_\beta(u)) = \phi'_\beta(\rho_\alpha(u))$ and hence $\rho_\beta(u) = \rho_\alpha(u)$. We thus have a well defined bundle isomorphism ρ, defined such that if $u \in \pi_I^{-1}(U_\alpha)$ then $\rho(u) = \rho_\alpha(u)$.

5. Clifford Algebras.

The algebras that we shall be concerned with are the real Clifford algebras or their complexifications [**4**]. We here summarise properties of the real Clifford algebras.

Let V be a real non-degenerate orthogonal space whose metric g can be diagonalised to have p plus and q minus signs. If $p + q = n$ then the Clifford algebra of V, $C(V,g)$, can be defined as that 2^n-dimensional algebra generated by V with the relations

$$xy + yx = 2g(x,y) \quad \forall x,y \in V \ . \tag{5.1}$$

The structure of the Clifford algebra is determined by p and q, $C(V,g)$ being isomorphic to some standard algebra $C_{p,q}(\mathcal{R})$. For n even the Clifford algebra is central simple (over $\mathcal{R}$). For n odd the Clifford algebra is either the direct sum of two isomorphic central simple algebras or isomorphic to the algebra of all complex matrices.

The involutory automorphism η satisfies

$$\eta x = -x \quad \forall x \in V \ . \tag{5.2}$$

The involutory anti-automorphism ξ satisfies

$$x^\xi = x \quad \forall x \in V \ . \tag{5.3}$$

We may naturally[1] identify elements of the Clifford algebra of V with elements of the exterior algebra of V with the products related such that

$$am = a \wedge m + i_{\tilde{a}} m \quad \forall a \in V \quad \forall m \in \Lambda(V) \ . \tag{5.4}$$

The interior operator $i_{\tilde{a}}$ is a graded derivation on the exterior algebra satisfying $i_{\tilde{a}} b = g(a, b)$ for all a, b in V. Although we can write down the general relation between the exterior and Clifford products it is usually sufficient to know that it exists, in practice using (5.4) and the following:

$$ma = a \wedge \eta m - i_{\tilde{a}} \eta m \quad \forall a \in V \quad \forall m \in \Lambda(V) \ . \tag{5.5}$$

The Clifford group Γ consists of all s such that

$$sxs^{-1} \in V \quad \forall x \in V \ . \tag{5.6}$$

The vector representation χ is defined by

$$\chi : \Gamma \longrightarrow O(p,q)$$

$$\chi(s).x = sxs^{-1} \ . \tag{5.7}$$

Let $\Gamma^{\pm}$ be the sub-group of Γ consisting of all s such that $\eta s = \pm s$ and Γ^+ be the sub-group of all s such that $\eta s = s$. The "twisted" vector representation ϕ is defined by

$$\phi : \Gamma^{\pm} \longrightarrow O(p,q)$$

$$\phi(s).x = \eta sxs^{-1} \ . \tag{5.8}$$

A norm homomorphism λ is defined by

$$\lambda : \Gamma^{\pm} \longrightarrow \mathcal{R}^*$$

[1]The Clifford and exterior algebras of V can both be defined as quotients of the tensor algebra of V. With this definition elements of the exterior algebra are equivalence classes of tensors, each class represented by a totally anti-symmetric tensor. Elements of the Clifford algebra are equivalence classes of tensors defined with a different equivalence relation. Each such class is represented by one and only one inhomogeneous sum of anti-symmetric tensors.

$$\lambda(s) = s^{\xi} s \ . \tag{5.9}$$

Another norm homomorphism is given by μ:

$$\mu : \Gamma^{\pm} \longrightarrow \mathcal{R}^*$$

$$\mu(s) = s^{\xi\eta} s \ . \tag{5.10}$$

Let ${}_{\pm}\Gamma^{\pm}$ be the sub-group of $\Gamma^{\pm}$ consisting of elements s with $\lambda(s) = \pm 1$, and ${}_{+}\Gamma^{\pm}$ the sub-group of all s with $\lambda(s) = +1$. Similarly we define ${}_{\pm}\Gamma^{+}$ and ${}_{+}\Gamma^{+}$. In an analogous way we define ${}^{\pm}\Gamma^{\pm}$ and ${}^{+}\Gamma^{\pm}$ with the norm μ. (Obviously ${}^{\pm}\Gamma^{+} =_{\pm} \Gamma^{+}$.) The following summarises the image of these groups under the "twisted" vector representation:

$$\begin{aligned} {}_{\pm}\Gamma^{\pm} &\longrightarrow O(p,q) \\ {}_{+}\Gamma^{\pm} &\longrightarrow O\uparrow(p,q) \\ {}^{+}\Gamma^{\pm} &\longrightarrow O^{+}(p,q) \\ {}_{\pm}\Gamma^{+} &\longrightarrow SO(p,q) \\ {}_{+}\Gamma^{+} &\longrightarrow SO^{+}(p,q) \ . \end{aligned} \tag{5.11}$$

The group $O\uparrow(p,q)$ is the orthochronous orthogonal group, the group that preserves the orientation of the negative-definite sub-space, whilst $O^{+}(p,q)$ preserves the "spacelike" semi-orientation.

If p and q are both even then the images of the various sub-groups under χ are the same as under ϕ: one may replace the map ϕ with χ in (5.11). If however p and q are both odd then $\chi({}_{+}\Gamma^{\pm}) = O^{+}(p,q)$ and $\chi({}^{+}\Gamma^{\pm}) = O\uparrow(p,q)$. When $p+q$ is odd then $\chi(\Gamma) = \chi(\Gamma^{+}) = \phi(\Gamma^{+})$. So the improper orthogonal transformations of V extend to outer automorphisms of the Clifford algebra.

One can define the "twisted Clifford group" $\tilde{\Gamma}$ by

$$s \in \tilde{\Gamma} \quad \text{iff} \quad \eta s x s^{-1} \in V \quad \forall x \in V \ . \tag{5.12}$$

It then follows that

$$\tilde{\Gamma} = \Gamma^{\pm} \ . \tag{5.13}$$

We sketch the proof. One has $\Gamma^{\pm} \subset \tilde{\Gamma}$, and any even or odd element (under η) of $\tilde{\Gamma}$ is in $\Gamma^{\pm}$. We need to show that all elements of $\tilde{\Gamma}$ are even or odd. Any orthogonal transformation can be written as a finite product of reflections and hence as $\phi(x_1 x_2 ... x_r)$ for some non-isotropic vectors $\{x_i\}$. Since $\phi(s) \in O(p,q)$ for all $s \in \tilde{\Gamma}$ it remains to show that the kernel of ϕ is $\mathcal{R}^*$. If $\phi(s) = 1$ and $s_{\pm} \equiv \frac{1}{2}(1 \pm \eta s)$ then s_{+} commutes with all elements of V whilst s_{-} anti-commutes with them. Hence $s_{+} \in \mathcal{R}^*$ and we want to show that $s_{-} = 0$. If a is odd such that

$ax = -xa$ for all $x \in V$ then it follows from (5.4) and (5.5) that $i_{\tilde{x}}a = 0 \ \forall x \in V$ and hence $a = 0$.

If n is even then all elements of Γ are either even or odd and so in this case $\tilde{\Gamma} = \Gamma$.

The group ${}_{\pm}\Gamma^{\pm}$ is also known as $Pin(p,q)$, with ${}_{\pm}\Gamma^{+}$ as $Spin(p,q)$ and ${}_{+}\Gamma^{+}$ as $Spin^{+}(p,q)$.

6. Clifford Bundles.

Let M be a real paracompact pseudo-Riemannian manifold. We can identify the vector space of $C(T_x^*M, g_x)$ with $\Lambda(T_x^*M)$, the product being given by (5.4). The Clifford bundle[5] $C(M)$ of M is the exterior bundle $\Lambda(M)$ of M, Clifford multipliction being defined on each fibre by (5.4). *We shall now restrict to the case of M even-dimensional; ensuring that the Clifford algebra is central simple over $\mathcal{R}$.* We can now make the following definition:

Definition: A spin structure for $C(M)$ is an equivalence class of bundles of irreducible representation modules for it, represented by $\mathcal{I}(M)$. A section of $\mathcal{I}(M)$ is called a spinor field on M. M is a spin manifold if and only if $C(M)$ admits a spin structure.

6.1

(We recall our earlier proviso that $\mathcal{I}(M)$ is assumed to have a D-linear structure.) We shall now relate this definition to more usual ones.

We firstly observe that the structure group of the Clifford bundle is reducible from $Aut\, C_{p,q}(\mathcal{R})$ to $O(p,q)$. This follows immediately from the existence of the pseudo-Riemannian structure and the fact that the Clifford bundle is the exterior bundle whose fibres are equiped with a Clifford structure. Thus the transition maps of the orthonormal frame bundle may be taken as transition maps for the Clifford bundle. From (4.1) we now see that a spin structure, as defined in (6.1), is in one to one correspondence with an equivalence class of coherent lifts from $O(p,q)$ to the Clifford group Γ. (Recall that any $s \in C_{p,q}^*(\mathcal{R})$ such that $sVs^{-1} \in V$ is in the Clifford group.) That is, if $\mathcal{I}(M)$ exists then there are trivialising maps $\{\phi_\alpha\}$, with associated transition maps $\{\tilde{f}_{\alpha\beta}\}$, such that (4.3) holds where the trivialising maps $\{\Phi_\alpha\}$ for $C(M)$ produce transition maps $\{f_{\alpha\beta}\}$ in $O(p,q)$. It is a standard result (see for example [**5**]) that the structure group of a bundle is reducible from G to H if G/H is diffeomorphic to $\mathcal{R}^n$. Now $\Gamma/{}_{\pm}\Gamma \simeq \mathcal{R}^+$, the multiplicative group of positive reals. This is diffeomorphic to the additive group of all reals, and hence any Γ-bundle is reducible to a ${}_{\pm}\Gamma$-bundle. So if $\{\tilde{f}_{\alpha\beta}\}$ are the Γ-valued transition maps of $\mathcal{I}(M)$ then there are Γ-valued maps $\{F_\alpha\}$ such that

$$F_\alpha(x)\tilde{f}_{\alpha\beta}(x)F_\beta(x)^{-1} = h_{\alpha\beta}(x) \qquad 6.2$$

for some $_{\pm}\Gamma$-valued transition maps $h_{\alpha\beta}$. Out of the norm homomorphism λ we construct a norm homomorphism ϑ:

$$\vartheta : \Gamma \longrightarrow \mathcal{R}^+$$

$$\vartheta(s) = \sqrt{|\lambda(s)|} \tag{6.3}$$

where the positive square root is taken. For $h_{\alpha\beta}(x) \in {}_{\pm}\Gamma$ we have $\vartheta\big(h_{\alpha\beta}(x)\big) = 1$. So if

$$\vartheta\big(F_\alpha(x)\big) \equiv \mu_\alpha(x) \tag{6.4}$$

then taking the norm of (6.2) produces

$$\vartheta\big(\tilde{f}_{\alpha\beta}(x)\big) = \mu_\beta(x)\mu_\alpha(x)^{-1} \ . \tag{6.5}$$

So we may define new trivialising maps for $\mathcal{I}(M)$, still satisfying (4.3), by

$$\phi'_\alpha(x) \equiv \mu_\alpha(x)\phi_\alpha(x) \ . \tag{6.6}$$

Since the maps μ_α take values in the centre of the algebra then (6.5) shows that the new transition maps $\tilde{f'}_{\alpha\beta}(x)$ are related to the old ones by

$$\tilde{f'}_{\alpha\beta}(x) = \frac{\tilde{f}_{\alpha\beta}(x)}{\vartheta\big(\tilde{f}_{\alpha\beta}(x)\big)} \ . \tag{6.7}$$

Hence $\tilde{f'}_{\alpha\beta}(x) \in {}_{\pm}\Gamma$.

We have actually shown that a spinor structure in the sense of (6.1) is in one to one correspondence with an equivalence class of coherent lifts of the transition maps of the orthonormal frame bundle into $_{\pm}\Gamma$. For the case in which M is both space and time orientable then we have lifts into $_{+}\Gamma^{+}$, otherwise known as $Spin^{+}(p,q)$, and (6.1) is then obviously equivalent to the standard definition. For the case in which M is not space-and-time-orientable then our definition (6.1) is equivalent to what is usually called a Pinor structure.

7. Some Generalisations.

The real central division algebras are the real numbers themselves, $\mathcal{R}$, and the quaternions $\mathcal{H}$. We have hitherto required any bundle of irreducible representation modules for those algebras whose Wedderburn decomposition involves the quaternions to have an $\mathcal{H}$-linear structure. We now consider relaxing this requirement, the transition maps not necessarily being $\mathcal{H}$-linear. We shall regard $A \simeq M_r(\mathcal{R}) \otimes \mathcal{H}$ as being a sub-algebra of M_{4r}, the irreducible representations of A being real vector spaces of dimension $4r$. If $\mathcal{I}(M)$ is a bundle carrying an irreducible representation of $\mathcal{A}(M)$ then we may choose local trivialising maps such that (4.3) holds. The transition maps are related as in (4.4) where now the transition maps $\tilde{f}_{\beta\alpha}(x)$ of $\mathcal{I}(M)$ are in $Gl(4r, \mathcal{R})$, not necessarily in A^*. Since all automorphisms of A are inner there is an $s_{\beta\alpha}(x) \in A^*$ such that

$$f_{\beta\alpha}(x).\Phi_\alpha(a_x) = s_{\beta\alpha}(x)\Phi_\alpha(a_x)s_{\beta\alpha}(x)^{-1} \quad . \qquad 7.1$$

We thus have $s_{\beta\alpha}(x)\Phi_\alpha(a_x)s_{\beta\alpha}(x)^{-1} = \tilde{f}_{\beta\alpha}(x)\Phi_\alpha(a_x)\tilde{f}_{\beta\alpha}(x)^{-1}$

and hence $s_{\beta\alpha}(x)^{-1}\tilde{f}_{\beta\alpha}(x)$ commutes with all A. If B is that sub-algebra of $M_{4r}(\mathcal{R})$ that commutes with A then $B \simeq \mathcal{H}$ $(M_{4r}(\mathcal{R}) \simeq M_r(\mathcal{R}) \otimes \mathcal{H} \otimes \mathcal{H})$. The transition maps of $\mathcal{I}(M)$ can be expressed as

$$\tilde{f}_{\beta\alpha}(x) = s_{\beta\alpha}(x)q_{\beta\alpha}(x) \quad \text{for} \quad s_{\beta\alpha} \in A^* \quad \text{and} \quad q_{\beta\alpha}(x) \in B \ . \qquad 7.2$$

Such a decomposition is not unique, there being the freedom to scale $s_{\beta\alpha}(x)$ by an element of R^*, $q_{\beta\alpha}(x)$ scaling with the inverse. Quaternion conjugation gives a norm on the quaternion algebra, the group of unit-norm quaternions being isomorphic to $SU(2)$. We can express $\tilde{f}_{\beta\alpha}(x)$ as in (7.2) with $q_{\beta\alpha}(x)$ chosen to be a unit-norm quaternion; there remaining the freedom to simultaneously change the sign of $s_{\beta\alpha}(x)$ and $q_{\beta\alpha}(x)$. Thus $\tilde{f}_{\alpha\beta}(x) \in \frac{A^* \times SU(2)}{Z_2}$. We obviously have the homomorphism h

$$h : A^* \times SU(2) \longrightarrow Aut\ A$$

$$(s, q) \longrightarrow \chi(s) \ .$$

This homomorphism passes to the quotient $\frac{A^* \times SU(2)}{Z_2}$ since Z_2 is in the kernel of χ. The transition maps for $\mathcal{I}(M)$ are related to those of $\mathcal{A}(M)$ by $h(\tilde{f}_{\alpha\beta}(x)) = f_{\alpha\beta}(x)$. The analogue of (4.1) is that such a bundle $\mathcal{I}(M)$ exists iff the transition maps for $\mathcal{A}(M)$ can be coherently lifted from $Aut\ A$ to $\frac{A^* \times SU(2)}{Z_2}$.

For $p-q = 4, 6 \bmod 8$, $C_{p,q}(\mathcal{R}) \simeq M_{2^{n/2-1}}(\mathcal{R}) \otimes \mathcal{H}$. Since the transition maps of the Clifford bundle can be taken in $O(p,q)$ then $C(M)$ has a bundle of irreducible representation modules, without $\mathcal{H}$-linear structure, iff there exist coherent lifts into $\frac{\Gamma \times SU(2)}{Z_2}$. Since $_\pm\Gamma$ is a normal sub-group of Γ, the quotient being $\mathcal{R}^+$, it follows that $\frac{\pm\Gamma \times SU(2)}{Z_2}$ is a normal sub-group of $\frac{\Gamma \times SU(2)}{Z_2}$ with quotient $\mathcal{R}^+$. Thus any $\frac{\Gamma \times SU(2)}{Z_2}$ bundle is reducible to a $\frac{\pm\Gamma \times SU(2)}{Z_2}$ bundle. Thus for p-q=4,6 mod 8 there exists a bundle of irreducible modules, without $\mathcal{H}$-linear structure, for $C(M)$ iff the transition maps can be coherently lifted from $O(p,q)$ to $\frac{\pm\Gamma \times SU(2)}{Z_2}$. The existence of such lifts corresponds to a generalised spinor structure [**6**] .

As an example of these generalised structures take M to be any orientable four-dimensional Riemannian manifold with orienting volume 4-form z. Then global sections of $C(M)$ are given by $P_\pm = \frac{1}{2}(1 \pm z)$. Since $C_{4,0}(\mathcal{R}) \simeq M_2(\mathcal{R}) \otimes \mathcal{H}$ we see that $P_\pm$ are a pair of primitive idempotents that decompose the Clifford bundle into sub-bundles of minimal left ideals. So in this (very special) case the Clifford bundle contains sub-bundles of irreducible representation modules. In general these bundles of representation modules do not have an $\mathcal{H}$-linear structure. A well known example of a Riemannian manifold without a spin structure is $CP(2)$ (regarded as a real manifold). In this case the bundle of irreducible Clifford modules without $\mathcal{H}$-linear structure exists, whilst none with an $\mathcal{H}$-linear structure exists.

The real Clifford algebra of an odd-dimensional vector space is not central simple, the centre being generated by any n-form. Thus automorphisms that do not leave n-forms fixed are not inner. In particular improper orthogonal transformations extend to outer automorphisms of the Clifford algebra. If we restrict to an orientable odd-dimensional manifold then the transition maps for the Clifford bundle lie in $SO(p,q)$, the special orthogonal transformations of T_x^*M extending to inner automorphisms of $C(T_x^*M,g)$.

If the orienting n-form is z satisfying $z^2 = 1$ then the idempotents $P_\pm = \frac{1}{2}(1 \pm z)$ decompose the Clifford bundle into two sub-bundles of central simple algebras. A bundle of irreducible representation modules for the reducible Clifford algebra is a bundle carrying an irreducible representation of either of these bundles of simple ideals. Thus we have already considered the matter of the existence of such a bundle.

For p-q=3,7 mod 8 $C_{p,q}(\mathcal{R}) \simeq M_{2^{(n-1)/2}}(\mathcal{R}) \otimes \mathcal{C}(\mathcal{R})$, the algebra of complex matrices. For an orientable pseudo-Riemannian manifold whose metric tensor has signature as above the transition maps of the Clifford bundle may be taken to be inner automorphisms. We may thus regard the Clifford bundle as a bundle of central simple algebras over $\mathcal{C}$. We can then argue as before that a bundle of irreducible modules exists iff we have coherent lifts of the SO(p,q)-valued transition maps into the Clifford group Γ. Now a little care is needed in normalising. In odd dimensions not all elements of Γ are even or odd whereas the norm λ is only defined on $\Gamma^\pm$. Any element of Γ can be written as the product of an element of the centre, here isomorphic to $\mathcal{C}$, and an element of Γ^+. Hence if $\tilde{f}_{\alpha\beta}(x) \in \Gamma$ then $\tilde{f}_{\alpha\beta}(x) = \lambda_{\alpha\beta}(x)s_{\alpha\beta}(x)$ for $\lambda_{\alpha\beta}(x) \in U(1)$ and $s_{\alpha\beta}(x) \in \Gamma^+$. Such a decomposition is unique up to the choice of sign of $\lambda_{\alpha\beta}(x)$. Thus there is a bundle carrying an irreducible representation of the Clifford bundle iff the SO(p,q)-valued transition maps can be coherently lifted into $\frac{\Gamma^+ \times U(1)}{Z_2}$. We could always choose new lifts in $\frac{\pm\Gamma^+ \times U(1)}{Z_2}$ and thus the existence of a bundle of irreducible representation modules is equivalent to the existence of a $Spin^{\mathcal{C}}$ structure[**7**] .

We finally briefly mention the case of a non-orientable odd-dimensional manifold. In odd dimensions the automorphism group of the centre of the Clifford algebra is isomorphic to Z_2; consisting of the identity and complex conjugation in the case of $\mathcal{C}$, or the identity and the involution "swap" that interchanges the simple components in the case of $\mathcal{R} \oplus \mathcal{R}$. If Ω is some Z_2 group of outer automorphisms of the Clifford algebra that induces these automorphisms of the centre then $S \in Aut\, C_{p,q}(\mathcal{R})$ can be written as $S = \chi(a)\omega$ for $a \in C^*_{p,q}(\mathcal{R})$, $\omega \in \Omega$. We have the semi-direct product composition $S_2S_1 = \chi\big(a_2\omega_2(a_1)\big)\omega_2\omega_1$. For the case in which the centre is $\mathcal{C}$ we regard $C_{p,q}(\mathcal{R}) \simeq M_{2^{(n-1)/2}}(\mathcal{R}) \otimes \mathcal{C}(\mathcal{R})$ as being a sub-algebra of $M_{2^{(n+1)/2}}(\mathcal{R})$. There is then a $t \in M_{2^{(n+1)/2}}(\mathcal{R})$ with $t^2 = 1$ and $tat^{-1} = a^* \ \forall a \in C_{p,q}(\mathcal{R})$. This element t generates a Z_2 sub-group $\tilde{\Omega}$ of $M_{2^{(n+1)/2}}(\mathcal{R})$. If $f \in M_{2^{(n+1)/2}}(\mathcal{R})$ such that $m \mapsto fmf^{-1}$, for $m \in C_{p,q}(\mathcal{R})$, defines an automorphism of $C_{p,q}(\mathcal{R})$ then $f = a\varpi$ for $a \in C^*_{p,q}(\mathcal{R})$ and $\varpi \in \tilde{\Omega}$. We again have a semi-direct product composition: $a_1\varpi_1a_2\varpi_2 = a_1(\varpi_1a_2{\varpi_1}^{-1})\varpi_1\varpi_2$. It follows that we have a bundle carrying an irreducible representation of the Clifford bundle iff we can coherently lift the transition maps from $Aut\, C_{p,q}(\mathcal{R})$ to $C^*_{p,q}(\mathcal{R})\, Ⓢ Z_2$ where the semi-direct product is with the group generated by t. The reducability of the Clifford bundle implies that this is equivalent to being able to coherently lift from O(p,q) into $\frac{\pm\Gamma^+ \times U(1)}{Z_2}\, Ⓢ Z_2$.

When the centre is $\mathcal{R} \oplus \mathcal{R}$ we cannot have a bundle of irreducible modules in the non-orientable case. The condition that we have a reducible bundle of representation modules, carrying a faithful representation of minimal dimension, is similar to that just mentioned. If in the orientable case we need coherent lifts from SO(p,q) to G then in the non-orientable case we need coherent lifts from O(p,q) to $G\, Ⓢ Z_2$. The group G is either $_\pm\Gamma^+$ or $\frac{\pm\Gamma^+ \times SU(2)}{Z_2}$.

8. Representing Spinor Fields With Differential Forms.

We assume that M is an even-dimensional spin manifold. Then we gave in section 4 a prescription for constructing a bundle carrying an irreducible representation of $C(M)$. (The bundle $C(M)$ being a special case of what was there denoted $\mathcal{A}(M)$.) If $\{U_\alpha\}$ is an open covering of M with $f_{\alpha\beta}$ transition maps for $C(M)$ then we used coherent lifts $\{\tilde{f}_{\alpha\beta}\}$ to quotient $X \equiv \bigcup_\alpha U_\alpha \times D^r$ to form the bundle $\mathcal{I}(M) = X/\varsigma$. The equivalence relation ς was given in (4.2). We shall here repeat this construction in a slightly different but equivalent way.

Assume that as in section 4 $\{\Phi_\alpha\}$ are local trivialising maps for $C(M)$ with resulting transition maps $\{f_{\alpha\beta}\}$, with coherent lifts $\{\tilde{f}_{\alpha\beta}\}$. If $C_{p,q}(\mathcal{R}) \simeq M_r(\mathcal{R}) \otimes D(\mathcal{R})$ then any minimal left ideal of $C_{p,q}(\mathcal{R})$ is an r-dimensional right D-module. Let I be any minimal left ideal (for example, that formed by the first 'column'). For each x in U_α we choose a minimal left ideal $I_{x\alpha}$ in $C(T^*_x, g_x)$ by requiring

$$\Phi_\alpha(I_{x\alpha}) = I \ . \qquad 8.1$$

As in section 2 we introduce $a_{\alpha\beta} \in C(T^*_x, g)$ such that

$$\Phi_\beta(a_{\alpha\beta}) = \tilde{f}_{\alpha\beta}(x) \ . \qquad 8.2$$

Then for all $u \in C(T_x^*, g_x)$ $\Phi_\alpha(u) = \Phi_\beta(a_{\alpha\beta} u a_{\alpha\beta}{}^{-1})$. So if $u \in I_\alpha$ then $a_{\alpha\beta} u a_{\alpha\beta}{}^{-1}$, and hence $u a_{\alpha\beta}{}^{-1}$, is in I_β. If now $Y_\alpha = U_\alpha \times I_\alpha$ and $Y = \bigcup_\alpha Y_\alpha$ then an equivalence relation ϱ is defined on Y by

$$(\alpha, x, \phi) \sim (\beta, y, \psi) \text{ iff } x = y \text{ and } \psi = \phi a_{\alpha\beta}{}^{-1} \quad . \qquad 8.3$$

Then $\mathcal{I}(M)$ is a bundle carrying an irreducible representation of $C(M)$ if $\mathcal{I}(M) \equiv Y/\varrho$.

The prescription we have given here is intentionally only a trivial variation of that given in section 4. That given here results in sections of $\mathcal{I}(M)$ (spinor fields) being equivalence classes of sections of $C(M)$ (differential forms). So a spinor field can locally be represented by a local differential form that lies in a minimal left ideal of the Clifford algebra at each point. It is perhaps worth emphasising that the Clifford elements $\{a_{\alpha\beta}\}$ that define the equivalence relation in (8.3) need not lie in the Clifford group (although the resulting bundle $\mathcal{I}(M)$ always admits local trivialisations such that the transition maps lie in the Pin group).

We believe, for reasons given in the introduction, that it is convenient to regard spinor fields as being sections of a bundle that carries an irreducible representation of the Clifford bundle. One can if one wishes then form such a bundle (when any exists) out of equivalence classes of certain differential forms: it being purely a matter of taste whether such a "concrete" spinor bundle is used rather than any other. When it comes to solving local spinor equations we have found it useful to represent our spinor fields by differential forms lying in minimal left ideals of the Clifford algebra rather than work with the components of the spinor in some more abstract basis.

Acknowledgement: We thank Graeme Segal for guiding us towards the proper approach to these problems.

References

1. J Milnor, Enseignement Math. **9**, p. 198.

2. H R Petry, Spin Structures on Lorentz Manifolds, Trieste **ISAS-44/84/EP**.

3. Y Ne'eman Dj Šijački, Phys. Letts **157B**, p. 267,275.

4. I R Porteous, Topological Geometry, Van-Nostrand-Reinhold, London.

7. M F Atiyah, R Bott, A Shapiro, Topology **3** (1964), p. 3.

5. S Kobayashi, K Nomizu, Principles of Differential Geometry, Interscience **vol 1**.

6. A Back, P G O Freund, M Forger, Phys. Letts **77B**, p. 181.

Department of Physics, University of Lancaster.

INEQUALITIES FOR SPINOR NORMS IN CLIFFORD ALGEBRAS

by

Gerald N. Hile
Department of Mathematics
2565 The Mall, University of Hawaii
Honolulu, HI 96822, USA

and

Pertti Lounesto
Institute of Mathematics
Helsinki University of Technology
SF-02150 Espoo, Finland

Abstract. In hypercomplex analysis one considers mappings from the euclidean space $\mathbf{R}^n$ to its Clifford algebra $\mathbf{R}_n$, where an inequality $|uv| \leq K_n |u||v|$ holds for u, v in $\mathbf{R}_n$. In this paper the smallest possible value of the constant K_n is determined. As a byproduct the authors present a more detailed description of the faithful matrix representations of Clifford algebras, which might also be useful for other purposes.

Introduction. Let $\mathbf{C}_n$ denote the complex universal Clifford algebra, with unity 1, generated as an algebra by the n elements $\mathbf{e}_1, \mathbf{e}_2, ..., \mathbf{e}_n$, and subject to the multiplication rules

(1) $\mathbf{e}_\alpha \mathbf{e}_\beta = -\mathbf{e}_\beta \mathbf{e}_\alpha$ for $\alpha, \beta = 1, 2, ..., n$ and $\alpha \neq \beta$,
(2) $\mathbf{e}_\alpha{}^2 = -1$ for $\alpha = 1, 2, ..., n$.

(See Ref [6].) The Clifford algebra $\mathbf{C}_n$ is spanned as a 2^n-dimensional complex linear space by the collection of linearly independent elements $\mathbf{e}_\mathrm{A}$, where the multi-index A ranges over all naturally ordered subsets of the first n positive integers $\{1,2,...,n\}$, and $\mathbf{e}_\mathrm{A} = \mathbf{e}_{(\alpha_1,\alpha_2,...,\alpha_k)}$ is defined as

(3) $$\mathbf{e}_{(\alpha_1,\alpha_2,...,\alpha_k)} = \mathbf{e}_{\alpha_1}\mathbf{e}_{\alpha_2}...\mathbf{e}_{\alpha_k} \text{ for } 1 \leq \alpha_1 < \alpha_2 < ... < \alpha_k \leq n .$$

A general element u of $\mathbf{C}_n$ has the form

(4) $$u = \Sigma_\mathrm{A}\, u_\mathrm{A} \mathbf{e}_\mathrm{A} ,$$

where the u_A's are complex scalars, and again A ranges over all ordered subsets of $\{1,2,...,n\}$. Associated with the element u is its *absolute value*, or *spinor norm*,

(5) $$|u| = \{\Sigma_\mathrm{A}\, |u_\mathrm{A}|^2\}^{1/2} .$$

It follows that there is a positive constant $K_n{}^\mathrm{C}$ such that

(6) $$|uv| \leq K_n{}^\mathrm{C}\, |u||v| \text{ , for all } u, v \text{ in } \mathbf{C}_n .$$

However, to our knowledge the smallest possible value of $K_n{}^\mathrm{C}$ has not previously been determined. We derive here the following best possible value of $K_n{}^\mathrm{C}$,

(7) $$K_n{}^\mathrm{C} = 2^{(n-[n/2])/2} = 2^{n/4} \text{ if } n \text{ is even}$$
$$= 2^{(n+1)/4} \text{ if } n \text{ is odd} .$$

Here $[n/2]$ is the greatest integer less or equal to $n/2$.

1. The main result of this paper

We have analogous estimates, although more complicated to state, for the real nondegenerate universal Clifford algebras $\mathbf{R}_{p,q}$. (See Ref. [6].) The algebra $\mathbf{R}_{p,q}$, for given nonnegative integers p and q, has as generators $\mathbf{e}_1, \mathbf{e}_2, ..., \mathbf{e}_n$, $(n = p + q)$, with the anticommutativivity condition (1) holding, but with (2) replaced by

(1.1) $$\mathbf{e}_\alpha{}^2 = +1 \quad \text{for } \alpha = 1, 2, ..., p$$
$$= -1 \quad \text{for } \alpha = p{+}1, ..., p{+}q = n.$$

The Clifford algebra $\mathbf{R}_{p,q}$ is 2^n-dimensional as a real linear space. In this paper we derive the best possible estimate

(1.2) $$|uv| \leq K_{p,q}\, |u||v| \quad \text{for } u, v \text{ in } \mathbf{R}_{p,q},$$

where

(1.3) $$K_{p,q} = 2^{h/2}, \quad h = q - r_{q-p}$$

and r_i is the *Radon-Hurwitz number* defined by the recursion formula $r_{i+8} = r_i + 4$ and the initial conditions

i	0	1	2	3	4	5	6	7
r_i	0	1	2	2	3	3	3	3.

Hence, an alternative way of writing (1.3) is (with $n = p + q$)

(1.4) $$K_{p,q} = 2^{n/4} \quad \text{if } q - p = 0, 6 \pmod 8$$
$$= 2^{(n-1)/4} \quad \text{if } q - p = 1, 3, 5 \pmod 8$$
$$= 2^{(n-2)/4} \quad \text{if } q - p = 2, 4 \pmod 8$$
$$= 2^{(n+1)/4} \quad \text{if } q - p = 7 \pmod 8.$$

To be more precise, in the Clifford algebras $\mathbf{R}_n = \mathbf{R}_{n,0}$ of the euclidean spaces $\mathbf{R}^n = \mathbf{R}^{n,0}$ these coefficients are

n	0	1	2	3	4	5	6	7
K_n	1	$\sqrt{2}$	$\sqrt{2}$	$\sqrt{2}$	$\sqrt{2}$	2	2	$2\sqrt{2}$

and in the Clifford algebras $\mathbf{R}_{0,n}$ of the anti-euclidean spaces $\mathbf{R}^{0,n}$

n	0	1	2	3	4	5	6	7
$K_{0,n}$	1	1	1	$\sqrt{2}$	$\sqrt{2}$	2	$2\sqrt{2}$	4 .

Also note that in these two cases the absolute value can be defined by the involutions, that is, by the reversion $|u|^2 = \mathrm{Re}(u^*u)$ for u in $\mathbf{R}_n$, and by the conjugation $|u|^2 = \mathrm{Re}(\bar{u}u)$ for u in $\mathbf{R}_{0,n}$. It follows that in these two cases with definite quadratic forms the square norm is invariant under the rotation group $\mathbf{SO}(n)$.

2. The Clifford algebras C_n

Our derivation of inequalities for the square norm makes use of matrix representations of the algebras $\mathbf{C}_n$. We let $\mathbf{C}(s)$ denote the complex algebra of all $s{\times}s$ matrices with complex entries. The notation ${}^2\mathbf{C}(s)$ refers to the complex algebra of all $2s{\times}2s$ matrices of the form

$$\begin{pmatrix} M & 0 \\ 0 & N \end{pmatrix},$$

where M and N are matrices from $\mathbf{C}(s)$, and 0 here denotes the $s{\times}s$ zero matrix. Thus ${}^2\mathbf{C}(s)$ is a proper subalgebra of $\mathbf{C}(2s)$.

As is well known (see [6]), we have the algebraic isomorpisms,

$$(2.1)\qquad \begin{aligned} \mathbf{C}_n &\simeq \mathbf{C}(2^{n/2}) && \text{if } n \text{ is even} \\ &\simeq {}^2\mathbf{C}(2^{(n-1)/2}) && \text{if } n \text{ is odd}. \end{aligned}$$

That is to say, there is a one-to-one correspondence, preserving algebraic operations, between elements u of $\mathbf{C}_n$ and matrices U in $\mathbf{C}(2^{n/2})$ or ${}^2\mathbf{C}(2^{(n-1)/2})$, depending on whether n is even or odd. The generators $\mathbf{e}_\alpha$ we will say to correspond to the matrices E_α in the appropriate matrix algebra;

$$(2.2)\qquad \mathbf{e}_\alpha \longleftrightarrow E_\alpha, \ \text{ for } \alpha = 1, 2, \ldots, n\,.$$

We also have the correspondence for an ordered subset $\mathrm{A} = (\alpha_1, \alpha_2, \ldots, \alpha_k)$ of

$\{1,2,...,n\}$,

(2.3) $\quad \mathbf{e}_A = \mathbf{e}_{(\alpha_1,\alpha_2,...,\alpha_k)} \longleftrightarrow E_A = E_{(\alpha_1,\alpha_2,...,\alpha_k)}$,

including also the correspondence $1 \longleftrightarrow I$, where I represents the identity matrix in the appropriate matrix algebra. The matrices E_α, according to (1) and (2), will have to satisfy corresponding conditions $E_\alpha E_\beta = -E_\beta E_\alpha$ for $\alpha \neq \beta$, and $E_\alpha^2 = -I$. We will require further conditions on these matrices, as listed in the following lemma:

Lemma 1. *The matrices* E_α, *in the correspondence* (2.2) *between generators of* $\mathbf{C}_n$ *and matrices in the appropriate matrix algebra given in* (2.1), *can be chosen so that they fulfill the conditions*

(a) $E_\alpha E_\beta = -E_\beta E_\alpha$ *for* $\alpha \neq \beta$, $\alpha, \beta = 1, 2, ..., n$
(b) $E_\alpha^2 = -I$ *for* $\alpha = 1, 2, ..., n$
(c) $E_\alpha^* = -E_\alpha$ *for* $\alpha = 1, 2, ..., n$ (E_α^* is the conjugate transpose of E_α)
(d) trace$(E_A) = 0$ *for* $A \neq \emptyset$, $A \subset \{1, 2, ..., n\}$
(e) *Exactly* $[n/2]$ *of the* E_α*'s are real, with the remainder purely imaginary.*

Before proving Lemma 1, we first show how it leads to our inequalities for the spinor norm.

Theorem 1. *In* $\mathbf{C}_n$ *we have the estimate* (6), *with* $K_n^{\mathbf{C}}$ *defined by* (7). *Moreover, this value of the constant* $K_n^{\mathbf{C}}$ *is the smallest possible.*

Proof: For a given complex matrix $M = (m_{\alpha\beta})$, let the matrix norm $\|M\|$, be defined by

(2.4) $\quad \|M\|^2 = \text{trace}(MM^*) = \sum_{\alpha,\beta} |m_{\alpha\beta}|^2.$

Then, as is well known, for complex matrices M and N one has the inequality

(2.5) $\quad \|MN\| \leq \|M\| \, \|N\|$.

By Lemma 1, there is one-to-one algebraic correspondence,

$$u = \Sigma_A u_A \mathbf{e}_A \longleftrightarrow U = \Sigma_A u_A E_A ,$$

between elements u of $\mathbf{C}_n$ and matrices U of the appropriate matrix algebra, such that

(2.6) $$E_\alpha E_\alpha^* = -E_\alpha^2 = I , \text{ for } \alpha = 1, 2, ..., n,$$

which implies

(2.7) $$E_A E_A^* = I , \text{ for } A \subset \{1, 2, ..., n\} .$$

Hence,

(2.8) $$\begin{aligned} \|U\|^2 &= \text{trace}(UU^*) = \text{trace}((\Sigma_A u_A E_A)(\Sigma_A \bar{u}_A E_A^*)) \\ &= \text{trace}\{\Sigma_A |u_A|^2 E_A E_A^* + \sum_{A \neq B} u_A \bar{u}_B E_A E_B^*\} . \end{aligned}$$

But if $A \neq B$, then conditions (a), (b), (c) of Lemma 1 imply that $E_A E_B^* = E_A(\pm E_B) = \pm E_C$ for some C, $C \neq \emptyset$, $C \subset \{1, 2, ..., n\}$; thus $\text{trace}(E_A E_B^*) = 0$ by condition (d). Therefore, in view of (2.7), we obtain from (2.8),

(2.9) $$\|U\|^2 = (\Sigma_A |u_A|^2)(\text{trace } I) = |u|^2 (\text{trace } I) .$$

Now for u, v in $\mathbf{C}_n$, with $u \longleftrightarrow U$, $v \longleftrightarrow V$, we have $uv \longleftrightarrow UV$, and therefore

(2.10) $$\|uv\| = (\text{trace } I)^{-1/2} \|UV\| \leq (\text{trace } I)^{-1/2} \|U\| \, \|V\| = (\text{trace } I)^{1/2} |u| \, |v| .$$

Since the constant K_n^{C} of (7) is $(\text{trace } I)^{1/2}$ in the appropriate matrix algebra (2.1), inequality (2.10) is the same as (6).

In order to see why the value of K_n^{C} in (7) is smallest possible, we observe that if one takes both M and N in (2.5) as the matrix with "1" in the upper left corner and zeros elsewhere, then equality holds in (2.5). Therefore if we choose both u and v in (2.10) as elements in $\mathbf{C}_n$ corresponding to this particular matrix (a primitive idempotent) in the appropriate matrix algebra, equality holds also in (2.10). □

Before proving Lemma 1, we state the following proposition whose verification is straightforward and easily shown:

Proposition 1. *Let* $E_1, E_2, ..., E_n$ *be complex square matrices fulfilling conditions* (a), (b), (c) *of Lemma* 1, *and let* P *be the product*

$$P = E_1 E_2 \dots E_n .$$

Then

$$\begin{aligned} P^2 &= +I \quad \text{if } n = 0, 3 \pmod 4 \\ &\quad -I \quad \text{if } n = 1, 2 \pmod 4 \\ P^* &= +P \quad \text{if } n = 0, 3 \pmod 4 \\ &\quad -P \quad \text{if } n = 1, 2 \pmod 4 \\ PE_\alpha &= -E_\alpha P \quad \text{if } n \text{ is even} \\ &\quad +E_\alpha P \quad \text{if } n \text{ is odd} \quad (\alpha = 1, 2, ..., n). \end{aligned}$$ □

Now we prove Lemma 1. First we check a few trivial cases;

$n = 0$; $\quad E_\varnothing = I = (1)$

$n = 1$; $\quad E_\varnothing = I = \begin{pmatrix} 1 & 0 \\ 0 & 1 \end{pmatrix}, \quad E_1 = \begin{pmatrix} i & 0 \\ 0 & -i \end{pmatrix}$

$n = 2$; $\quad E_\varnothing = I = \begin{pmatrix} 1 & 0 \\ 0 & 1 \end{pmatrix}, \quad E_1 = \begin{pmatrix} i & 0 \\ 0 & -i \end{pmatrix}, \quad E_2 = \begin{pmatrix} 0 & 1 \\ -1 & 0 \end{pmatrix}, \quad E_{12} = \begin{pmatrix} 0 & i \\ i & 0 \end{pmatrix}.$

Clearly conditions (a)–(e) of Lemma 1 all hold for these matrix representations. We proceed fo higher values of n by induction arguments.

First suppose n is even, and assume we have matrices $E_1, E_2, ..., E_n$ in $\mathbf{C}(2^{n/2})$ satisfying conditions (a)–(e). We construct matrices $F_1, F_2, ..., F_n, F_{n+1}$ in ${}^2\mathbf{C}(2^{n/2})$ according to the formulas

(2.11) $\quad F_\alpha = \begin{pmatrix} E_\alpha & 0 \\ 0 & -E_\alpha \end{pmatrix}$ for $\alpha = 1, 2, ..., n$; $\quad F_{n+1} = i^{(n+2)/2} \begin{pmatrix} P & 0 \\ 0 & -P \end{pmatrix}$,

where P is the product matrix $P = E_1 E_2 ... E_n$ of Proposition 1. Then $(F_{n+1})^2 = -I$, $(F_{n+1})^* = -F_{n+1}$, and $F_{n+1} F_\alpha = -F_\alpha F_{n+1}$ for $\alpha = 1, 2, ..., n$. Thus it is easily seen that conditions (a), (b), (c) hold for the collection $\{F_\alpha\}$. Moreover, by the induction hypothesis (e), F_{n+1} is imaginary, which establishes (e) for the case n+1. It remains to check condition (d).

If $A \subset \{1, 2, ..., n\}$, then (with $|A|$ = number of elements in A)

$$F_A = \begin{pmatrix} E_A & 0 \\ 0 & (-1)^{|A|}E_A \end{pmatrix};$$

thus $A \neq \emptyset$ implies $\text{trace}(F_A) = 0$ since $\text{trace}(E_A) = 0$. If $(n+1) \in A$, $A \subset \{1, 2, ..., n, n+1\}$, set $B = A \backslash \{n+1\}$; then

$$F_A = F_B F_{n+1} = i^{(n+2)/2} \begin{pmatrix} E_B P & 0 \\ 0 & (-1)^{|A|}E_B P \end{pmatrix}.$$

If $B \neq \{1, 2, ..., n\}$, then $E_B P = \pm E_C$ for some C, $C \neq \emptyset$, $C \subset \{1, 2, ..., n\}$, and hence $\text{trace}(E_B P) = 0$, $\text{trace}(F_A) = 0$. If $B = \{1, 2, ..., n\}$, then $|A| = n+1$ (an odd number), and hence $(-1)^{|A|} = -1$, $\text{trace}(F_A) = 0$.

Finally, in order to guarantee that the collection $\{F_A\}$ is linearly independent, thereby forming a linear basis for ${}^2\mathbf{C}(2^{n/2})$, it is sufficient to observe that the product $F_1F_2...F_nF_{n+1}$ of all the generators is not a scalar multiple of the identity matrix (see [6], ch. 13, Thm. 13.10 and its proof). But the trace condition (d) rules out this possibility, the trace of the identity matrix being positive.

Now, in order to establish Lemma 1 for the next even number, $n+2$, we add to the collection $\{F_\alpha\}$ of (2.11) the additional matrix

$$(2.12) \qquad F_{n+2} = \begin{pmatrix} 0 & I \\ -I & 0 \end{pmatrix}$$

obtaining now $n+2$ generators in $\mathbf{C}(2^{(n+2)/2})$. Then (a), (b), (c) are clear for this larger collection and, since F_{n+2} is real, (e) also holds. We check now (d). If $A \subset \{1, 2, ..., n+2\}$, $A \neq \emptyset$, $(n+2) \notin A$, then we have already shown that $\text{trace}(F_A) = 0$. If $(n+2) \in A$, then the diagonal blocks of F_A are 0 (as in (2.12)), and again $\text{trace}(F_A) = 0$.

Finally, we observe that the product $F_1F_2...F_{n+2}$ is not a scalar multiple of the identity, being a nonsingular matrix with zero trace. □

3. The Clifford algebras $\mathbf{R}_{p,q}$

First we discuss matrix representations of the real Clifford algebras $\mathbf{R}_{p,q}$. We let $\mathbf{R}(s)$ denote the real algebra of all $s{\times}s$ matrices with real entries, and we define ${}^2\mathbf{R}(s)$ analogously to ${}^2\mathbf{C}(s)$. We let $\mathbf{H}$ denote the quaternion ring, with generators i and j, and with $k = ij$. In a likewise analogous manner we define the matrix algebras $\mathbf{H}(s)$ and ${}^2\mathbf{H}(s)$. All these algebras now are understood as real algebras, with multiplication only by real scalars. With this understanding, according to [6], ch. 13, we have the following algebraic isomorphisms (where $n = p+q$):

$$\begin{aligned}(3.1)\qquad \mathbf{R}_{p,q} &\simeq \mathbf{R}(2^{n/2}) && \text{if } q-p = 0, 6 \pmod 8\\ &\simeq \mathbf{C}(2^{(n-1)/2}) && \text{if } q-p = 1, 5 \pmod 8\\ &\simeq \mathbf{H}(2^{(n-2)/2}) && \text{if } q-p = 2, 4 \pmod 8\\ &\simeq {}^2\mathbf{H}(2^{(n-3)/2}) && \text{if } q-p = 3 \pmod 8\\ &\simeq {}^2\mathbf{R}(2^{(n-1)/2}) && \text{if } q-p = 7 \pmod 8)\,.\end{aligned}$$

Thus we have a one-to-one algebraic correspondence, say $u \longleftrightarrow U$, between elements u of $\mathbf{R}_{p,q}$ and matrices U of the appropriate real algebra as defined by (3.1). Likewise, we have correspondences

$$(3.2)\qquad \mathbf{e}_\alpha \longleftrightarrow E_\alpha\,, \qquad \mathbf{e}_A \longleftrightarrow E_A\,,$$

as in (2.2) and (2.3), relating generators $\{\mathbf{e}_\alpha\}$ and corresponding matrices $\{E_\alpha\}$. We require further conditions on the matrices $\{E_\alpha\}$ as described in the next lemma:

Lemma 2. *The matrices* E_α *in the correspondence* (3.2) *between generators of* $\mathbf{R}_{p,q}$ *and matrices in the appropriate matrix algebra, as listed in* (3.1), *can be chosen so as to fulfill the conditions*

(a) $E_\alpha E_\beta = -E_\beta E_\alpha$ *for* $\alpha \neq \beta,\ \alpha, \beta = 1, 2, \ldots, n$

(b) $E_\alpha{}^2 = +I$ *for* $\alpha = 1, 2, \ldots, p$

$\qquad -I$ *for* $\alpha = p+1, 2, \ldots, p+q$

(c) $E_\alpha E_\alpha{}^* = I$ *for* $\alpha = 1, 2, \ldots, n$

(d) *If* $A \subset \{1, 2, \ldots, n\},\ A \neq \emptyset$, *then*

(*i*) $\operatorname{trace}(E_A) = 0$, *when* $q-p = 0, 6, 7 \pmod 8$

(*ii*) $\operatorname{trace}(E_A) = 0$ *or* $E_A{}^* = -E_A$, *when* $q-p = 1, 2, 3, 4, 5 \pmod 8$.

Remark: If M is real or complex matrix, then M^* is the transpose or conjugate transpose of M, respectively. If $M = (m_{\alpha\beta})$ is a matrix of quaternions, we define M^* again as the conjugate transpose of M, $M^* = (m_{\beta\alpha})$, where for a quaternion $q = q_0 + q_1 i + q_2 j + q_3 k$ its conjugate is $q = q_0 - q_1 i - q_2 j - q_3 k$, with $|q|$ defined as $|q|^2 = qq = q_0^2 + q_1^2 + q_2^2 + q_3^2$. The matrix norm of a quaternion matrix is again defined by (2.4), with (2.5) continuing to hold.

Before giving the proof of Lemma 2, we use it to prove our inequality on the spinor norm:

Theorem 2. *In* $\mathbf{R}_{p,q}$ *we have the estimate* (1.2), *with* $K_{p,q}$ *defined by* (1.4). *Moreover, this value of* $K_{p,q}$ *is the smallest possible.*

Proof: Proceeding as in the proof of Theorem 1, we arrive at (2.8), where the values u_A are now real scalars instead of complex. Condition (c) of Lemma 2 implies (2.7); thus (2.8) may be written as

$$\|U\|^2 = |u|^2(\text{trace } I) + \tfrac{1}{2}\sum_{A\neq B} u_A u_B \text{ trace}(E_A E_B{}^* + E_B E_A{}^*) \,. \tag{3.3}$$

But conditions (b) and (c) of Lemma 2 imply that $E_\alpha{}^* = E_\alpha$ for $\alpha = 1, 2, ..., p$, and $E_\alpha{}^* = -E_\alpha$ for $\alpha = p+1, ..., p+q$; hence, if $A \neq B$ then $E_A E_B{}^* + E_B E_A{}^* = \pm(E_C + E_C{}^*)$ for some $C, C \neq \emptyset, C \subset \{1, 2, ..., p+q\}$. Therefore condition (d) of Lemma 2 and (3.3) combine to yield again (2.9), and thus also (2.10). We observe that $(\text{trace } I)^{1/2}$ is the required $K_{p,q}$ of (1.4). The argument that this value of $K_{p,q}$ is best possible is the same as the argument for $K_n{}^C$ in the proof of Theorem 1. □

We now prove Lemma 2. Again we check a few trivial cases:

$p = 0$, $q = 0$: $E_\emptyset = I = (1)$

$p = 1$, $q = 0$: $E_\emptyset = I = \begin{pmatrix} 1 & 0 \\ 0 & 1 \end{pmatrix}$, $E_1 = \begin{pmatrix} 1 & 0 \\ 0 & -1 \end{pmatrix}$

$p = 0$, $q = 1$: $E_\emptyset = (1)$, $E_1 = (i)$

$p = 0$, $q = 2$: $E_\emptyset = (1)$, $E_1 = (i)$, $E_2 = (j)$, $E_{12} = (k)$

$p = 0$, $q = 3$: $E_\emptyset = I = \begin{pmatrix} 1 & 0 \\ 0 & 1 \end{pmatrix}$, $E_1 = \begin{pmatrix} i & 0 \\ 0 & -i \end{pmatrix}$, $E_2 = \begin{pmatrix} j & 0 \\ 0 & -j \end{pmatrix}$, $E_3 = \begin{pmatrix} k & 0 \\ 0 & -k \end{pmatrix}$.

Conditions (a)–(d) of Lemma 2 are easily checked for these matrix representations.

Next we show that if Lemma 2 is true for $\mathbf{R}_{p,q}$, where $p \geq 1$, then it must also be true for $\mathbf{R}_{q+1,p-1}$. Let $E_1, E_2, ..., E_{p+q}$ be the matrix generators for $\mathbf{R}_{p,q}$ fulfilling the conditions of Lemma 2. We define new matrices $F_1, F_2, ..., F_{p+q}$ according to

$$F_\alpha = E_1 E_{p+\alpha} \text{ for } \alpha = 1, 2, ..., q, \quad F_{q+1} = E_1,$$
$$F_{q+\alpha} = E_1 E_\alpha \text{ for } \alpha = 2, ..., p.$$

Then $F_\alpha^2 = +I$ for $\alpha = 1, 2, ..., q+1$, and $F_\alpha^2 = -I$ for $\alpha = q+2, ..., q+p$. Conditions (a)–(d) are easily verified for the collection $\{F_\alpha\}$, with each F_A being $\pm E_B$ for some B. As in the proof of Lemma 1, condition (d) guarantees that the product $F_1 F_2 ... F_{p+q}$ is not a scalar multiple of the identity; thus the collection $\{F_A\}$ is linearly independent. Note also that $(p\text{-}1) - (q+1) = 6 - (q\text{-}p)$ (mod 8). This observation is in agreement with (3.1), which indicates that if $q'-p' = 6 - (q\text{-}p)$ (mod 8), and if $p'+q' = p+q$, then the matrix algebras for $\mathbf{R}_{p',q'}$ and $\mathbf{R}_{p,q}$ are the same.

Now we show that if Lemma 2 is true for $\mathbf{R}_{p,q}$ with $p \geq 4$, then it is also true for $\mathbf{R}_{p-4,q+4}$. Let $E_1, E_2, ..., E_{p+q}$ be the matrix generators for $\mathbf{R}_{p,q}$ fulfilling the conditions of Lemma 2, and define matrices $F_1, F_2, ..., F_{p+q}$ by

$$F_\alpha = E_\alpha \text{ for } \alpha = 1, 2, ..., p-4 \text{ and } \alpha = p+1, ..., p+q,$$
$$F_\alpha = E_{p-3} E_{p-2} E_{p-1} E_p E_\alpha \text{ for } \alpha = p-3, p-2, p-1, p\,.$$

Then $F_\alpha^2 = +I$ for $\alpha = 1, 2, ..., p-4$, and $F_\alpha^2 = -I$ for $\alpha = p-3, ..., p+q$. Again conditions (a)–(d) are easily checked, with each F_A being $\pm E_B$ for some B. The difference $q-p$ is invariant modulo 8 under the transformation $(p,q) \to (p-4, p+4)$; since also $p+q$ is invariant, (3,1) indicates that the matrix algebras for $\mathbf{R}_{p,q}$ and $\mathbf{R}_{p-4,q+4}$ are the same, as we just verified.

Finally, we show that if Lemma 2 is true for $\mathbf{R}_{p,q}$, then it is also true for $\mathbf{R}_{p+1,q+1}$. First consider the cases $q-p = 0, 1, 2, 4, 5, 6$ (mod 8). Assume the lemma is true for given values p,q, and let $E_1, E_2, ..., E_{p+q}$ be the matrix generators of the lemma. We define a family $F_1, F_2, ..., F_{p+q+2}$ of matrices twice the size according to

$$F_\alpha = \begin{pmatrix} E_\alpha & 0 \\ 0 & -E_\alpha \end{pmatrix}, \ \alpha = 1, 2, ..., n\,; \quad F_{n+1} = \begin{pmatrix} 0 & I \\ I & 0 \end{pmatrix}, \ F_{n+2} = \begin{pmatrix} 0 & I \\ -I & 0 \end{pmatrix}$$

Then $F_\alpha{}^2 = +I$ for $\alpha = 1, 2, ..., p$ and $\alpha = n+1$, and $F_\alpha{}^2 = -I$ for $\alpha = p+1, ..., p+q$ and $\alpha = n+2$. Again conditions (a)–(d) for the collection $\{F_\alpha\}$ are easily verified. The cases $q-p = 3, 7 \pmod 8$ are slightly different. We first write each F_α in the form

$$E_\alpha = \begin{pmatrix} G_\alpha & 0 \\ 0 & H_\alpha \end{pmatrix}, \qquad \alpha = 1, 2, ..., p+q = n ,$$

where G_α and H_α are square matrices of half the size of E_α. We define the family $\{F_\alpha\}$ of matrices twice the size of the matrices $\{E_\alpha\}$ by

$$F_\alpha = \begin{pmatrix} G_\alpha & 0 & 0 & 0 \\ 0 & -G_\alpha & 0 & 0 \\ 0 & 0 & H_\alpha & 0 \\ 0 & 0 & 0 & -H_\alpha \end{pmatrix}, \quad \alpha = 1, 2, ..., n$$

$$F_{n+1} = \begin{pmatrix} 0 & I & 0 & 0 \\ I & 0 & 0 & 0 \\ 0 & 0 & 0 & I \\ 0 & 0 & I & 0 \end{pmatrix}, \qquad F_{n+2} = \begin{pmatrix} 0 & I & 0 & 0 \\ -I & 0 & 0 & 0 \\ 0 & 0 & 0 & I \\ 0 & 0 & -I & 0 \end{pmatrix}.$$

Then conditions (a)–(d) can be easily checked for $\{F_\alpha\}$.

By the notation $(p,q) \to (r,s)$ we mean that the truth of Lemma 2 for $\mathbf{R}_{p,q}$ implies the truth for $\mathbf{R}_{r,s}$. Thus far, we have verified the implications

(3.4) $\qquad (p,q) \to (q+1,p-1), \quad (p,q) \to (p-4,q+4), \quad (p,q) \to (p+1,q+1);$

we have also shown that the lemma is true for the cases $(p,q) = (0,0), (0,1), (0,2), (0,3), (1,0)$. Beginning with these elementary cases and applying (3.4), we can readily show that the lemma is true for the values $(p,0)$ and $(0,q)$ where $0 \le p,q \le 7$; from $(p,q) \to (p+1,q+1)$ we infer its validity for all (p,q) such that $-7 \le q-p \le 7$. Then the implications

$$(p,q) \to (p+4,q+4) \to (p,q+8),$$
$$(p,q) \to (p+4,q+4) \to (q+5,p+3) \to (q+1,p+7) \to (p+8,q)$$

establish the lemma for all (p,q). □

REFERENCES

The absolute value of the Clifford algebras, or the spinor norm, has been studied by E. Artin [1], H. Bass [2], H. Zassenhaus [7] and I. Porteous [6]. See also Ref. [8]. This norm has been employed in hypercomplex analysis by F. Brackx, R. Delanghe and F. Sommen [3] and also by B. Goldschmidt [5]. For physical applications of the Clifford algebras, see various articles in the Proceedings of the "Workshop on Clifford Algebras" edited by J. S. R. Chisholm and A. K. Common [4].

[1] E. Artin: *Geometric Algebra*. Interscience, New York, 1957.

[2] H. Bass: Clifford algebras and spinor norms over a commutative ring. *Amer. J. Math.* **96** (1974), 156-206.

[3] F. Brackx, R. Delanghe, F. Sommen: *Clifford Analysis*. Pitman Books, London, 1982.

[4] J. S. R. Chisholm, A. K. Common (ed.): *Clifford Algebras and their Applications in Mathematical Physics*. Reidel, Dordrecht, 1986.

[5] B. Goldschmidt: Existence and representations of solutions of a class of elliptic systems of partial differential equations of first order in the space. *Math. Nachr.* **108** (1982), 159-166.

[6] I. Porteous: *Topological Geometry*. Van Nostrand Reinhold, London, 1969. Cambridge University Press, Cambridge, 1981.

[7] H. Zassenhaus: On the spinor norm. *Arch. Math. (Basel)* **13** (1962), 434-451.

[8] Correspondence under the pseudonym R. Lipschitz. *Ann. Math.* **69** (1959), 247-251.

THE IMPORTANCE OF SPIN *

A.O. Barut**

International Centre for Theoretical Physics, Trieste-Miramare, Italy

ABSTRACT

The spin properties of fundamental particles manifested as a magnetic moment are reviewed. The possible dominance of magnetic forces at short distances is discussed, which leads to a magnetic interpretation of strong and weak interactions. The construction of mesons, baryons and heavy leptons and their internal quantum numbers from the two basic absolutely stable spinors, electron and neutrino, is given.

* Presented at the "Conference on Spinors in Physics and Geometry" in honor of Paolo Budinich, Sept. 1986.

** Permanent address: University of Colorado, Physics Dept., Boulder CO 80309, U.S.A.

1. INTRODUCTION

I should like to talk about spinning elementary particles rather than spinors. We have heard many talks at this meeting on the remarkable geometrical and group theoretical properties of spinors as mathematical objects. I shall elaborate on the equally remarkable, if not more, physical role played by spin in particle physics. Without the existence of the electron spin we would not have the spinors, hence this meeting here today.

2. THE MAGNETISM OF ELEMENTARY PARTICLES

In atomic physics the spin manifests itself as a magnetic dipole moment and gives to the electron a certain internal structure; it is no longer a pure point particle. Very important spin effects are summarized in the Pauli principle which can also be roughly viewed as an attractive force in the singlet state and as a repulsive force in the triplet state of two identical particles. In addition, there are spin-orbit and spin-spin forces between the magnetic moments which are of electromagnetic nature, but these forces are weak compared to the Coulomb force. They are however known with very high precision from the fine - and hyperfine - structure of atomic energy splittings. In relativistic atomic physics the Dirac equation describes the magnetic moment of the electron in a more organic manner in the sense that the <u>minimal</u> U(1) - gauge coupling of the Dirac equation to the electromagnetic field $A_\mu(x)$ implies automatically the normal magnetic moment $g = 2$ for the electron. Dirac has discovered, as he states, his equation by chance. The mathematical and physical richness underlying this very simple equation, $(\gamma^\mu i\partial_\mu - m)\psi = 0$, does not cease to amaze us sixty years later, and cannot be just accidental. The Dirac matrices γ^μ ,best viewed as new internal dynamical variables, attribute to the electron that certain elusive internal structure called "the zitterbewegung" which carries the magnetic moment, as we shall see. There is no mathematical mystery if we accept the spin as some discrete internal quantum number arising from the representations of the little group SU(2). But if we ask about the physical origin of spin there is a great deal of mystery how and why this peculiar internal structure is realized and how it differs from an extended top, and how the forces depend on this structure.

When we come to nuclear and particle physics the spin effects or the spin forces become more important. Again we always impose the Pauli principle. The spin-orbit and spin-spin forces (so called tensor forces) between the nucleons in a nucleus are empirically of the form of magnetic moment interactions[(1)], but much larger in magnitude than the magnetic moments of the nucleons would indicate. Recall that the magnetic moments of the proton and the neutron are some 2,000 times smaller than that of the electron, 2.79 μ_N and -1.91 μ_N respectively, where μ_N is the nuclear magneton, $\mu_N = \frac{e\hbar}{2M_pc} = \frac{e\hbar}{2mc}\left(\frac{m}{M_p}\right) = \mu_o(m/M_p)$, μ_o = Bohr magneton, M_p = proton mass, m = electron mass. The fact that the magnetic moments of the proton and neutron differ from the Dirac value 1 μ_N already indicates that these are not simply descirable by a Dirac equation. We have of course since then ample evidence in high energy processes that p and n are composite particles. The deviation of the magnetic moment from the Dirac value is called the <u>anomalous magnetic moment</u> which is thus 1.79 μ_N for proton p and -1.91 μ_N for neutron n ; because n , being neutral has no minimal coupling to A_μ, hence no normal magnetic moment. By the way, the electron also acquires an effective anomalous magnetic moment a due to its interactions with its own self field, $a = (\alpha/2\pi)\mu_o$, which is of the same order of magnitude as the full magnetic moment of the nucleons. We shall come back to this anomalous magnetic moment later. The tensor forces of nuclear physics are therefore some effective spin forces due to the exchange and direct forces between the constituents of the nucleons.

In particle physics one has thought for a long time that the spin forces are small and unimportant, especially at high energies. However, recent experiments indicate that there are very large unexpected spin effects[(2)]. In the case of composite particles, one goes of course to what one thinks to be the more elementary constituents of nucleons and tries to find out the spin interactions between them. If these are the quarks and gluons interacting via the minimal coupling of QCD, then it is very difficult to understand the large polarization asymmetries between protons[3].

In hadron spectroscopy using valence quarks as constituents, spin-spin and spin-orbit forces have been found to be necessary in the phenomenological potential picture[4]. Their nature is thought to be non electromagnetic. One could associate new gluonic type dipole moments to quarks. Quarks have of course the ordinary magnetic moments as well; sometimes even an anomalous magnetic moment is assigned to them. As we shall see, magnetic forces at short distances are of the same order as strong forces. But the potential picture for quark spectroscopy has serious problems at large distances due to confinement and van der Waals forces[5], and at short distances due to the increase of spin-forces[6].

We can best summarize the situation in nuclear and particle physics by saying that although the spin effects are very important their origin and their precise form is far from being clear.

After this brief overview of the magnetism of elementary particles I should like to return to the magnetic dipole moment of the electron, presumably the best known case, and that of the neutrino.

3. MAGNETIC DIPOLE MOMENTS OF ELECTRON AND NEUTRINO

The magnetic moment of the electron is actually a magnetic form factor $f_2(q^2)$, a function of the square of the momentum transfer four vector, q^2. The magnetic dipole moment per se is the value $a = f_2(0)$ at zero momentum transfer, i.e. the static value of the form factor. It is true that this value for the electron is measured with an incredible accuracy

$$a_{exp} = \frac{1}{2}(g-2)_{exp} = 1159\ 652\ 797(128) \times 10^{-12} .$$

The best QED-perturbation theory value is

$$a_{th} = \frac{1}{2}(g-2)_{th} = 1\ 159\ 652\ 193(4) \times 10^{-12}$$

so that there is still a discrepancy which can be as large as [7].

$$a_{th} - a_{exp} = 604\,(149) \times 10^{-12}.$$

The neutrino, like the neutron, can only have an anomalous magnetic moment. In general, the origin of the anomalous magnetic moment of a particle might be the composite nature of the particle. Or it may be intrinsic in which case we would have besides the minimal Dirac coupling, also a Pauli coupling of the form $a\bar{\Psi}\sigma_{\mu\nu}\Psi F^{\mu\nu}$ in the action. Now for simplicity (and in QED for reason of renormalizability) one prefers to have the Dirac coupling only, which is derived from a gauge principle. But the Pauli coupling can also be derived from a gauge principle [8]. As to the renormalizability, nonperturbative methods allow us to deal with the Pauli coupling as well. We should therefore not dismiss the Pauli coupling at least for the neutrino, for it may bring a simplification to the weak interaction theory as we shall see.

A two-component neutrino cannot have a Pauli coupling, but a four-component neutrino can have a magnetic moment even if it is massless [9]. A four component massless neutrino in a magnetic field is not diagonal in its chiral left and right states, but as soon as it leaves the magnetic field the two L and R components become uncoupled. Furthermore, while crossing sufficiently strong magnetic fields only one of the spin (or helicity) components of a magnetic dipole gets through, the other is reflected or trapped.

An upper limit to the magnetic moment of the neutrino of about can be given from different considerations. At this value, the electron-neutrino scattering cross section becomes equal to the observed value, or to the Weinberg-Salam cross section [10]. In other words, a neutrino with a magnetic moment provides a purely electromagnetic model of neutral currents. Also at this value of the magnetic moment the tunnelling out of electron and neutrino allows one to calculate the lifetime for neutron or muon. Hence the Fermi coupling constant can be reduced to a magnetic interaction and thereby the universality of the weak interactions[11]. The very small magnetic moment of the neutrino makes it difficult to make a test with laboratory magnetic fields. But very large magnetic fields occur in neutron stars (10^{12} gauss), and in the sun over long distances. The solar neutrino problem[12] (1/3 of the expected solar neutrino seem to arrive to earth) might find an explanation with a neutrino magnetic moment. One could look for a correlation of the neutrino flux with solar magnetic flare, and evaluate the deflection or trapping of neutrinos between the solar centre and corona. Of course, very large magnetic fields also occur near elementary particles. For example, at 2.8 fermis from the electron a localized magnetic moment of μ_o produces 10^{17} gauss. In the next section I discuss another striking effect of such large magnetic fields at short distances between elementary particles.

4. THE MAGNETISM AT SHORT DISTANCES

We have mentioned that magnetic forces are of the order of α^2 smaller than the electric forces at atomic distances (10^{-8} cm). If one extrapolates them however down to nuclear distances (10^{-13} cm) they become $1/\alpha^4$ times larger than the electric forces. At these distances one must use of course relativistic dynamics. This softens a bit the form of the magnetic potentials at short distances. But then we must take into account the effect of the anomalous magnetic moment which gives rise to singular potentials at short distances even relativistically. The static value of the anomalous magnetic moment for the electron that we have quoted $\frac{\alpha}{2\pi}\mu_o$ refers to the free electron. For a strongly bound electron the situation is quite different. Anomalous magnetic moment in fact depends on the value of the external field, more generally on the wave function of the state itself. To see this and for later purposes I now discuss the non-perturbative origin of the anomalous magnetic moment.

For an electron in a fixed external field A_μ^{ext} and a radiation field A_μ satisfying coupled Maxwell-Dirac equation, we can express $A_\mu(x)$ as the Lienard-Wiechert potential of the electron current

$$A_\mu(x) = e\int dy\, D(x-y)\bar{\psi}(y)\gamma_\mu\psi(y), \tag{1}$$

where $D(x-y)$ is the causal Green's function. Consequently, the electron sees besides A_μ^{ext} also the self-field $A_\mu(x)$ in (1) and satisfies the non-linear equation

$$\left\{\gamma^\mu\left(i\partial_\mu - eA_\mu^{ext}\right) - m\right\}\psi(x) = e^2\gamma^\mu\psi(x)\int dy\, D(x-y)\bar{\psi}(y)\gamma_\mu\psi(y). \tag{2}$$

It has been shown elsewhere [(13)] that this nonlinear equation reproduces all the QED effects like spontaneous emission, vacuum polarization and Lamb-shift. Physically, eq. (2) means that we are treating the action of the field produced by the electron (the self field (1)) back on the electron in a self-consistent way, as in classical radiation theory. As in any other self-consistent process with feedback the equation is nonlinear, meaning that the radiative effects in a state ψ depend on ψ itself. Among these radiative effects is also the emergence of the anomalous magnetic moment. For example, if A_μ^{ext} is a magnetic field, or a Coulomb field, then part of the Lamb-shift can be attributed to an anomalous radiative magnetic moment for the electron.

Now the nonlinear equations of the type (2) can have two types of solutions. A solution in which the right hand side gives a small shift to the solutions of the left ahand side, such as precisely the Lamb-shift. There can be a second type of solution where the right hand side dominates. These are the "soliton-type" solutions of nonlinear equations. Such solutions have been interpreted as the self-focusing or self-organizing solutions of nonperturbative electrodynamics [(14)]. Intuitively, for highly localized states of electrons and positrons the magnetic fields produced by them is so large and localized that it is sufficient to keep the particles going around in that localized orbits, hence the name self-focusing solutions.Equivalently, for orbiting highly localized electrons and positrons, the anomalous magnetic moment increases, the normal magnetic moment decreases, so that total magnetic moment remains constants. Then one can indeed show that for such large anomalous magnetic moments, electrons and positrons can form massive new resonance states, called "magnetic resonances". In fact this is a mechanism of generation of mass, from almost massless constituents. Mass of particle is the motion of its constituents[15].

5. COMPOSITE MODEL OF PARTICLES

We can make therefore the hypothesis that electrons and neutrinos can form, at short distances, (almost pointlike it turns out) new resonances of very high mass which we can identify with mesons and heavy leptons.

It is reasonable to assume that electrons and electron neutrinos are the only absolutely stable particles, hence can be taken to be the basic building blocks of all other particles which are then composite and generally unstable. In fact the only other candidates for stable particles, besides e and ν_e, are at present, proton and ν_μ (perhaps ν_τ). The search for stability of p and ν_μ is under way experimentally.

The idea of making the real, absolutely stable particles as building blocks of matter[16] is very attractive, because these particles being undestructible and unmodifiable will remain so under all interaction, hence there is an end of subdividing matter into smaller and smaller entities. Also empirically we observe all new unstable "particles" decaying in succession until the stable constituents e, ν_e, γ are reached. It is a theorem of S-matrix theory, that an unstable particle is a resonance in the channel of its decay product S. Hence we can go backwards, and construct all resonances from their decay products, thus eventually from stable particles e, ν_e, γ. Roughly speaking a particle is into what it decays

Electron and neutrino are of course very akin and may be considered as two states of one bigger entity. In fact, the conformally invariant Dirac equation is necessarily 8-dimensional [18] containing two types of Dirac particles, one charged and one neutral which can be identified with e and ν_e [19]. This unification is at a deeper level. In the following we shall treat e and ν_e as separate particles and unify everything else.

6. CONSTRUCTION OF PARTICLE MULTIPLETS FROM THE BASIC SPINORS ELECTRONS AND NEUTRINOS

The question therefore arises if the seemingly very complex world of mesons, baryons, and heavy leptons can be built up from just two particles, e and ν_e. The answer is yes, perhaps unexpectedly. Is is sufficient to have one charged and one neutral spinor. I shall now outline this scheme which includes, in a mathematically equivalent way, the classification of hadrons based on quark model, but more, and discuss the origin of the internal quantum numbers [20].

The "Aufbau" principle consists of building from e, ν_e first simple two or three body systems, then using these together with e, ν_e again to form more complex systems which are however less stable, and so on. A hierarchy of particles with decreasing stability is established. Stability criterion turns out to be very important in understanding the approximately conserved quantum numbers like "strangeness", "charm", etc.

Thus, at the first level A we put the two absolutely stable spinors: e,

ν_e (and their antiparticles). Their numbers provide two absolutely conserved quantum numbers which we identify with charge Q and lepton number L. All charge in the world is due to the electron. This simple starting point explains two of the most fundamental laws of physics: the quantization of charge and the conservation of charge.

In the second level B we form the two next most stable particles

$$p = (e^+ e^+ e^-) \quad , \quad \nu_\mu = (\nu_e \nu_e \bar{\nu}_e)$$

Their number is conserved for all processes lasting shorter than their lifetimes. We identify these with baryon number B and N_{ν_μ}. As I have indicated the lifetimes of p and ν_μ have not yet been determined.

If we start with equal numbers of e^+ and e^- and form $p = (e^+e^+e^-)$ then we are left with e^- and p, and no e^+ and $\bar{p}$. Since p and $\bar{p}$ annihilate, only one of these species will eventually survive and become abundant. This process explains the asymmetry between matter and antimatter in the world in a simple way.

At the third level C we have again the next two most stable particles after the level B

$$n = (p e^- \bar{\nu}_e) \quad , \quad \mu = (\nu_\mu e^- \bar{\nu}_e) . \tag{3}$$

with a lifetime τ_C. In this case we know that n and μ indeed decay into the assumed constituents. These two particles are of the same nature, with an almost universal decay rate (modified by their different phase spaces) and with the same building principle from levels A and B. Their number N_n and N_μ are not absolutely conserved, but again for all processes which last shorter than τ_C. Thus N_n is conserved in nuclear physics and we can identify N_μ with the "strangeness" quantum number[(16)].

The six most important particles are in the levels A, B, C. From here on the remaining particles are less and less stable and their lifetimes cluster around 10^{-8}, 10^{-10}, 10^{-12}, 10^{-16}, 10^{-18}, up to 10^{-24} sec, which is about the limit of observability. There seems to be a certain quantization of lifetimes

[17]. I now give a general rule for all the particle states. Let ℓ ($\bar{\ell}$) denote the leptons, and b ($\bar{b}$) some basic baryons which we can take to be any number of $\ell = (e, \nu_e, \mu, \nu_\mu, \ldots)$ and $b = (p, n, \Lambda^0, \Lambda^+ \ldots)$. Then we form the the mesons and remaining baryon as

$$M = (\ell \bar{\ell}) , \quad B = (b \ell \bar{\ell}) . \tag{4}$$

I shall not go here into the details of mixing effects, like $M = (\ell\bar{\ell}) \oplus (\ell'\bar{\ell}')$. If one takes two, three or four ℓ's and b's one can show quite rigorously that one obtains the same classification as the SU(2), SU(3) and SU(4)-quark models. For example at the SU(4)-level we have a correspondence to the integrally charged quark model with colour, although "colour" as such has not been introduced[21]. A more direct transformation of the leptonic model (4) to quark model can be given using the notion of T and V- rishons[22].

Different branching ratios of unstable particles in decay correspond to different rearrangements of the constituents. Rearrangement of the constituents also occurs during scattering of the composite particles. The observed approximate SU(2) (isospin) or SU(3)-symmetries can in fact be reduced to the symmetries under the exchange of constituents. We have become accustomed to using the language of continuous Lie groups for internal symmetries. In practice, however, we use first of all Lie algebras rather than groups, and then only low dimensional representations. Such low dimensional representations coincide with those of finite subgroups of Lie groups. Consequently all the results of internal symmetries can be reconstructed using finite permutation groups[23] and charge conservation, hence physically by the exchange of the constituents.

7. ON THE ORIGIN OF SPIN

An elaboration of the enormous importance of spin would be incomplete without some reflection on the origin of spin. The mathematics of spin in quantum theory in the form of Pauli and Dirac spinors, a finite dimensional spin space, the operators acting on it, etc. is no longer a mystery. But what is a mystery is perhaps what all this means physically as to the internal structure of the particle. It is clear that we have new internal degrees of freedom besides the usual phase space $(\bar{p},\bar{q})$. What is the best way of visualizing the internal structure? I like a picture that emerges from the Dirac equation. The charge of the electron performs a rapid oscillatory motion around a fictitious centre of mass. Thus the motion of the charge is helical as the electron moves (called "zitterbewegung" by Schrödinger). The velocity and momentum of the electron become independent dynamical variables. The momentum refers to the centre of mass. Position and velocity commute, but position and momentum do not commute[(24)]. The orbital angular momentum of the internal oscillations becomes identical with the spin with the correct gyromagnetic ratio[(25)] g = 2. It has recently been found that such a picture can be realized in classical relativistic mechanics for a particle which has precisely the orbit as a zitterbewegung[(26)]. Thus the electron is neither a point,nor a small extended ball, but the internal oscillations endows the electron with a structure which is expressed in the spin properties of the particle. A notion of "antiparticle" emerges already here in the classical problem. The quantization of this classical system gives precisely the Dirac electron, thereby a solution is obtained to the old problem of finding the classical analog of the Dirac electron. I hope that a picure might give us a clue to another physical mystery, namely the origin of the Pauli principle. As I mentioned, the Pauli principle can be expressed as an additional force between two electrons, repulsive when spins are parallel, attractive when spins are antiparallel. Thus it should be connected with the forces between two internal oscillations, like forces between current loops.

8. CONCLUSIONS

The spin is not just an inessential property of the particles to cause small spin-orbit and spin-spin effects which seem to break, annoyingly, some nice symmetries. Rather the magnetic effects may be elevated to equal status as electric effects, the latter responsible for atomic and molecular structure, the former for nuclear and hadronic structures. The two absolutely stable spinors, the electron and the neutrino, can be taken to be the fundamental building blocks of matter. This simple picture explains the quantization and conservation of charge, the matter-antimatter asymmetry,the origin of approximate quantum numbers. The group structure of the multiplets can rigorously be established. What is more difficult, however, is the dynamical problem of calculating the masses of mesons and hadrons. If the magnetic interactions suffice to do that,we would have an already unified theory of strong, weak and electromagnetic interactions, unified in electromagnetism[(27)] : what we have called strong and weak interactions are just new manifestations of the electromagnetic interactions. No new particles and no new forces are postulated. The only parameter in the theory being the mass m and charge e of the electron and the magnetic moment of the neutrino. The strong interaction scale is introduced via the magnetic moment e/2m of the electron, and weak interaction scale by the magnetic moment of the neutrino. The dynamical problem is not fully solved. The difficulties with relativistic two and three-body problems with magnetic forces are enormous. But enough simplified models have been solved and rather striking mass formulas for mesons, proton, and even Z^0 have been obtained by relativistic semiclassical arguments (28., 14.) that I feel that the model and the underlying idea "makes sense" and fits in, "the sounds will come later".

REFERENCES

1) A. DeShalit and H. Feshbach, Theoretical Nuclear Physics, New York, Wiley 1974; p. 13.

2) For a review see A.D. Krisch, Comments Nucl. & Particle Physics 11, 93 (1983); Ann. Review of Nuclear and Particle Science, Vol. 31, 107 (1981).

3) For a recent discussion see G. Preparata and J. Soffer, Physics Letters B, 180, 281 (1986);
H.J. Lipkin, Nature 324, 14 (1986).

4) See for example, H. Schnitzer, Phys. Letters 149 B, 408 (1984); 134,253 (1984).

5) A.O. Barut and R. Rączka, Modern Physics A, 2, 265 (1987).

6) A.O. Barut, Proc. Kaziemierz Conference on Elementary Particle, Univ. of Warsaw Press, 1986, (ed. Z. Ayduk), p. 293.

7) M.A. Samuel, Phys. Rev. Lett. 57, 3133 (1986).

8) A.O. Barut and J. McEwan, Phys. Lett. 135 B, 171 (1984).

9) A.O. Barut and J. McEwan, Lett. Math. Phys. 11, 67 (1986).

10) Z.Z. Aydin, A.O. Barut and I.H. Duru, Phys. Rev. D 28, 2872 (1983);
Phys. Rev. D 32, 3051 (1985).

11) A.O. Barut and G. Strobel, KNAM - Revista de Fisica, 4, 151 (1982).

12) J.N. Bahcall et al, Rev. Mod. Phys. 54; 767 (1982).

13) A.O. Barut and J.Kraus, Foundations of Physics 13, 189 (1983);
A.O. Barut and J.-F. van Huele, Phys. Rev. A 32, 3887 (1985).

14) A.O. Barut, Suppl. Hadr. Journal, Vol. I, 314 (1985);
A.O. Barut, Found. of Physics, 17, 549 (1987).

15) A.O. Barut, Lett. Math. Phys. 10, 195 (1985)

16) A.O. Barut, Surveys in High Energy Physics, 1, 113 (1980); American Institute of Physics Conference Proceedings, 71, 73 (1980).

17) A.O. Barut, in "Quantum Theory and Structure of Space-Time", Vol. IV, München C. Hanser Verlag, 1981 (edit. L. Castell), p. 152.

18) A.O. Barut and R. Haugen, Nuovo Cim. 18 A, 495 and 511 (1973); Lett. N.C. 7, 625 (1973);

P. Budinich, P. Furlan and R. Rączka, Nuovo Cim. 52 A, 191 (1979).

19) A.O. Barut and B.W. Xu, Physics Letters 101 B, 437 (1981).

20) A.O. Barut, in Quantum Theory and Structure of Space-Time, Vol. V, München, C. Hanser Verlag, 1983 (ed. L. Castell), p. 122.

21) A.O. Barut and S. Basri, Lett. N. G. 35, 200 (1982); Intern. J. Theor. Phys. 22, 691 (1983).

22) A.O. Barut and D.Y. Chung; Lett. N. C. 33, 225 (1983).

23) A.O. Barut, Physica 114 A, 221 (1982).

24) A.O. Barut and A.J. Bracken, Phys. Rev. D23, 2454 (1981).

25) A.O. Barut and A.J. Bracken, Phys. Rev. D24, 3333 (1981).

26) A.O. Barut and M. Zang , Phys. Rev. Lett. 52, 2009 (1984).

27) A.O. Barut, Unification based on Electromagnetism
Ann. der Physik, 43, 83 (1986) - contains detailed bibliography.

28) A.O. Barut, Lett. N. C. 34, 1 (1985).

THE THEORY OF WORLD SPINORS

Yuval Ne'eman
Sackler Faculty of Exact Sciences*
Tel Aviv University, Tel Aviv, Israel**
and Center for Particle Theory***
University of Texas, Austin, Texas, USA

* Wolfson Chair Extraordinary in Theoretical Physics

** Supported in part by the US-Israel Binational Science Foundation

*** Supported in part by the US DOE Grant DE-FG05-85ER40200

Abstract

We review the confusion that caused the long-held belief that the real linear groups and the diffeomorphisms do not have bivalued spinorial representations. We then describe these infinite unitary representations of the $s\ell(n,\mathbb{R})$ for n=2,3,4 and the Dirac-like equation for such a spinor "manifield" in flat space and in Einstein-Cartan or Metric-Affine Gravity. We present the transition to world spinor manifields for the Riemannian and the Affine case and for theories gauging $\overline{GL}(4,\mathbb{R})$ holonomically and anholonomically. We discuss applications to hadron spectroscopy, to the coupling of hadrons to gravity and to the theory of the superstring.

1. The Confusion over the Spinorial Representations of the Linear Groups

For more than fifty years after the discovery of the electron and the subsequent introduction of the spinor representations of $\overline{SO}(3)$ (=SU(2)) and of the induced Poincaré group into the world of physics, the existence of spinor representations of the corresponding linear groups $\overline{GL}(3,\mathbb{R})\supset\overline{SO}(3)$, $\overline{GL}(4,\mathbb{R})\supset\overline{SO}(1,3)$ was unknown or unnoticed throughout the physics community. Almost every textbook on General Relativity, upon reaching the subject of spinors, contains a sentence such as "there are no representations of GL(4,$\mathbb{R}$), or even 'representations up to a sign', which behave like spinors under the Lorentz subgroup" (this excerpt is from one of the best-known texts, perhaps rightly considered as the most complete and the most "physical" treatise on Gravity available to date). Even a review on Supergravity published as late as 1985 still contains the same piece of Folklore!

On one occasion, an article by two of our most distinguished colleagues purportedly proves an important point in Quantum Gravity, building the proof on the premise that spinorial representations do not exist for the linear groups...

I have related elsewhere[1,2] why Cartan is sometimes given as the original reference for the erroneous statement. With all our modern disregard of authority, it was hard for me to assume that Cartan must have been wrong, when I first realized that I was looking at the matrices of a spinorial representation of $\overline{GL}(3,\mathbb{R})$. I checked and found the relevant discussion in the "Leçons sur la Theorie des Spineurs"[3]. In the last section of that work, Cartan indeed studies the construction of a Dirac-like linear equation on a Riemannian manifold and concludes that "it is impossible to construct a Riemannian covariant derivative for the equation [which indeed should be invariant under the diffeomorphisms and under their linear GL(4,$\mathbb{R}$) subgroup - Y.N] using a finite number of components", a very correct statement. That phrase, limiting the negative theorem to the case of a finite number of components, appears to have been lost in the transmission, like a scribe's error in a Biblical manuscript. Cartan's other relevant discussion is in his study (in the same text) of the 2-dimensional unimodular groups SU(2), SL(2,$\mathbb{C}$) and SL(2,$\mathbb{R}$). He explains that SU(2) is simply-connected and thus has no double-covering; on the contrary, SU(2) itself is isomorphic to $\overline{SO}(3)$, the double-covering of SO(3). SL(2,$\mathbb{C}$) is a non-compact group whose maximal compact subgroup is SU(2). The topology of a non-compact group G is that of its maximal compact subgroup K, as can be observed in the Iwasawa decomposition[4],

$$G = KAN \qquad (1)$$

(A is abelian, N is nilpotent), i.e. every matrix of G can be written as a product of matrix representations of K, A and N. Those of A and N are topologically trivial or contractible to a point so that the topology of G is that of K. In this case too, it is $SL(2,\mathbb{C})$ itself which is isomorphic to the double-covering $\overline{SO}(1,3)$ of another group, the Lorentz group SO(1,3). Now tackling $SL(2,\mathbb{R})$ Cartan is well aware of the availability of an infinite layer of coverings for the compact subgroup SO(2), but his discussion of the representations of $SL(2,\mathbb{R})$ centers on the fact that $SL(2,\mathbb{R})$ itself is the double-covering of SO(1,2) or of the direct homographic group for one real variable

$$z' = \frac{az+b}{cz+d}\ , \quad ad - bc > o$$

And yet Cartan does provide a construction of multiple-coverings for that group. Bargmann[5], in his study of all representations of $SL(2,\mathbb{R})$ included the coverings, whose representations appear both in the $C_q^{\frac{1}{2}}$ continuous class and in the discrete $D_k^{\pm}$.

Considering that this conference in Trieste is at the same time a commemorative occasion honouring our distinguished colleagues A. Lichnerowitz and P. Budinich, I am very happy to have saved Cartan's reputation and thus preserved the prestige of Age and Authority... I shall use this opportunity to wish them both an active and pleasant continuation of their creative careers, at least until the ripe age of 120 (following the tradition in which I was educated...).

Returning to the $G:=GL(n,\mathbb{R})$ for $n>2$ the semi-simple compact subgroups $\mathcal{O}_s:=SO(n)$ all have spinorial representations and can always be double-covered; so all $GL(n,\mathbb{R})$ have a double-covering $\overline{G}:=\overline{GL}(n,\mathbb{R})$ whose compact subgroup is $\overline{\mathcal{O}}_s=\overline{SO}(n)$. The double-valued representations of SO(n) become single-valued on $\overline{SO}(n)$. This is precisely the definition of a spinor. These representations of G(or $S:=SL(n,\mathbb{R})$) whose SO(n) reduction yields double-valued representations are thus also double-valued for G or S and become single-valued for $\overline{G}$ or $\overline{S}$. In all these cases, the double-coverings of the linear groups taken as matrix groups <u>exist only in infinite matrices</u>.

The diffeomorphisms ("$\mathcal{D}$iff"), an infinite group, are generally represented non-linearly over some linear subgroup. In world-tensors[6], the linear subgroup is G. Before it was realized that there existed $\overline{G}$ spinors, theories <u>gauging $GL(4,\mathbb{R})$</u> itself were forced to represent it non-linearly for the spinorial matter fields, now inducing the representation over the Poincaré subgroup[7,8]. This is still useful, I think, in the understanding of how the mere existence of a spin 2 massless state induces the dynamics of Einstein Gravity[9] over the special relativistic Hilbert space. The Hilbert space is a reducible representation of the Poincaré group, and General Covariance is realized non-linearly over it, through the non-linear application of graviton states[10]. Note that I have been using $SL(n,\mathbb{R})$ and $GL(n,\mathbb{R})$ indiscriminately. The matrix representations are the same. The topology is, of course, different, the full-linear (like its orthogonal $O(n)$ compact subgroup) including twice the number of connected components, as compared to the unimodular group-manifolds.

Gelfand and Graev[11] had studied the $SL(n,\mathbb{R})$ and in particular $SL(3,\mathbb{R})$, constructing the Principal Series of unitary irreducible representations ("unirreps"). However, they did not deal exhaustively with the other classes and did not address the issue of the double-covering at all. Neither did others who tackled this problem in the Fifties and Sixties. I think the issue of an $SL(n,\mathbb{R})$ spinor was first raised in 1965, when with Dothan and Gell-Mann we were interested in using $SL(3,\mathbb{R})$ for Regge sequences of hadronic levels[12]. It had occurred to us (and independently to A. Barut, A. Bohm and C. Fronsdal[13]) that the rich and apparently unending spectrum of hadron excitations should be described by the (infinite) unitary representations of some non-compact group, like the spectrum of the harmonic oscillator (described by $SU(3,1)$), hydrogen atom ($SO(4,1)$ or $SO(4,2)$) or spinning top[14] ($SL(4,\mathbb{R})$). Helped by Feynman, we developed a method of constructing "ladder" representations, a type of multiplicity free (in terms of the reduction over the compact subgroup) representation in which the sequence of representations of the compact subgroup forms a one-dimensional array: the non-compact raising and lowering operators connect the states corresponding to a given representation of the compact subgroup in the reduction to only one "higher" irreducible representation and at most one "lower" one. For the $|\Delta J|=2$ systematics of the Regge sequences (and of deformed nuclei, in Nuclear Physics), two ladder unirreps of $SL(3,\mathbb{R})$ fitted beautifully ($\{J\}$ is the range of values of J)

$$\mathcal{D}^{\ell}(\sigma;|J_{min}|), \sigma\epsilon\mathbb{R}, \ |J_{min}| = 0,1; \{|J|\} = |J_{min}| + 2|\mathbb{Z}|$$

but what about the nucleon $N(938), |J_{min}|=\frac{1}{2}$ and $\Delta(1238), (|J_{min}|=\frac{3}{2})$ trajectories? It seemed possible to construct them, but our ignorance of the principles governing the topology of groups made us abandon the problem. Similarly, with

Dothan[19)], we constructed the ladder representations of SL(4,ℝ) for the spinning top, but did not find the spinors.

Note that this was the first time Regge sequences were treated as an approximation to an infinite set of states, similar to the harmonic oscillator, etc. Originally they had been viewed as resulting from (pionic) Yukawa potentials, and these had a very short series of poles. After SL(3,ℝ) it became common to search for an infinite sequence as the better model for Regge excitations, and the Dual Models and Strings continued this approach.

Interestingly, we did suspect[12)] some connection with gravity, but rather through its <u>inertial</u> component. We conjectured that the current-algebra-like generators producing this SL(3,ℝ) in hadrons and in nuclei were in fact the quadrupole-vibration rates. In a way this was a very early interpretation of hadrons as extended structures, and moments of inertia are then directly involved. Following our suggestion, Biedenharn and co-workers later tried out the model on nuclei[15)], with modest success. A more extensive application has since emerged[16)] with a claim to an improved understanding of the algebraic framework.

Still on the track of an algebraic derivation of the Regge sequences, I suggested to D. W. Joseph in 1969 (we were together at Caltech for a term) that we try and construct $\overline{SL}(3,\mathbb{R})$ representations with $J_{min} = \frac{1}{2}$ or $\frac{3}{2}$. It seemed to work but after returning to our respective home universities, Joseph reported that the $J_{min} = \frac{1}{2}$ was OK but the $J_{min} = \frac{3}{2}$ had developed a singularity, i.e. it did not exist. Having in mind the N and Δ trajectories, and knowing too little about representations that might not be of the ladder type (they do exist for $J_{min} = \frac{3}{2}$, as we shall see later), I lost hope and interest, and suggested to Joseph that he go ahead and publish his results in some mathematical journal. He did try but the referees were not familiar with this material and insisted on mathematical purity. He abandoned[17)] it. It was all rediscovered three years later by Biedenharn and collaborators[15)], including the puzzle of the $J_{min} = \frac{3}{2}$ which was again settled by Ogievetsky and Sokachev[18)].

With all this, the association with the issue of curved space spinors had not yet occurred to me. It was only when the discovery of Supergravity had rekindled my interest in gravity and while I was reading about Gravity as a gauge theory that I realized[1,19,20)] that the $J_{min} = \frac{1}{2}$ unirrep of $\overline{SL}(3,\mathbb{R})$ really described the Hilbert space states of a $\overline{GL}(4,\mathbb{R})$ spinor!

Hehl and collaborators were reviewing the case for torsion[21)]. Their review also contained the sacerdotal statement about the inexistence of GL(4,$\mathbb{R}$) spinors, and they then remarked that this was an obstacle for an attempt to gauge GL(4,$\mathbb{R}$) since it would make it difficult to represent matter fields in such a theory. They were interested in this gauge group at the quantum level in a Palatini approach, later reproducing Einstein's metric Riemannian manifold in the equations of motion ("Metric-Affine" Gravity[22]). I wrote to Hehl and suggested a solution for that matter field, as an anholonomic GL(4,$\mathbb{R}$) spinor field. It was the start of a rewarding collaboration on both the Metric Affine[23] and the Poincaré Gauge Theory[24].

Dj. Šijački, then a student of Biedenharn, had pursued the Regge program in his thesis and studied a covariant version[25] involving GL(4,$\mathbb{R}$). In 1975 he published the definitive work on $\overline{SL}(3,\mathbb{R})$, an exhaustive catalogue[26] of all unirreps, including the double covering. In general, mathematicians have not touched the issue, and B. Speh's comprehensive work[27] on SL(3,$\mathbb{R}$) and SL(4,$\mathbb{R}$) contains everything except for the unirreps of the double covering.

For SL(4,$\mathbb{R}$), Sijacki may soon be publishing a similar complete classification[28]. Meanwhile, we had tried an Affine gravity[29,30] theory, leading to a natural two-graviton "strong-gravity" theory. For this purpose, we constructed the multiplicity-free set, and completed it in 1985[31]. With A. Cant[32] I constructed the field equations for an anholonomic GL(4,$\mathbb{R}$) spinor. I then developed the transition to world spinors[33] and with Šijački we used the results first for hadrons[34], renewing and "modernizing" the 1965 program, with a very good fit to the entire hadron spectrum, using just a very few GL(4,$\mathbb{R}$) unirreps. We then completed the formalism for world spinors[35].

2. Unitary Representations of $\overline{SL}(n,\mathbb{R})$ for n=2,3,4

A. $\overline{SL}(2,\mathbb{R})$

Bargmann[10] calculated the complete classification of the irreducible unitary representations of SL(2,$\mathbb{R}$). He listed 3 classes, defined by σ, the eigenvalue of the quadratic Casimir operator and some characteristic value of m the helicity eigenvalue in the SO(2) subgroup (see Fig. 1).

(a) the principal series

$$\mathcal{D}^P(\sigma;\overline{m}):\ 0<\sigma\varepsilon\mathbb{R},\quad \overline{m}=0;\quad \{m\} = \mathbb{Z} + \overline{m} \tag{2a}$$

(b) the supplementary series

$$\mathcal{D}^S(\sigma;\overline{m}):\ \frac{1}{4}<\sigma<\infty\ ,\ \overline{m} = \frac{1}{2};\quad \{m\} = \mathbb{Z} + \overline{m} \tag{2b}$$

(c) the discrete

c1 $\mathcal{D}^d(m_{min})$: $\{m_{min}\} = \frac{1}{2}|\mathbb{Z}|$, $\sigma = m_{min}(1-m_{min})$; $\{m\} = m_{min} + |\mathbb{Z}|$ (2c1)

c2 $\mathcal{D}^d(m_{max})$: $\{m_{max}\} = -\frac{1}{2}|\mathbb{Z}|$, $\sigma = -m_{max}(1 + m_{max})$; $\{m\} = m_{max} - |\mathbb{Z}|$ (2c2)

The representations of the covering groups $\overline{SL}(2,\mathbb{R})$ and $\overline{\overline{SL}}(2,\mathbb{R})$ ... $\overline{SL}^n(2,\mathbb{R})$ extend the supplementary series (b),

$\mathcal{D}^S(\sigma;\bar{m})$: $\bar{m} = 1/2c$, c the order of the covering (2d)

B. $\overline{SL}(3,\mathbb{R})$

For $\overline{SL}(3,\mathbb{R})$, there are two Casimir operators, with eigenvalues σ and δ. The compact subgroup $\overline{SO}(3)$ representations are given by $|J|$, the angular momentum. Šijacki listed[26] four classes (see Fig. 2).

(a) the principal series

$\mathcal{D}^P(\sigma,\delta;\bar{J},\bar{K})$: $\sigma = -3 + i\sigma_2$, $\delta = -1 + i\delta_2$; $\sigma_2,\delta_2 \varepsilon \mathbb{R}$

$\{\bar{J},\bar{K}\} = o,\frac{1}{2},1$; $\{\bar{J}-\bar{K}\} = o,1$; $\bar{J} \geq \bar{K}$ (3a)

The $|J|$ eigenvalues recur (with different $|K|$) r times:

a1: $(\bar{J}=o,\ \bar{K}=o)$: $\{J^r\}=o,2^2,3,4^3,5^2,6^4,7^3,\ldots$

a2,3: $(\bar{J}=1,\ \bar{K}=o,1)$: $\{J\}=1,2,3^2,4^2,5^3,6^3,7^4,\ldots$

a4: $(\bar{J}=\frac{1}{2},\ \bar{K}=\frac{1}{2})$: $J^r = \frac{1}{2},\frac{3}{2}^2,\frac{5}{2}^3,\frac{7}{2}^4,\ldots$

thus including a4, a (spinorial) representation of the double-covering.

(b) the supplementary series

$\mathcal{D}^S(\sigma,\delta_1;\bar{J},\bar{K})$: $\sigma \varepsilon \mathbb{R}$, $\{\bar{J},\bar{K}\}$ and $\{J^r\}$ as in the principal series, with

b1: $1 > \delta_1 \varepsilon \mathbb{R}$

b2,3: $1 > \delta_1 \varepsilon \mathbb{R}$ (3b)

b4: $\frac{1}{2} > \delta_1 \varepsilon \mathbb{R}$

(c) the discrete series

$\mathcal{D}^d(\sigma,\bar{J})$: $\sigma \varepsilon \mathbb{R}$; $\{\bar{J}\} = \frac{3}{2},2,\frac{5}{2},3,\ldots$ (3c)

$\{J^r\}$: $\bar{J},\bar{J}+1,(\bar{J}+2)^2,(\bar{J}+3)^2,(\bar{J}+4)^3,(\bar{J}+5)^3,\ldots$

(d) the ladder (degenerate) series

$\mathcal{D}^\ell(\sigma;\bar{J})$: $\{\bar{J}\}=o,\frac{1}{2},1$; $\sigma \varepsilon \mathbb{R}$; $\sigma(\bar{J}=\frac{1}{2})=o$ (3d)

$\{|J|\} \equiv \bar{J} \pmod 2$, $|J| \geq \bar{J}$

In the applications, $\mathcal{D}^\ell(o;\frac{1}{2})$ and $\mathcal{D}^d(\sigma;\frac{3}{2})$ are the most interesting ones. $\mathcal{D}^\ell(o;\frac{1}{2})$ is the one we constructed with D.W. Joseph in 1969.

C. $\overline{SL}(4,\mathbb{R})$

For $\overline{SL}(4,\mathbb{R})$, some of the work was done by Kihlberg[36] in 1966, working on an extended structure for hadrons. Speh classified everything but the double-covering[27]. We may soon have the complete picture[28]. Meanwhile we list the complete classification for the multiplicity-free set[31] (see Fig. 3).

(a) the principal series

$\mathcal{D}^P(e_2;\overline{J}_1,\overline{J}_2)$: $e_2 \varepsilon R$

($\overline{J}_1,\overline{J}_2$ denote the "co-ordinates" of the lowest $\overline{SO}(4)$ sub-multiplet)

$$\overline{J}_1=o,\ \overline{J}_2=o\ ;\ \{|J_1|+|J_2|\} \equiv o \pmod 2 \tag{4a}$$

$$\overline{J}_1=1,\ \overline{J}_2=o\ ;\ \{|J_1|+|J_2|\} \equiv 1 \pmod 2$$

(b) the supplementary series

$$\mathcal{D}^S(e_1;\overline{J}_1,\overline{J}_2)\ :\ o<|e_1|>1,\quad \overline{J}_1=1,\ \overline{J}_2=o \tag{4b}$$

(c) the discrete series

$$\mathcal{D}^d(\overline{J},o)\ \&\ \mathcal{D}^d(o,\overline{J})$$
$$\{\overline{J}\} = \tfrac{1}{2},1,\tfrac{3}{2},2,\ldots$$
$$|J_1|-|J_2| \geq \overline{J} \tag{4c}$$
$$|J_1|+|J_2| \equiv \overline{J} \pmod 2$$

(d) the ladder series

$$\mathcal{D}^{\ell}(J)\ ,\quad \{\overline{J}\} = o,\tfrac{1}{2}.$$
$$|J_1| = |J_2| = \overline{J} + |Z| \tag{4d}$$

The spinors exist only in the discrete series, for these multiplicity-free representations.

I shall not describe the construction and properties of the representations, since they are covered in detail in Professor Dj. Šijački's contribution to this conference.

D. Some Construction Tips

1. A very useful decomposition is that of Harish-Chandra's[37] "sub-quotient theorem. Using (1), one identifies M, the centralizer of A in K,

$$K \supset M\ ,\quad [M,A] = o \tag{5}$$

and defines the subgroup MAN. Using the representations of MAN, one then induces those of $G \supset MAN$. For example, taking[1] $SL(3,\mathbb{R})$, K is SO(3) and M is then a four element group including the three inversions in the 2-3, 3-1, 1-2 planes and the identity. This is obvious as only diagonal 3x3 matrices will commute with the entire A, and to get det(M)=1 for real eigenvalues, there can only be (+1,-1,-1), (-1,+1,-1), (-1,-1,+1) and (+1,+1,+1). If we now go to the covering group $\overline{SL}(3,\mathbb{R})$, $\overline{K}$ is SU(2) and M is represented by $\exp(\pm i\pi\sigma_i)$ (the σ_i are the Pauli matrices), yielding (±) the Pauli matrices and the identity as a finite group. It is over this group (the "Weyl group") that one then induces the entire $\overline{SL}(3,\mathbb{R})$ representation.

2. For the multiplicity-free representations, one can construct the representations from those of a semi-direct-product algebra in which all the non-compact operators of the Lie algebra of G have been contracted. This operation, the inverse of the Wigner-Inönu contraction is achieved by using a "decontraction formula" first introduced in 1965[12,13),

$$N_u = pY_u + \frac{i}{2(C_2)^{\frac{1}{2}}} \left[\sum_a K_a^2 , Y_u\right] , \quad p\varepsilon R \tag{5}$$

where N_u is the non-compact generator in G, K_a the set of compact generators, C_2 the eigenvalue of $\sum_a K_a^2$, Y_u an Abelian wave function with the same behavior under K as that of N,

$$[K_a, K_b] = if_{abc} K_c \tag{5a}$$

$$[K_a, K_u] = if_{auv} N_v \tag{5b}$$

$$[K_a, Y_u] = if_{auv} Y_v \tag{5c}$$

$$[Y_y, Y_v] = o \tag{5d}$$

eq. (5) then ensures that

$$[N_u, N_v] = if_{uva} K_a \tag{5e}$$

We have used the method in the case of $\overline{SL}(4,\mathbb{R})$, for example[29,31). For G=SL(n,R), the Y_u and N_u will always be a symmetric tensor under the K=SO(n).

3. The study of the algebra commutators and the resulting raising and lowering action of the N_u leads to an identification of <u>the relevant Hilbert spaces</u>. These are distinguished by the lowest K representation k_o and the way in which the lowering $N_u^{(-)}$ have to destroy the k_o states. Once these are determined, one has to identify <u>all possible kernels that can be used to get a Hermitian scalar product</u>. Non-unitary representations may exist fulfilling the previous conditions but not the last one. For the unitary representations, all these stages are essential.

3. Structure of $\overline{SL}(4,\mathbb{R})$

A. The Center

SO(3,3) is a non-compact group whose compact subgroup is K=SO(3)xSO(3). The complete double-covering is $\overline{K}$=SU(2)xSU(2). However, instead of having two completely independent SU(2) groups in the representations, we may constrain the representations to be either of integer spin or of half-integer simultaneously in both SU(2) groups. Each SU(2) has a 2-element center (the 2x2 matrices +1 and -1), and rather than dividing them both out and getting back SO(3)xSO(3), since

$$SO(3) = SU(2)/\mathbb{Z}(2) \tag{6a}$$

to get SO(4) we divide

$$SO(4) = [SU(2)_I \times SU(2)_{II}]/\mathbb{Z}(2)_{diagonal} \tag{6b}$$

where

$$\begin{aligned} &\mathbb{Z}(2)_I : (\pm 1)_I , \; \mathbb{Z}(2)_{II} : (\pm 1)_{II} \\ &\mathbb{Z}(2)_{diagonal} : (+1_I \;\&\; +1_{II}, -1_I \;\&\; -1_{II}) \end{aligned} \tag{6c}$$

and

$$\overline{K} = SU(2) \times SU(2) = \overline{SO}(4) \tag{6d}$$

At the level of G=SL(4,$\mathbb{R}$), we find accordingly that SL(4,$\mathbb{R}$) is the double covering of SO(3,3), and $\overline{SL}(4,\mathbb{R})$ its quadruple covering. This is described in the exact sequences

$$\begin{array}{ccccccccc} & & & & 1 & & 1 & & \\ & & & & \downarrow & & \downarrow & & \\ 1 & \longrightarrow & \mathbb{Z}(2)_{diag} & \longrightarrow & \mathbb{Z}(2)\times\mathbb{Z}(2) & \longrightarrow & \mathbb{Z}(2) & \longrightarrow & 1 \\ & & & & \downarrow & & \downarrow & & \\ 1 & \longrightarrow & \mathbb{Z}(2)_{diag} & \longrightarrow & \overline{SL}(4,\mathbb{R}) & \longrightarrow & SL(4,\mathbb{R}) & \longrightarrow & 1 \\ & & & & \downarrow & & \downarrow & & \\ & & & & SO(3,3) & & SO(3,3) & & \\ & & & & \downarrow & & \downarrow & & \\ & & & & 1 & & 1 & & \end{array} \tag{6e}$$

B. The Algebra of SL(4,$\mathbb{R}$)

$s\ell(4,\mathbb{R})$ is given by the commutation relations (a,b=o,..3)

$$[Q_{ab}, Q_{cd}] = ig_{bc}\, Q_{ad} - ig_{ad}\, Q_{cb} \tag{7}$$

where g_{ab} can be either δ_{ab} or the Minkowski metric η_{ab}= diag (1,-1,-1,-1). We shall use the latter. Define the Lorentz generators (of SL(2,$\mathbb{C}$) as the anti-symmetric

$$M_{ab} = Q_{[ab]} \tag{7a}$$

and the relativistic generators of shear as the symmetric

$$T_{ab} = Q_{(ab)} \tag{7b}$$

Using the commutators with (i,j=1,2,3) the angular momentum $J_i = \frac{1}{2}\varepsilon_{ijk} M_{jk}$ subalgebra su(2) we observe that the T_{ab} split into J=2 (the T_{jk}), J=1 (the $C_j = T_{oj}$) and J=o (the T_{oo}). We can now construct the generators if the su(2)xsu(2) compact subalgebra,

$$J_i^{I/II} = \frac{1}{4}\varepsilon_{ijk} M_{jk} +/- \frac{1}{2} T_{oi} = \frac{1}{2}(J_i +/- C_i) \tag{7c}$$

The commutation relations can then be rewritten as (we denote $L_i = M_{oi}$)

$$[J_i, J_j] = i\varepsilon_{ijk} J_k \tag{8a}$$

$$[J_i, C_j] = i\varepsilon_{ijk} C_k \tag{8b}$$

$$[C_i, C_j] = i\varepsilon_{ijk} J_k \tag{8c}$$

$$[J_i, L_j] = i\varepsilon_{ijk} L_k \tag{8d}$$

$$[L_i, L_j] = -i\varepsilon_{ijk} J_k \tag{8e}$$

$$[L_i, C_j] = -i(T_{ij} + \delta_{ij} T_{oo}) \tag{8f}$$

$$[L_i, T_{jk}] = = i(\delta_{ij} C_k + \delta_{ik} C_j) \tag{8g}$$

$$[C_i, T_{jk}] = -i(\delta_{ij} L_k + \delta_{ik} L_j) \tag{8h}$$

$$[L_i, T_{oo}] = -2iC_i \tag{8i}$$

$$[C_i, T_{oo}] = -2iL_i \tag{8j}$$

$$[J_i, T_{jk}] = i\varepsilon_{ij\ell} T_{\ell k} + i\varepsilon_{ik\ell} T_{j\ell} \tag{8k}$$

$$[J_i, T_{oo}] = o \tag{8ℓ}$$

$$[T_{ij}, T_{k\ell}] = -i(\delta_{ik}\varepsilon_{j\ell m} + \delta_{i\ell}\varepsilon_{jkm} = \delta_{jk}\varepsilon_{i\ell m} + \delta_{j\ell}\varepsilon_{ikm}) J_m \tag{8m}$$

$$[T_{ij}, T_{oo}] = o \tag{8n}$$

(a) is su(2); (abc) make up su(2)xsu(2); (ade) is $s\ell(2,\mathbb{C})$; (akm) is $s\ell(3,\mathbb{R})$

4. Covariance and Equivalence

A. Algebraic implications

Gravitation involves two invariance principles - General Covariance and Equivalence. Covariance refers to the passive application of the infinite Diffeomorphism group $\mathcal{D}iff: x^\mu \to x'^\mu = f^\mu(x^\mu)$, x^μ the "curvilinear" (holonomic) co-ordinate. The x^μ dependence implies some limited resemblance to a local gauge. Indeed, the covariant derivative

$$D_\mu \phi^\Xi = [\delta^\Xi_\Lambda \partial_\mu - \Gamma^\tau_{\sigma\mu}\{\textstyle\sum^\sigma{}_\tau\}^\Xi_\Lambda]\phi^\Lambda \tag{9}$$

with $\sum^\sigma{}_\tau$ the generator matrices of GL(4,**R**) in the representation ϕ^Ξ (the index Ξ standing for some tensor $\phi^{\zeta\eta\kappa\cdot\cdot}{}_\lambda$) is the same as for gauging G:=GL(4,R), even though the structure constants of *Diff* are completely different from those of gauged GL(4,R). The resemblance is due to *Diff* being <u>represented</u> (for tensor fields) <u>non-linearly over its</u> $G\subset$*Diff* <u>linear subgroup</u>, with the infinitesimal variation

$$x^\mu \to x^\mu + \xi^\mu(x) \quad , \quad \delta\phi^\Xi = \partial_\mu \xi^\lambda \{\textstyle\sum^\mu{}_\lambda\}\phi^\Xi - \xi^\lambda \partial_\lambda \phi^\Xi \tag{10}$$

where the last term is the "alibi" transformation. The "alias" again involves GL(4,**R**) generators.

Equivalence is a local gauge principle, involving both passive and <u>active</u> transformations. In Einstein's original version and in Einstein-Cartan gravity, the gauge group is the Lorentz group L:=SO(1,3). In Metric-Affine and in Affine gravity, this is either S:=SL(4,**R**) or G. In the limit of a vanishing gravitational field the appropriate covariant derivative reduces to the ordinary derivative. In the "crosswise" Einsteinian coupling in which the conserved (energy-momentum) translation current ($\phi_{\mu\nu}$) is coupled to the gauge field of the (homogeneous) Lorentz transformations, this ensures universality and the equivalence between inertial and gravitational mass.

To summarize: covariance is realized through the fields carrying linear representations of G. Equivalence requires invariance under SL(2,**C**)$\subset$SL(4,**R**), i.e. the (J,L) system in (8).

B. <u>Why SL(4,**R**) in Tensors and SL(2,**C**) in Spinors are Represented Non-Unitarily</u>

There is, however, a dilemma. If we use intrinsic unitary representations of S or G, the $s\ell(2,\mathbf{C})$ Lorentz subgroup will also be represented unitarily, e.g. by infinite discrete representations à la Gelfand-Naimark. Assuming a proton (or a pion) to be a $J=\frac{1}{2}$ (or 0) state, we require a Lorentz boost acting on it to increase its linear momentum or velocity, but not to change its nature. However, should the proton (or pion) field be carrying a unitary representation of $s\ell(2,\mathbf{C})$, the Lorentz boost would transform the $J=\frac{1}{2}$(o) state partly to $J=\frac{3}{2}$(1). In relativistic quantum field theory <u>we therefore use finite and thus non-unitary representations</u> of the Lorentz group and GL(4,**R**). In this way, deriving the $s\ell(2,\mathbf{C})$ Lorentz generators from a Noether theorem, we have (for a Dirac spinor field, for instance)

$$M_{ab} = \int d^3x\{(\bar{\psi}\gamma^o\sigma_{ab}\psi + x_{[a}\bar{\psi}\gamma^o\partial_{b]}\psi)$$
$$+ (ix_{[a}\bar{\psi}\partial_{p]}\psi) \tag{11}$$
$$+ \text{hermitian conjugate}.$$

However, since $s\ell(2,C)$ is non-compact, the finite σ_{ab} representation is non-unitary: the matrices

$$\sigma^+_{oi} = -\sigma_{oi}\ ,\ \sigma^+_{ij} = +\sigma_{ij} \tag{12}$$

This makes the $M_{ij} = J_{ij}$ angular momentum have both intrinsic (spin) and orbital components, whereas in the $M_{oi} = L_i$ the intrinsic pieces just cancel and the boost acts only orbitally (i.e. kinetically, by imparting momentum). Note that in an extended-body treatment of the proton or pion (rather than a field theory) the question does not arise and it is evidently assumed that a Lorentz boost can indeed connect to a higher spin state. In this article, we are interested in fields, and our new infinite spinor fields should not be excited intrinsically by Lorentz boosts. Note that the excitations are allowed as far as the T_{ij} are concerned, since this assumes the presence of a gravitational field (since T_{ij} is not a symmetry of special relativity) with J=2 and it can indeed connect a $J=\frac{1}{2}$ state to $J=\frac{5}{2}$, etc....

C. The Deunitarizing Automorphism A

Fulfillment of this requirement is attained through the following deunitarizing automorphism of the $GL(4,\mathbb{R})$ (or any $GL(n,\mathbb{R})$, $n\geq4$). The commutation relations (8) are invariant under the exchange

$$A: C_i \to iL_i\ ,\ L_i \to iC_i \tag{13}$$

with all other commutators unchanged. We can now perform the following sequence:

(step A) use physical $Q_{ab}(C^{ph}_i, L^{ph}_i)$ as in (8)

(step B) apply A: $Q_{ab} \to Q^A_{ab}(C^A_i = L^{ph}_i, L^A_i = iC^{ph}_i)$ (14)

(step C) represent the "unphysical" Q^A_{ab} $\overline{SL}(4,R)^A$ group as in (4) by unitary infinite-dimensional representations. Thus, the L^A will be represented by unitary infinite-dimensional irreducible representations, while the C^A_i will be given by the infinite set of finite matrices of $so(4)$ for each level in the reduction of S over its compact subgroup

(step D) apply $A^{-1}: \mathcal{D}(Q^A_{ab}) \to \mathcal{D}(Q^{ph}_{ab})$, the physical generators $L^{ph}_i = iC^A_i$ are now represented by an infinite reducible sum of finite non-hermitian matrices just as in ordinary S tensors (and in Lorentz spinors). (14)

5. The Lorentz-Invariant Spinor Equation in $\overline{GL}(4,\mathbb{R})$ Theories

A. The equation

The first $\overline{GL}(4,\mathbb{R})$ spinor equation was a holonomic one (for a "world spinor"), suggested by J. Mickelsson[38]. However, it was based on a global $\overline{GL}(4,\mathbb{R})$, a symmetry which persists in the (flat) limit of a vanishing gravitational field, thus violating the Equivalence Principle. It is possible that this could be adjusted by introducing in the Lagrangian a mechanism of spontaneous symmetry breakdown. The Higgs-type fields should then reduce the equation into an infinite sum of Lorentz-invariant special relativistic equations[7,8,29,30].

To avoid a violation of the Equivalence Principle, we have selected a different route. With A. Cant[32], we have first written an SL(2,$\mathbb{C}$)-invariant equation, acting on a $\overline{GL}(4,\mathbb{R})$ spinor. Such a spinor is then first a reducible structure, fitting special relativity. We have indeed exploited this feature directly, in applying the equation to the systematics of hadron excitations[34]. The same equation fits a $\overline{GL}(4,\mathbb{R})$ spinor in an Einstein-Cartan-Sciama-Kibble theory, with the Lorentz group acting locally on the tetrad indices.

The same spinor "manifield" can be used as a matter field in all $\overline{GL}(4,\mathbb{R})$ gauge theories, whether the Metric-Affine (Anholonomic) theory[22,23] or the (equivalent, I think) holonomic Hamiltonian theory of Komar[39]. In all these theories the matter Lagrangian is the special-relativistic one, with all derivatives replaced by $\overline{GL}(4,\mathbb{R})$ covariant derivatives (the $Q_a{}^b$ is that of eq. (7))

$$\delta^\mu_a \partial_\mu \Psi^u \rightarrow h^\mu_a (\delta^u_v \partial_\mu - \omega_\mu{}^e{}_f \{Q_e{}^f\}^u{}_v)\Psi^v \tag{15}$$

(u,v are the infinite-component indices) ω is the $\overline{GL}(4,\mathbb{R})$ connection. In Komar's theory, h^μ_a is also replaced by δ^μ_a.

As a linear Dirac -like equation, it reads, in flat space,

$$(i\{\Gamma^\mu\}^u{}_v \partial_\mu - \delta^u_v \kappa)\ \Psi^v = 0 \tag{16}$$

Since the equation is only Lorentz invariant, and using (13) and (14),

$$\kappa = \mathscr{f}(J^2 - L^2_{ph}) = \mathscr{f}(J^2 + C^2_A) \tag{17}$$

In addition, requiring only Lorentz invariance implies that the Γ^μ make a Lorentz 4-vector, or using the deunitarizing automorphism A as prescribed in (14), under the $s\ell(2,C)$ of $\{J\&L^{ph}\}$, the latter given by the finite, so(4)-like matrices of J and -iC in (4), the Γ^μ are in ($\mathcal{D}_F$ stands for the non-unitary finite rep.)

$$\Gamma^{\mu} \sim \mathcal{D}_F(\tfrac{1}{2},0) \oplus \mathcal{D}_F(0,\tfrac{1}{2}) \qquad (18)$$

This point is very important[34]. In Fig. 4 we show how, due to this feature, the (flat-space) Γ^{μ} matrices do not connect any states but those on the immediate row on both sides of the diagonal, in a (J^{I},J^{II}) plane (the $J^{I,II}$ being defined by eq. (7c)).

In curved space, i.e. in the presence of the gravitational field, the replacement (15) should be enacted in eq. (16), and the gravitational connection field $(\omega^{e}{}_{f}\, Q_{e}{}^{f})$ connects all states.

B. The Application to Hadrons: Assignments

This is a revival of the 1965 algebraic treatment of hadron Regge sequences[12] based on SL(3,ℝ). This time[34] we can deal with all spinor states, and we can do it covariantly. Basically, relativistic fields have been very useful phenomenologically. Covariance is apparently a very powerful constraint. Looking at the band structure of hadron resonances (now much richer than in 1965), only our $\overline{SL}(4,\mathbb{R})$ spinor and tensor manifields present an appropriate covariant field representation. Dynamically, it is highly probable that the whole systematics are due to QCD, just as the whole of Chemistry is due to the Schrödinger equation with an electromagnetic potential. In neither case can one go very far in actual calculation, and assuming that the resulting states arrange themselves so as to fit a phenomenological covariant description appears highly plausible. Indeed, the phenomenological fit is very satisfactory. In Table I and Fig. 5 we give as an example the assignment of all N states to $\mathcal{D}^{d}(\frac{1}{2},0) \oplus \mathcal{D}^{d}(0,\frac{1}{2})$ (See Fig. 3) of equations (4c). Note that following the irrelevance of the usual CPT treatment on the infinite- component case[40], we have not assigned one half of the states to the antiparticles as in the Dirac spinor and have used both representations for independent states, assuming all anti-particles to go into a conjugate reducible representation. It is also possible alternatively to follow the Dirac assignments and have two independent such coupled pairs of representations, with the particles in $(\frac{1}{2},0)$ for the $\frac{1}{2}^{+}$ "based" system and their anti-particles filling $(0,\frac{1}{2})$ in one reducible spinor manifield; and the $\frac{1}{2}^{-}$ "based" system in $(0,\frac{1}{2})$ with its antipartcles in $(\frac{1}{2},0)$, for the other manifield.

Note that as a result of (18), the proliferation of states is somewhat subdued and may affect the Hagedorn thermodynamics (we have not yet checked this point).

The $J=\frac{1}{2}$ original SU(3) octet states can be assigned to similar representations, one such for each octet state. For the $\Delta(1238\ J=\frac{3+}{2})$ and $\Delta(1700)\ J=\frac{3-}{2}$ we have suggested using a Rarita-Schwinger like spinor-vector manifield ψ_μ, i.e. affixing a $(\frac{1}{2},\frac{1}{2})$ external index onto the $\mathcal{D}^d(\frac{1}{2},o)\oplus\mathcal{D}^d(o,\frac{1}{2})$ of eq. (4c) and (16). Table II gives the actual phenomenological assignments. We have not yet calculated in detail the infinite Rarita-Schwinger like equations.

For mesons (see Table III) considering their quark-antiquark structure, we have required the $(\frac{1}{2},\frac{1}{2})$ as lowest state, i.e. $\mathcal{D}^\ell(\frac{1}{2})$ of (4d). Note that should the existence of "glueballs" really be established, they would have to be assigned to $\mathcal{D}^\ell(o)$ of (4d). Both meson manifields obey a Klein-Gordon like infinite equation.

In ref.[34] we have discussed mass formulae based on (17) and other predictions. There is considerable predictive power in that classification (see Fig. 6).

C. Hadrons: Black Holes and "Gravitational Radiation Lasers"

It is possible that a more direct connection be established between QCD (or its approximations, such as the Skyrme model) and our Covariance-based phenomenology. Another and perhaps more direct application is the use of those manifields in calculations relating to hadrons in strong gravitational field. The fall of protons onto a Black Hole should be restudied using our spinor manifields, since the gravitational field can certainly pump-up and fill the higher states in the nucleon manifield. One may even think of "gravitational radiation lasers" in this context.

6. Infinite Frames and World Spinors

A. $\aleph_0$-ads

We now have to address the issue of transforming a special-relativity [or anholonomic $s\ell(2,\mathbb{C})$, or alternatively anholonomic $s\ell(4,\mathbb{R})$] infinite spinor obeying (16) into a world-spinor, transforming bi-valuedly under the Diffeomorphism group. For tensors, we remember that the ϕ^{ab} anholonomic tensor (a scalar under Diff, tensor under local $s\ell(2,\mathbb{C})$) can be factorized into a world-tensor and tetrads (e^a_μ, the inverse of h^a_μ in (15)):

$$\phi^{ab\cdots}(x) = \phi^{\mu\nu\cdots}(x)\ \ e^a_\mu(x)\ \ e^b_\nu(x)\ldots \tag{19}$$

and a local-$s\ell(2,\mathbb{C})$-scalar becomes a world-scalar via (η_{ab} is the Minkowski metric $g_{\mu\nu}$ the Einstein metric)

$$\phi^a(x)\ \eta_{ab}\ \phi^b(x) = \phi^\mu(x)\ e^a_\mu(x)\ \eta_{ab}\ e^b_\nu(x)\ \phi^\nu(x) = \\ = \phi^\mu(x)\ g_{\mu\nu}(x)\ \phi^\nu(x) \tag{20}$$

Similarly, one may define "infinite tetrads" on $\aleph_o$-ads,

$$\Psi^U(x) = E^U{}_M(x)\ \Psi^M(x) \tag{21}$$

where Ψ^M is a (holonomic) world-spinor manifield, the index M (like U) describing a countable infinity of components. Note that though both Ψ^U and Ψ^M are $\overline{SL}(4,\mathbb{R})$ spinors, the Ψ^U in the case of flat space and of Einstein-Cartan theory is enacted upon by the $s\ell(2,\mathbb{C})$ subgroup only, whereas Ψ^M supports $\overline{SL}(4,\mathbb{R})$ and through it, the non-linear action of Diff.

Note that <u>in the case of Metric Affine Gravity</u>, Ψ^U also carries an $s\ell(4,\mathbb{R})$ (anholonomic) representation, and <u>the $E^U{}_M$ is itself a single-valued SL(4,R) x-dependent transformation matrix</u>.

B. <u>The G_{MN} metric</u>

We can also transform a Lorentz scalar into a world-scalar,

$$(\Psi^+(x))^U\ \{B\}_{UV}\ (\Psi(x))^V = (\Psi^+(x))^M\ E^U_M(x)\ \{B\}_{UV}\ E^V_N(x)\ (\Psi(x))^N \tag{22}$$

where $\{B\}$ is the Γ^o(infinite matrix) in (16). We may now recombine and define an infinite representation of the metric,

$$E^U_M(x)\ \{B\}_{UV}\ E^V_N(x) = G_{MN}(x) \tag{23}$$

Using constant transition matrices K for $SL(2,\mathbb{C})$ and R for $\overline{SL}(4,\mathbb{R})$, we may write

$$e^a_\mu(x) = K^a_U\ E^U_M(x)\ R^M_\mu \tag{24}$$

and for a gravitino

$$\psi^\alpha_\mu(x) = K^\alpha_U E^U_M(x)\ R^M_\mu \tag{25}$$

where K^α_U relates finite and infinite spinor indices and is itself therefore an "even" transition coefficient of $SL(2,\mathbb{C})$, whereas K^a_U and R^M_μ are "odd" and relate spinors to vectors in the two groups. Equations (24), etc. imply

$$D_\mu E = o\ ,\ D_\mu G = o \tag{26}$$

C. <u>Transformations</u>

We may write for an infinitesimal transformation ($\varepsilon(x)$ is the local Lorentz parameter, and $x^{\mu'} = x^\mu + \xi^\mu$),

$$\delta E^U_M = -\frac{i}{2}\varepsilon_b{}^a(x)\ \{M_a{}^b\}^U_V\ E^V_M(x) + \partial_\mu\xi^\rho\}\{Q^\mu_\rho\}^N_M\ E^U_N(x) \tag{27}$$

$\{Q_\rho{}^\mu\}^N_M$ are the $\overline{SL}(4,\mathbb{R})$ matrices calculated in (8) except that they are given in the holonomic co-ordinates. We can relate the two systems (if we deal with a

Metric Affine anholonomic theory)

$$\{Q_\nu{}^\mu\}^M_N = h^\mu_a(x)(H^M_U(x)\ \{Q_b{}^a\}^U_V\ E^V_N(x))\ e^b_\nu(x) \qquad (28)$$

(H^M_U is the inverse of E^U_M) the left hand side in the constant $\overline{SL}(4,\mathbb{R})$ matrix $\sum_\nu{}^\mu$ of equations (9) and (10) (for $\phi^\Lambda \to \psi^M$), up to a unitary locally-dependent transformation which can be absorbed by the Diff gauge,

$$\delta\Psi^M(x) = \partial_\mu \xi^\nu\ \{\textstyle\sum^\mu{}_\nu\}^M_N\ \psi^N(x) - \xi^\lambda\ \partial_\lambda \psi^M \qquad (29)$$

$\psi^M(x)$ can enter plain Einstein gravity and will not modify in that case the metricity (or Riemannian nature) of the theory. However, it can also serve as matter field in a non-Riemannian Affine theory of Gravity[29] and will thus contribute to the Shears in the Hypermomentum currents. These will be the currents of T_{ab} in (8).

In a Metric-Affine theory[22,23], the Shear currents will contribute off-mass-shell.

We can now also use the E^U_M and H^M_U with the Mickelsson[38] world spinor, transforming it into an anholonomic frame of a Metric-Affine theory.

7. $\overline{GL}(10,\mathbb{R})$ Anholonomic Spinors for Superstrings

We have recently applied[41,42] the method to suggest a possible solution to the problem (untouched until 1986) of embedding the superstring in a generic curved manifold. Several problems arise:

(a) the existence of curvature in the embedding manifold generates "radiative corrections" in the equations fixing the "critical" dimensionality in which the conformal anomaly is removed from the quantum theory. This was the main reason because of which the problem was not tackled until recently. In one very recent approach this is answered by imposing the vanishing of all radiative corrections as a condition on the embedding manifold. Indeed, it appears to amount to the imposition of (classical) Riemannian conditions.

(b) the theory of the superstring is described by frames $X^m(\zeta)$ that are general-invariant in the 2-dimensions of the string's world sheet (co-ordinates $\zeta^\alpha, \alpha=1,2$),

$$L_B = g^{\alpha\beta}(\zeta)\ \partial_\alpha X^m(\zeta)\ \partial_\beta X^n(\zeta)\ \eta \qquad (30)$$

(the Lagrangian is also conformally invariant in ζ^α). L_B is the bosonic part of the Lagrangian, η_{mn} the Minkowski metric in 10 dimensions. Thus, the embedding manifold appears as a ζ-local symmetry in the tangent frame to the world sheet at ζ. Inducing curvature in that tangent cannot be performed by the conventional tetrad formalism of (19)-(20), since it is based on taking the tangent to x^μ by putting up a frame $Y^m(x)$,

$$e^m_\mu(x') = \frac{\partial Y^m(x)}{\partial x^\mu}\Big|_{x=x'} \qquad (31a)$$

to be compared with

$$\frac{\partial X^m}{\partial \zeta^\alpha} \qquad (31b)$$

of eq. (30).

Replacing the flat tangent to ζ^α by a "curved tangent" appears ill-defined.

(c) the superstring involves spinor frames for supersymmetry, an essential ingredient in the removal of tachyons and in fixing the dimensionality of the embedding. Even if $X^m(\zeta)$ in (30) is replaced by $X^\mu(\zeta)$, the spinor frames $\Theta^{Au}(\zeta)$ in

$$[\overline{\Theta}^{1u}(\zeta)\ (\gamma^m)_{uv}\frac{\partial}{\partial\zeta^\alpha}\ \Theta^{1v}(\zeta)][\overline{\Theta}^{2u'}(\zeta)\ (\gamma^n)\frac{\partial}{\partial\zeta^\alpha}\ \Theta^{2v'}(\zeta)]+\ldots \qquad (32)$$

[A=1,2 is a spinor index on the sheet; u,v a spinor index under $\overline{SO}(1,9)$ in the embedding] require the introduction of tetrads in $\gamma^m = \gamma^\mu e^a_\mu$, with the above difficulties of (b).

To answer (b) and (c), we have introduced curvature in the embedding by <u>constructing in the tangent to ζ^α</u> infinite $\overline{GL}(4,\mathbb{R})$ frames that can carry Diff(10,$\mathbb{R}$),

$$X^m(\zeta) \to X^M(\zeta) \quad , \quad \Theta^u(\zeta) \to \Theta^U(\zeta) \qquad (33)$$

selecting representations whose lowest $\overline{SO}(1,9)$ level is identical with X^m and Θ^u. The deunitarizing automorphism A exists in 10 dimensions as well, so that the $\overline{SO}(1,9)$ subgroup representations are finite and non-unitary and coincide with the unitary representations of $\overline{SO}(10)$. To preserve the critical dimension (a), we have embedded

$$\overline{GL}(10,\mathbb{R}) \subset \overline{GQ}(10,\mathbb{R})$$

in a hyperexceptional supergroup and proved that it preserves the supersymmetry even in the curved manifold. We refer the reader to ref.[41,42] for details.

References

1. Y. Ne'eman, Ann. Inst. Henri Poincaré, A28, 369 (1978).
2. Y. Ne'eman, Foundations of Physics 13, 467 (1983).
3. E. Cartan, Leçons sur la Theorie des Spineurs", Hermann & Co., Paris (1948), Part II, sect. 177, pp. 89-91.
4. K. Iwasawa, Ann. Math. (Princeton) 50, 507 (1949).
5. V. Bargmann, Ann. Math. (Princeton), 48, 568 (1947).
6. B. S. De Witt, in Relativity Groups and Topology, B.S. DeWitt and C.M. DeWitt eds., Gordon and Breach, New York (1964), pp. 587-826.
7. V. I. Ogievetsky and I. V. Polubarinov, Sov. Phys. JETP 21, 1093 (1965).
8. C. J. Isham, A. Salam and J. Strathdee, Ann. Phys. (N.Y.) 62, 98 (1971).
9. S. Weinberg, Phys. Rev. 135, B1049 (1964).
10. Y. Ne'eman, in Differential Geometric Methods in Theoretical Physics, (Proc. Clausthal 1986 Conf.), H. D. Doebner ed., Lect. Notes in Math., Springer Verlag, Berlin-Heidelberg-N.Y., in press. Available as Tel Aviv Un. report TAUP N179-86.
11. I. M. Gelfand and M. I. Graev, Izv. Akad. Nauk SSR, Ser. Matem. 17, 189 (1953); Eng. version Amer. Math. Soc. Transl. 2, 147 (1956).
12. Y. Dothan, M. Gell-Mann and Y. Ne'eman, Phys. Lett. 17, 148 (1965).
13. A. Barut and A. O. Bohm, Phys. Rev. 139B, 1107 (1965).
14. Y. Dothan and Y. Ne'eman, in Resonant Particles (Proc. 2nd Athens, Ohio Conf.) B. A. Munir ed., Univ. of Ohio Pub., Athens (1965), p. 17. Reprinted in Symmetry Groups in Nuclear and Particle Physics, F. J. Dyson ed., W. A. Benjamin, N.Y. (1966), p. 287.
15. L. C. Biedenharn, R. Y. Cusson, M. Y. Han and O. L. Weaver, Phys. Lett. B42, 257 (1972).
16. G. Rosensteel and D. J. Rowe, Ann. of yHys. (N.Y.), 96, 1 (1976).
17. D.W. Joseph, "Representations of the Algebra of SL(3,ℝ) with |ΔJ|=2", Univ. of Nebraska preprint, unpub. (referred to in refs. 15 and 26).
18. V. I. Ogievetsky and E. Sokatchev, Teor. Mat. Fiz., 23, 214 (1975); Eng. version Theoret. Math. Phys. (USSR), 23, 462 (1975).
19. Y. Ne'eman, in GR8 Proc. VIII Intern. Conf. on General Relativity and Gravitation, Univ. of Waterloo, Waterloo (Canada),(1977) p. 262
20. Y. Ne'eman, in Proc. Nat. Acad. Sci., USA, 74, 4157 (1977).
21. F. W. Hehl, P.v.d.Heyde, G. D. Kerlick and J. M. Nester, Rev. Mod. Phys., 48, 393 (1976).

22. F. W. Hehl, G. D. Kerlick and P. v.d. Heyde, Phys. Lett. 63B, 446 (1976); Z. Naturforsch, A31, 111, 524, 823 (1976);
E. A. Lord, Phys. Lett. A65, 1 (1978);
F. W. Hehl, G. D. Kerlick, E. A. Lord and L. L. Smalley, Phys. Lett. B70, 70 (1977);
F. W. Hehl and G. D. Kerlick, GRG9, 691 (1978);
F. W. Hehl, E. A. Lord and L. L. Smalley, GRG13, 1037 (1981).
23. F. W. Hehl, E. A. Lord and Y. Ne'eman, Phys. Lett. B71, 432 (1977); Phys. Rev. D17, 418 (1978).
24. F. W. Hehl, Y. Ne'eman, J. Nitsch and P. v.d. Heyde, Phys. Lett. B78, 102 (1978).
25. Dj. Šijački, Ph.D Thesis, Duke University (1974).
26. Dj. Šijački, Jour. Math. Phys. 16, 298 (1975).
27. B. Speh, Mat. Ann. 258, 113 (1981).
28. Private communication from Dj. Šijački.
29. Y. Ne'eman and Dj. Šijački, Ann. of Phys. (NY), 120, 292 (1979).
30. Y. Ne'eman and Dj. Šijački, Proc. Nat. Acad. Sci., USA, 76, 561 (1979).
31. Dj. Šijački and Y. Ne'eman, Jour. Math. Phys. 26,2457 (1985).
32. A. Cant and Y. Ne'eman, Jour. Math. Phys., 26, 3180 (1985).
33. Y. Ne'eman, "World Spinors in Riemannian Gravity", report at the Fibre Bundle Workshop, Beersheva (1985), unpub.
34. Y. Ne'eman and Dj. Šijački, Phys. Lett. 157B, 267 (1985).
35. Y. Ne'eman and Dj. Šijački, Phys. Lett. 157B, 275 (1985).
36. A. Kihlberg, Arkiv. f. Fysik, 32, 241 (1966) and erratum (1980).
37. Harish-Chandra, Proc. Nat. Acad. Sci., USA, 37, 170, 362, 366, 691 (1951).
38. J. Mickelsson, Comm. Math. Phys., 88, 551 (1983).
39. A. Komar, Phys. Rev. D27, 2277 (1983); 30, 305 (1984);
Jour. Math. Phys. 26, 831 (1985);
P. G. Bergmann and A. B. Komar, Jour. Math. Phys. 26, 2030 (1985).
40. D. T. Stoyanov and I. T. Todorov, Jour. Math. Phys. 9, 2146 (1968).
41. Y. Ne'eman and Dj. Šijački, Phys. Lett. B174, 165 (1986).
42. Y. Ne'eman and Dj. Šijački, Phys. Lett. B174, 171 (1986).

Table I

Assignment of N states[a)]

$\mathcal{D}^{disc}(\frac{1}{2},0)$	J^P	N	$\mathcal{D}^{disc}(0,\frac{1}{2})$	J^P	N
$(\frac{1}{2},0)$	$\frac{1}{2}^+$	N(940)	$(0,\frac{1}{2})$	$\frac{1}{2}^-$	N(1535)
$(\frac{3}{2},1)$	$\frac{1}{2}^+$	N(1440)	$(1,\frac{3}{2})$	$\frac{1}{2}^-$	N(1650)
	$\frac{3}{2}^-$	N(1520)		$\frac{3}{2}^+$	
	$\frac{5}{2}^+$	N(1680)		$\frac{5}{2}^-$	N(1675)
$(\frac{5}{2},2)$	$\frac{1}{2}^+$	N(1710)	$(2,\frac{5}{2})$	$\frac{1}{2}^-$	N(2090)
	$\frac{3}{2}^-$	N(1700)		$\frac{3}{2}^+$	N(1720)
	$\frac{5}{2}^+$	N(2020)		$\frac{5}{2}^-$	N(2200)
	$\frac{7}{2}^-$	N(2190)		$\frac{7}{2}^+$	N(1990)
	$\frac{9}{2}^+$	N(2220)		$\frac{9}{2}^-$	N(2250)
$(\frac{7}{2},3)$	$\frac{1}{2}^+$	N(2100)	$(3,\frac{7}{2})$	$\frac{1}{2}^-$	
	$\frac{3}{2}^-$	N(2080)		$\frac{3}{2}^+$	
	$\frac{5}{2}^+$			$\frac{5}{2}^-$	
	$\frac{7}{2}^-$			$\frac{7}{2}^+$	
	$\frac{9}{2}^+$			$\frac{9}{2}^-$	
	$\frac{11}{2}^-$	N(2600)		$\frac{11}{2}^+$	
	$\frac{13}{2}^+$	N(2700)		$\frac{13}{2}^-$	

a) Underlined are uncertain.

Table II

Assignment of Δ states.

$\mathcal{D}^{\text{disc}}(\frac{1}{2},0)_\mu$	J^P	Δ	$\mathcal{D}^{\text{disc}}(0,\frac{1}{2})_\mu$	J^P	Δ
$(1,\frac{1}{2})$	$\frac{1}{2}^-$	Δ(1620)	$(\frac{1}{2},1)$	$\frac{1}{2}^+$	Δ(1550)
	$\frac{3}{2}^+$	Δ(1232)		$\frac{3}{2}^-$	Δ(1700)
$(2,\frac{3}{2})$	$\frac{1}{2}^-$	Δ(1900)	$(\frac{3}{2},2)$	$\frac{1}{2}^+$	Δ(1910)
	$\frac{3}{2}^+$	Δ(1600)		$\frac{3}{2}^-$	Δ(1940)
	$\frac{5}{2}^-$			$\frac{5}{2}^+$	Δ(1905)
	$\frac{7}{2}^+$	Δ(1950)		$\frac{7}{2}^-$	
$(3,\frac{5}{2})$	$\frac{1}{2}^-$	Δ(2150)	$(\frac{5}{2},3)$	$\frac{1}{2}^+$	
	$\frac{3}{2}^+$	Δ(1920)		$\frac{3}{2}^-$	
	$\frac{5}{2}^-$	Δ(1930)		$\frac{5}{2}^+$	
	$\frac{7}{2}^+$			$\frac{7}{2}^-$	Δ(2200)
	$\frac{9}{2}^-$			$\frac{9}{2}^+$	Δ(2300)
	$\frac{11}{2}^+$	Δ(2420)		$\frac{11}{2}^-$	
$(4,\frac{7}{2})$	$\frac{1}{2}^-$		$(\frac{7}{2},4)$	$\frac{1}{2}^+$	
	$\frac{3}{2}^+$			$\frac{3}{2}^-$	
	$\frac{5}{2}^-$	Δ(2350)		$\frac{5}{2}^+$	
	$\frac{7}{2}^+$	Δ(2390)		$\frac{7}{2}^-$	
	$\frac{9}{2}^-$	Δ(2400)		$\frac{9}{2}^+$	
	$\frac{11}{2}^+$			$\frac{11}{2}^-$	
	$\frac{13}{2}^-$	Δ(2750)		$\frac{13}{2}^+$	
	$\frac{15}{2}^+$	Δ(2950)		$\frac{15}{2}^-$	

Table III

Assignment of SU(3) meson states.

$\mathcal{D}^{ladd}(\frac{1}{2})$	J^P	π	K	η	η'	$\mathcal{D}^{ladd}(\frac{1}{2})$	J^P	ρ	K^*	ω	ϕ
$(\frac{1}{2},\frac{1}{2})$	0^-	$\pi(140)$	K(493)	$\eta(549)$	$\eta'(958)$	$(\frac{1}{2},\frac{1}{2})$	0^+	$\delta(980)$	$\kappa(1350)$	S(975)	$g_S(1240)$
	1^+	B(1235)	Q(1280)	H(1190)			1^-	$\rho(770)$	$K^*(892)$	$\omega(783)$	$\phi(1020)$
$(\frac{3}{2},\frac{3}{2})$	0^-	$\pi(1300)$	K(1400)		L(1440)	$(\frac{3}{2},\frac{3}{2})$	0^+			$\varepsilon(1300)$	
	1^+	A(1270)	Q(1400)	D(1285)	E(1420)		1^-				
	2^-		L(1580)				2^+	$A_2(1320)$	$K^*(1430)$	f(1270)	f'(1525)
	3^+						3^-				
$(\frac{5}{2},\frac{5}{2})$	0^-	$\pi(1770)$	K(1830)			$(\frac{5}{2},\frac{5}{2})$	0^+				S(1730)
	1^+				D(1530)		1^-	$\rho(1600)$	$K^*(1650)$		$\phi(1680)$
	2^-	A(1680)	L(1770)				2^+				f(1810)
	3^+						3^-	g(1690)	$K^*(1780)$	$\omega(1670)$	
	4^-						4^+				
	5^+						5^-				

Fig. I Unitary Irreducible Representations of SL (2 ℝ) and Coverings

(a) Principal Series

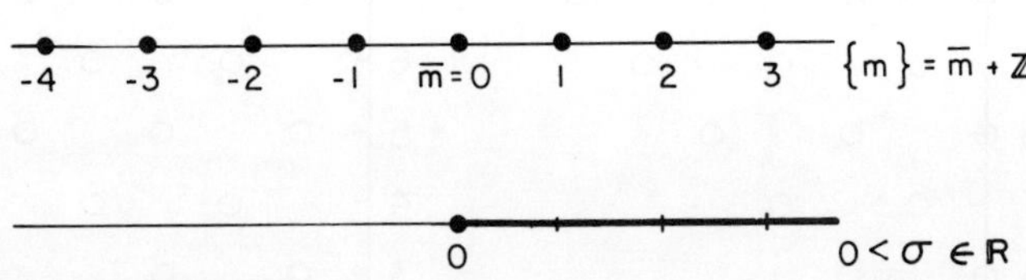

(b) Supplementary Series

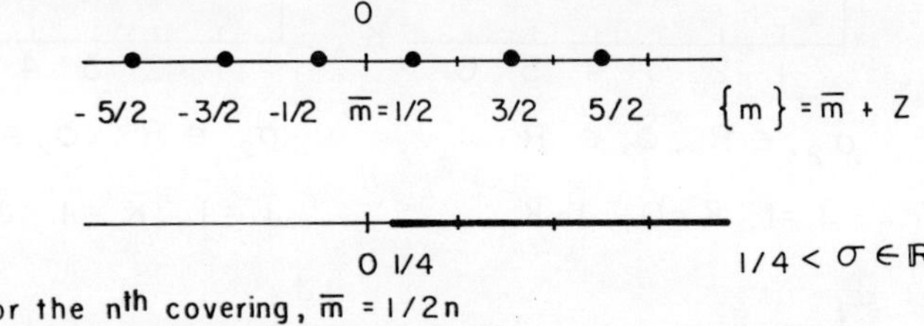

For the nth covering, $\overline{m} = 1/2n$

(c) Discrete Series (in the drawn example, $\overline{m} = -1/2$)

c.1 : bounded below

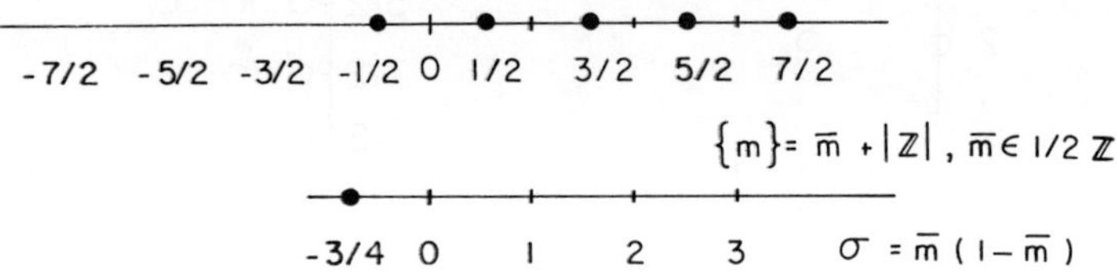

c.2 : bounded above

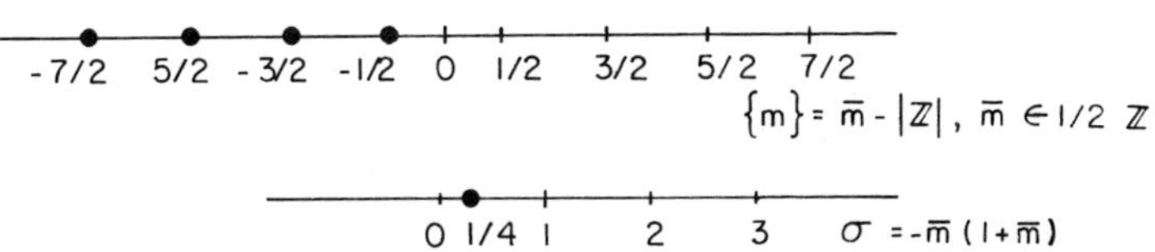

Fig. 2 Unitary Irreducible Representations of $\overline{SL}(3,\mathbb{R})$

A. Principal and Supplementary Series

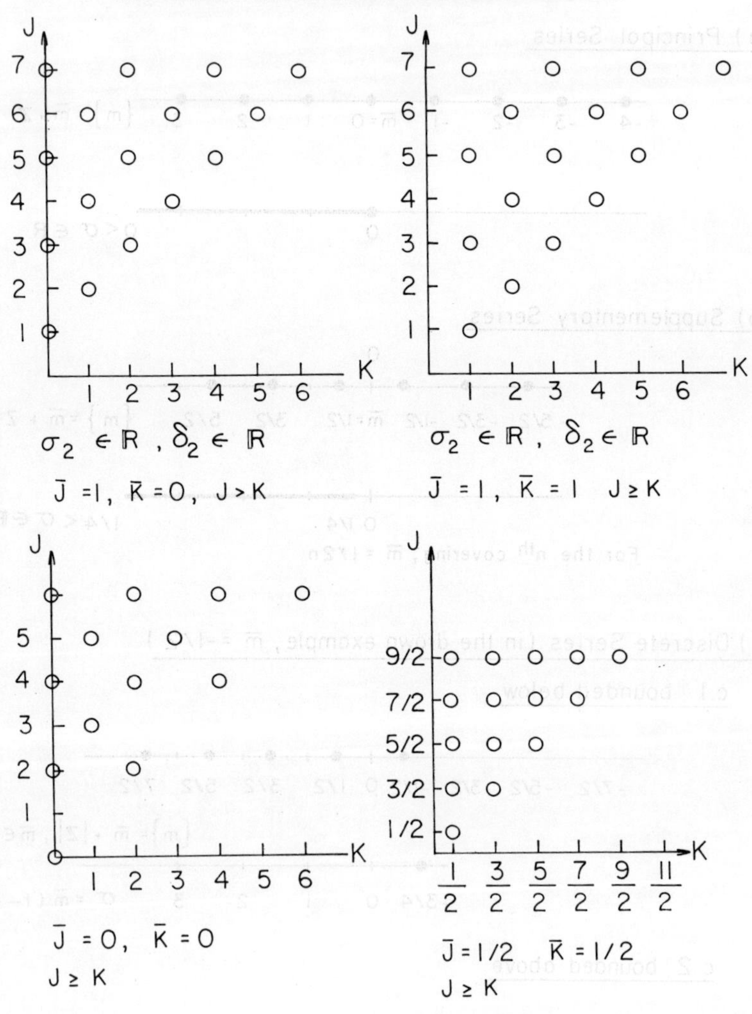

Fig. 2 Unitary Irreducible Representations of $\overline{SL}(3, \mathbb{R})$

B. Discrete and Ladder Series

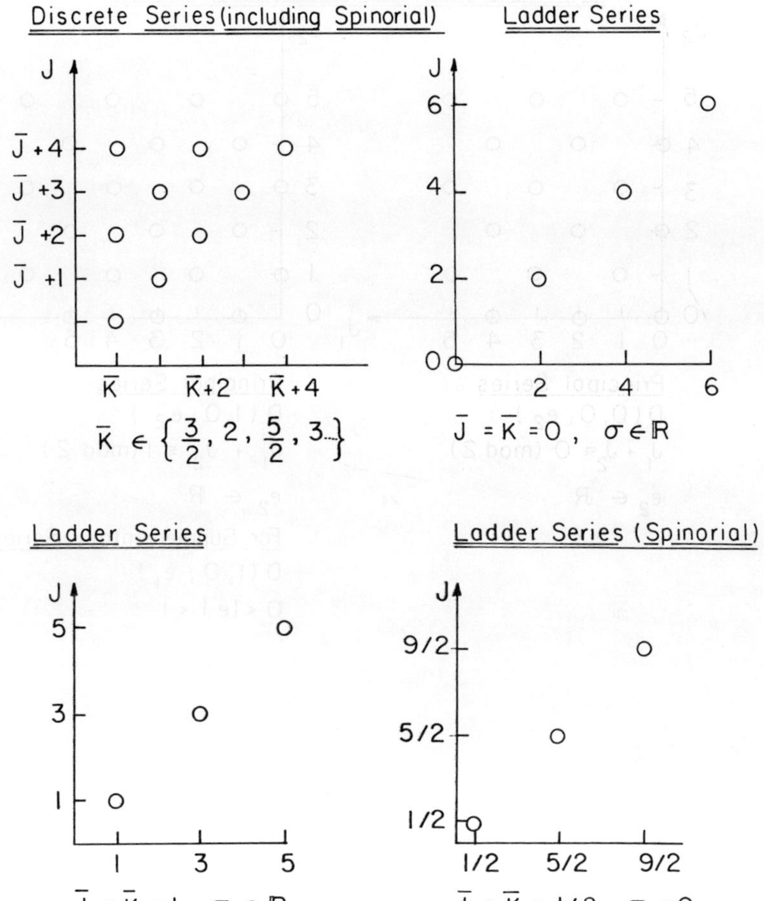

Fig.3 Multiplicity-free Unitary Irreducible Representations of $\overline{SL}(4,\mathbb{R})$

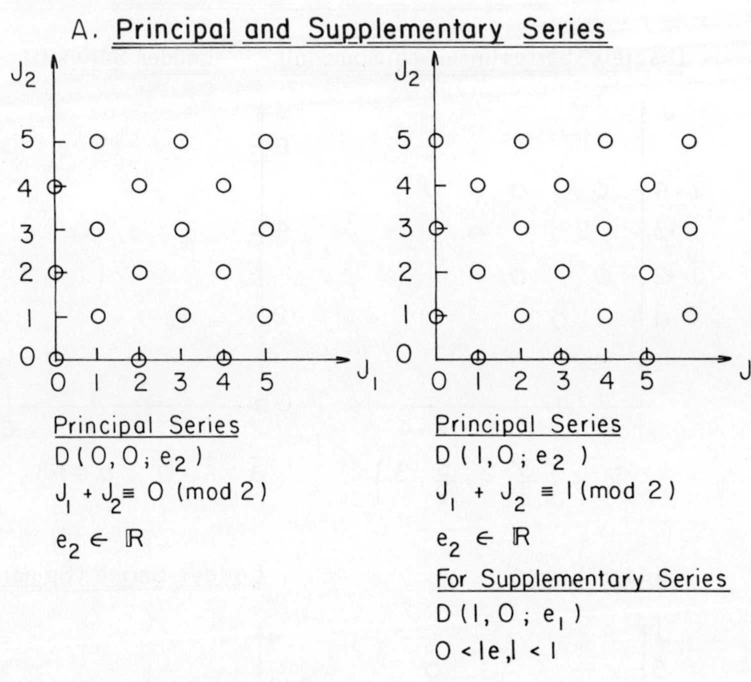

Fig.3 Multiplicity-free Unitary Irreducible Representations of $\overline{SL}(4,\mathbb{R})$

B. Discrete Series (including spinorial)

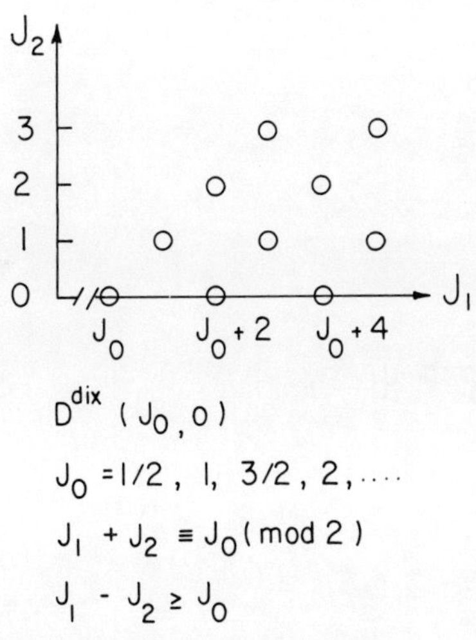

$D^{dix}(J_0, 0)$

$J_0 = 1/2,\ 1,\ 3/2,\ 2, \ldots$

$J_1 + J_2 \equiv J_0 \pmod 2$

$J_1 - J_2 \geq J_0$

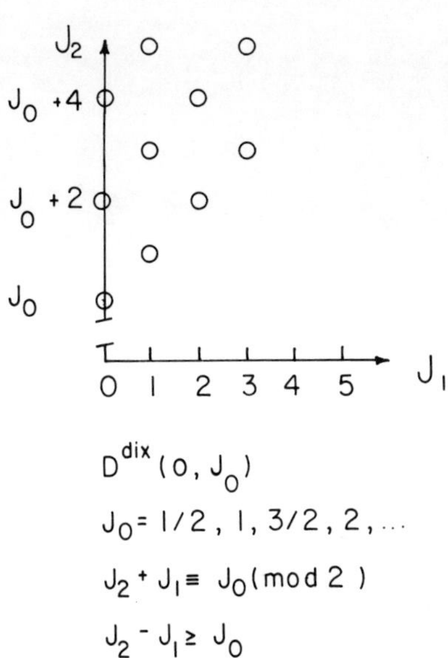

$D^{dix}(0, J_0)$

$J_0 = 1/2,\ 1,\ 3/2,\ 2, \ldots$

$J_2 + J_1 \equiv J_0 \pmod 2$

$J_2 - J_1 \geq J_0$

Fig.3 Multiplicity-free Unitary Irreducible Representations of $\overline{SL}(4,\mathbb{R})$

C. Ladder Series

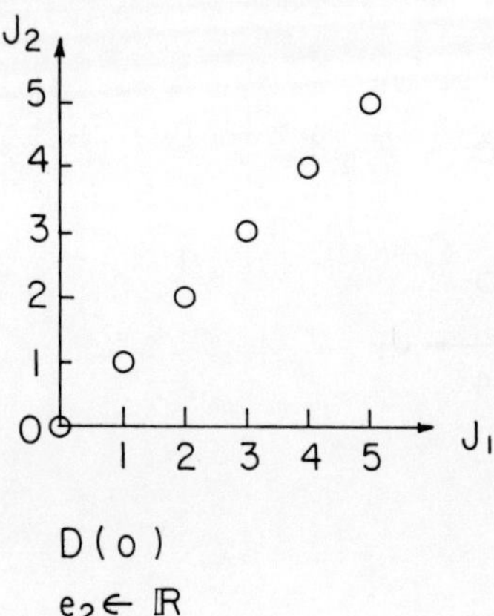

$D(0)$

$e_2 \in \mathbb{R}$

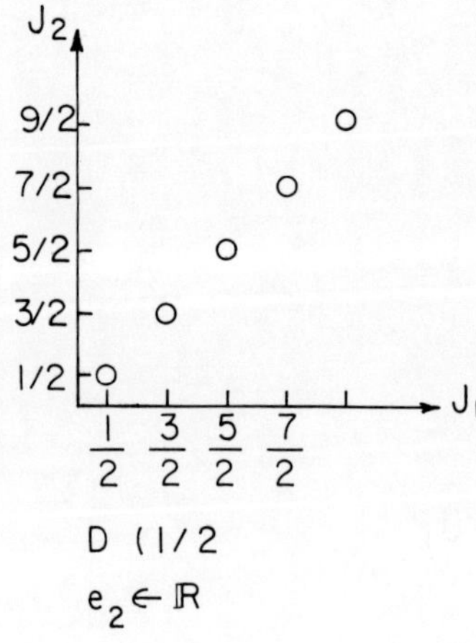

$D(1/2$

$e_2 \in \mathbb{R}$

Lorentz invariant equation

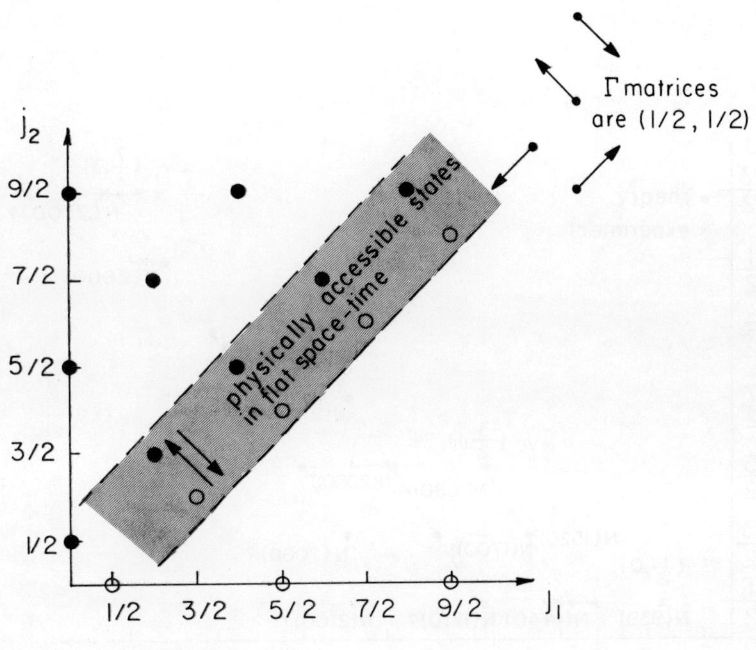

○ $D^{disc.}(1/2, 0)$

● $D^{disc.}(0, 1/2)$

Fig . 4

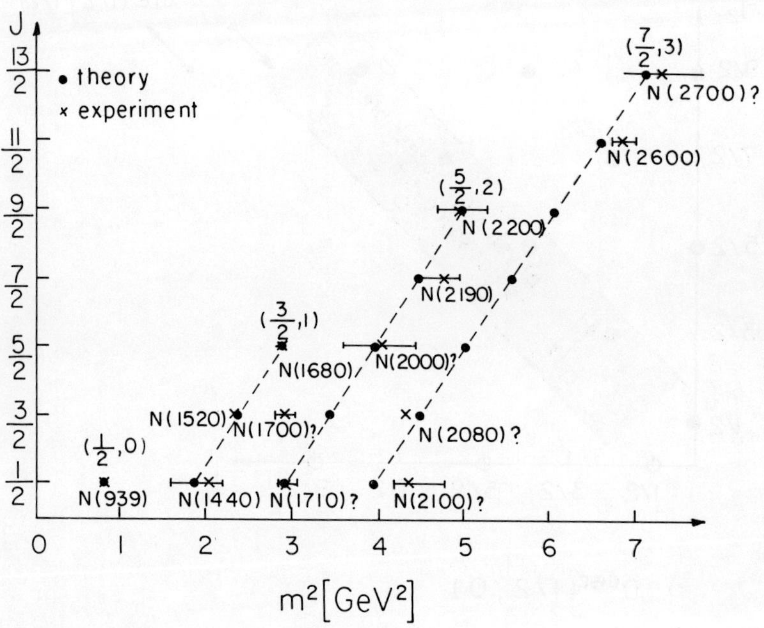
J
13/2
11/2
9/2
7/2
5/2
3/2
1/2
• theory
× experiment
(7/2, 3)
N(2700)?
N(2600)
(5/2, 2)
N(2200)
N(2190)
(3/2, 1)
N(1680)
N(2000)?
N(1520)
N(1700)?
N(2080)?
(1/2, 0)
N(939)
N(1440)
N(1710)?
N(2100)?
0
1
2
3
4
5
6
7
m²[GeV²]

Fig. 5

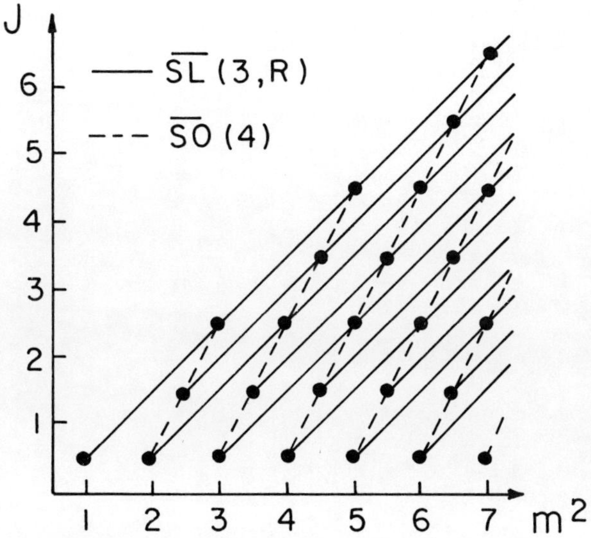

Mass-spectrum $m^2(J, n)$

$$\vec{J} = \vec{j}_1 + \vec{j}_2 \qquad n = j_1 + j_2 - J$$

Fig : 6

FINAL LIST OF PARTICIPANTS

Name/Institution	Country
1. AICHELBURG Peter Institute of Physics University of Vienna Boltzmanngasse 5 1090 Vienna	Austria
2. BARUT A.O. University of Colorado Boulder Co 80309	USA
3. BASHIR Mohammed Ali Dept. of Mathematics King Saudi University Riyadh	Saudi Arabia
4. BLAU Matthias Institute of Physics University of Vienna Boltzmanngasse 5 1090 Vienna	Austria
5. BOŽIC Mirjana Institute of Physics P.O. Box 56 Belgrade	Yugoslavia
6. BUDINICH Paolo Director SISSA/ISAS	Italy
7. BUGAJSKA Krystina Institut für Theoretische Physik Technische Universität Leibnizstrasse 10 D-3392 Clausthal-Zellerfeld	FRG
8. BURSTALL E. School of Mathematics Rice University Houston, Texas 77001	USA
9. CAHEN Michel Dept. de Mathématiques Université Libre Campus Plaine Blvd du Triomphe 1050 Bruxelles	Belgium

Name/Institution	Country
10. CAIANIELLO Edoardo Università di Salerno 84100 Salerno	Italy
11. CATENACCI Roberto Dipartimento di Matematica Università di Pavia 27100 Pavia	Italy
12. COQUEREAUX Robert CNRS Luminy 13288 Marseille	France
13. CRUMEYROLLE Albert 17, Av. de Lauragais 31400 Toulouse	France
14. DABROWSKI Ludwik SISSA/ISAS	Italy
15. DOEBNER Heinz Dieter Institut für Theoretische Physik Technische Universität Leibnitzstrasse 10 D-3392 Clausthal-Zellerfeld	FRG
16. EZIN Jean Pierre U.E.R. de Mathématiques Faculté de Sciences Université Nationale du Bénin Cotonou	Rep. Pop. Bénin
17. FALTAS M.S. Department of Mathematics Alexandria University Moharrem Bay Alexandria	Egypt
18. FRANC A. Dép. de Mathématiques Université Libre Campus Plaine Blvd du Triomphe 1050 Bruxelles	Belgium

Name/Institution	Country
19. FURLAN Giuseppe (Director) Dip. di Fisica Teorica University of Trieste	Italy
20. FURLAN Paolo Dip. di Fisica Teorica University of Trieste	Italy
21. FURUTANI K. ICTP (present)	Italy/Japan
22. GAMEDZE T. (present) ICTP	Italy/Swaziland
23. GIGENA Salvador Technische Universität Berlin 1000 Berlin (present)	FRG
24. GUTT Simone Département de Mathématiques Université Libre Campus Plaine Blvd du Triomphe 1050 Bruxelles	Belgium
25. HUGHSTON Lane P. Lincoln College Oxford OXI 3RD	U.K.
26. KYRIAZIS Athanasios Math. Institute University of Athens Athens 106 79	Greece
27. KREUZER Maximilian Institut für Theoretische Physik Universität Wien Boltzmanngasse 5 1040-Vienna	Austria
28. LANDI Giovanni SISSA/ISAS	Trieste

Name/Institution	Country
29. LEMAIRE L. Département de Mathématiques Université Libre Campus Plaine Blvd du Triomphe 1050 Bruxelles	Belgium
30. LICHNEROWICZ André Collège de France Paris	France
31. LOUNESTO Perrti Helsinki University of Technology 02150 Espoo	Finland
32. MAATTA E. Dept. of Mathematics University of Oulu Linnanmaa 90570 Oulu 57	Finland
33. MARGERIN C. Dépt. de Mathématiques Ecole Polytechnique 91128 Palaiseau Cédex	France
34. MOESGEN Karl J. Comandante Jimenez 293 Lima	Peru
35. MORAWIECZ Paweł Inst. of Theoretical Physics University of Wroclaw	Poland
36. MORENO Carlos Plaza de la Republica del Ecuador 3 Madrid 28016	Spain
37. NE'EMAN Yuval Tel Aviv University Ramat Aviv Tel Aviv	Israel
38. NIEDERLE Jiri Academy of Sciences Prague	Czechoslovakia

	Name/Institution	Country
39.	PLYMEN Roger mathematics Department The University Manchester M13 9PL	U.K.
40.	RACZKA Richard Inst. Of Nuclear Research and Nuclear Energy Hoza 69 00-681 Warsaw	Poland
41.	RATTO A. (present) ICTP	Italy/U.K.
42.	RASHID Muneer Ahmad Mathematics Dept. Ahmadu Bells University Zaria	Nigeria
43.	RAWNSLEY John Mathematics Institute University of Warwick Coventry CV4 7AL	U.K.
44.	REBHAN Anton Inst. für Theoretische Physik Universität Wien 1040 Vienna	Austria
45.	REINA Cesare Dip. di Fisica Via Celoria 16 20133 Milano	Italy
46.	RIGOLI M. ICTP	Italy
47.	SIJACKI Djordie Inst. of Physics P.O. Box 57 11001 Belgrade	Yugoslavia
48.	TEKIN Dereli D. Physics Department University of Lancaster Lancaster	UK

Name/Institution	Country
49. TODOROV I.T. Institute for Nuclear Research and Nuclear Energy Blvd. Lenin Sofia	Bulgaria
50. TRAUTMAN Andrzej (Director) Institute of Physics University of Warsaw Warsaw	Poland
51. TRIBUZY R. (present) ICTP	Italy/ Brazil
52. TUCKER Robin W. Dept. of Physics University of Lancaster Lancaster	U.K.
53. URBANEK Peter Institut für Theoretische Physik Universität Wien 1090 Vienna	Austria
54. VALLI G (present) ICTP	Italy/U.K.
55. VILLAROEL J. University of Salamanca Salamanca	Spain
56. WESTBURY Bruce 74 Kenmare Road Liverpool L15 3HQ	U.K.
57. WIECZOREK Eberhard Institut für Hochenenergiephysik 1615 Zeuthen	GDR